Lecture Notes in Physics

The Lecture Notes in Physics

The series Lecture Notes in Physics (LNP), founded in 1969, reports new developments in physics research and teaching – quickly and informally, but with a high quality and the explicit aim to summarize and communicate current knowledge in an accessible way. Books published in this series are conceived as bridging material between advanced graduate textbooks and the forefront of research to serve the following purposes:

• to be a compact and modern up-to-date source of reference on a well-defined topic;

• to serve as an accessible introduction to the field to postgraduate students and nonspecialist researchers from related areas;

• to be a source of advanced teaching material for specialized seminars, courses and schools.

Both monographs and multi-author volumes will be considered for publication. Edited volumes should, however, consist of a very limited number of contributions only. Proceedings will not be considered for LNP.

Volumes published in LNP are disseminated both in print and in electronic formats, the electronic archive is available at springerlink.com. The series content is indexed, abstracted and referenced by many abstracting and information services, bibliographic networks, subscription agencies, library networks, and consortia.

Proposals should be sent to a member of the Editorial Board, or directly to the managing editor at Springer:

Dr. Christian Caron
Springer Heidelberg
Physics Editorial Department I
Tiergartenstrasse 17
69121 Heidelberg/Germany
christian.caron@springer-sbm.com

James W. LaBelle
Rudolf A. Treumann
(Eds.)

Geospace Electromagnetic Waves and Radiation

 Springer

Editors

Professor James W. LaBelle
Department of Physics
Wilder Laboratory 6127
Dartmouth College
Hanover, NH 03755
USA
E-mail:
jlabelle@aristotle.dartmouth.edu

Professor Rudolf A. Treumann
Universität München
Sektion Geophysik
Theresienstraße 41
80333 München
Germany
E-mail:
art@mpe.mpg.de

J.W. LaBelle and R.A. Treumann, *Geospace Electromagnetic Waves and Radiation*, Lect. Notes Phys. 687 (Springer, Berlin Heidelberg 2006), DOI 10.1007/b11580119

ISSN 0075-8450

ISBN 978-3-642-06760-0   e-ISBN 978-3-540-33203-9

Springer is a part of Springer Science+Business Media
springer.com
© Springer-Verlag Berlin Heidelberg 2006
Softcover reprint of the hardcover 1st edition 2006

Preface

The "Ringberg Workshop on High Frequency Waves in Geospace" convened at Ringberg Castle, Bavaria, from July 11 to 14, 2004. Approximately 30 attendees from 11 countries gathered at the castle for a program of invited talks and posters focussed on outstanding problems in high-frequency waves, defined broadly as waves exceeding a few kHz in frequency. Thirteen invited presentations comprise the contents of this volume. These articles provide introductions to current problems in geospace electromagnetic radiation, guides to the associated literature, and tutorial reviews of the relevant space physics. As such, this volume should be of value to students and researchers in electromagnetic wave propagation in the environment of the Earth at altitudes above the neutral atmosphere, extending from the ionosphere into outer space.

The contributions are broadly grouped into three parts. Part I, entitled "High-Frequency Radiation" focusses on radiation processes in near-Earth plasmas. Benson et al. present a tutorial review of Z-mode emissions, which so far have received relatively little attention and are the subject of few such reviews despite their abundant presence in geospace. Hashimoto et al. continue with another tutorial review on the terrestrial continuum radiation, the relatively weak radio emissions that fill the entire outer magnetosphere and provide information about the magnetospheric plasma boundaries and the state of the magnetospheric plasma density. Louarn reviews the ideas relevant to the generation of Auroral Kilometric Radiation (AKR), by far the most powerful and significant of the high-frequency radiations in the magnetosphere. Fleishman introduces the topic of diffusive synchrotron radiation, a mechanism not widely appreciated by geophysicists, but which may play a role in several magnetospheric, heliospheric, and even astrophysical settings. Pottelette and Treumann end this chapter with a discussion of the latest ideas about the relationship between auroral acceleration processes and radiation processes such as AKR, a subject which has been transformed in the last decade due to observations with the FAST and CLUSTER satellites.

Part II of this monograph, entitled "High-frequency waves," focusses on wave physics. Sonwalkar presents a lengthy and comprehensive review of whistler-mode propagation in the presence of density irregularities. James' paper deals with recent results from the OEDIPUS-C sounding rocket, combined with recent innovations in antenna theory, which lead to the provocative but significant conclusion that field strengths measured by many previous observations of auroral hiss using dipole antennas may need to be revised downward. Lee et al. present a novel theoretical method for analyzing mode-coupling and mode-conversion of high-frequency waves, with applications to geophysical plasmas. Yoon et al. treat the subject of mode-conversion radiations, which are replete in the Earth's environment, both in the ionosphere, magnetosphere, and solar wind. Vaivads et al. conclude this part with a review of high-frequency waves related to magnetic reconnection as the generator region of high-frequency waves and radiation in geospace, a very important and hot topic, especially in the light of recent CLUSTER satellite observations.

Part III of the monograph is devoted to new analysis techniques and instrumentation transforming research on high-frequency waves. Pécseli and Trulsen discuss novel ideas on the forefront of linking wave observations to theoretical models. Santolík and Parrot apply sophisticated wave propagation analysis tools to the study of AKR. Finally, Kletzing and Muschietti discuss wave particle correlators, describing the physics that can be investigated with them and including results from a recent state-of-the-art wave-particle correlator flown in the Earth's auroral ionosphere.

This monograph would not have been possible without the assistance of the many referees. Special thanks are due to M. André, R.E. Ergun, J.R. Johnson, E.V. Mishin, R. Pottelette, O. Santolík, V.S. Sonwalkar, and A.T. Weatherwax. We thank Dr. Axel Hörmann and his team for creating the gracious, welcoming environment at Ringberg Castle, which allowed a creative workshop to take place and thereby inspired this volume. We also thank the International Space Science Institute Bern for support. Finally, the editors at Springer, especially Dr. Christian Caron, deserve thanks for supporting the timely publication of this work and helping to assure its high quality.

Hanover, New Hampshire, and Munich
June 2005

James LaBelle
Rudolf Treumann

Contents

Part II   High-Frequency Waves

8 Mode Conversion Radiation
in the Terrestrial Ionosphere and Magnetosphere

9 Theoretical Studies of Plasma Wave Coupling:
A New Approach

10 Plasma Waves Near Reconnection Sites

Part III   High-Frequency Analysis Techniques and Wave Instrumentation

11 Tests of Time Evolutions in Deterministic Models, by Random Sampling of Space Plasma Phenomena

H.L. Pécseli, J. Trulsen

12 Propagation Analysis of Electromagnetic Waves: Application to Auroral Kilometric Radiation

O. Santolík, M. Parrot

List of Contributors

Allan T. Weatherwax
Siena College, Dept. Physics
515 Loudon Rd.
Loudonville, NY 12211-1462, USA
aweatherwax@siena.edu

Andris Vaivads
Swedish Institute of Space Physics
Box 537, Uppsala, SE 751 21, Sweden
andris@irfu.se

Bodo W. Reinisch
University of Massachusetts
Lowell, MA 01854-0000, USA
bodo.reinisch@uml.edu

Craig Kletzing
University of Iowa/Phys. &
Astronomy
Iowa City, IA 52242-0000, USA
craig-kletzing@uiowa.edu

Donald L. Carpenter
Stanford University
Star Lab. Electr. Engng. Dept.
Stanford, CA 94305-9515
dlc@nova.stanford.edu

D.-H. Lee
Kyung Hee University
Dept. Astron. Space Sci.
Yongin, Kyunggi 449-701, Korea
dhlee@khu.ac.kr

E.-H. Kim
Kyung Hee University
Dept. Astron. Space Sci.
Yongin, Kyunggi 449-701, Korea
ehkim@khu.ac.kr

Gregory D. Fleishman
26 Polytekhnicheskaya
St. Petersburg 194021
Russian Federation
gregory@sun.ioffe.rssi.ru

Hans L. Pecseli
University of Oslo/Inst. Physics
Box 1048, Blindern, N-0316
Oslo, Norway
hans.pecseli@fys.uio.no

Hiroshi Matsumoto
Res. Inst. Sustainable Humanosphere
Kyoto University
Uji, Kyoto 611-0011, Japan
matsumot@rish.kyoto-u.ac.jp

H. Gordon James
Communications Research Center
Canada
Ottawa, Ontario KH2 882, Canada
gordon.james@crc.ca

James LaBelle
Dartmouth College
Dept. Phys. Astron.
Wilder Laboratory
Hanover, New Hampshire 03755
USA
jlabelle@aristotle.dartmouth.edu

James L. Green
NASA/Goddard Space Flight Center
MC 630, Bldg. 26,
Greenbelt, MD 20771-1000, USA
James.L.Green@nasa.gov

Jan Trulsen
University of Oslo
Institute of Theoretical Astrophysics
Box 1029 Blindern
N-0315 Oslo, Norway
jan.trulsen@astro.uio.no

Kozo Hashimoto
Res. Inst. Sustainable Humanosphere
Kyoto University
Uji, Kyoto 611-0011, Japan
kozo@rish.kyoto-u.ac.jp

K. Kim
Ajou University
Dept. Molecular Sci. Techn.
Suwon, Kyunggi 443-749, Korea
kkim@au.ac.kr

K.-S. Kim
Kyung Hee University
Dept. Astron. Space Sci.
Yongin, Kyunggi 449-701, Korea
kskim@khu.ac.kr

Laurent Muschietti
University of California
Space Sciences Laboratory
Berkeley, CA 94720-7450, USA
laurent.muschietti@ssl.ucb.edu

Marilia Samara
Dartmouth College
Dept. Phys. Astron., Wilder
Laboratory
Hanover, NH 03755, USA
marilia@aristotle.dartmouth.edu

Mats André
Swedish Institute of Space Physics
Box 537, Uppsala, SE 751 21
mats.andre@irfu.se

Michel Parrot
LPCE/CNRS
3a av. de Recherche Science
45071 Orleans, Cedex 02, France
mparrot@cnrs-orleans.fr

Ondřej Santolík
Charles Univ. Prague
Faculty of Math. & Phys.
Prague 8, CZ-18000
Czech Republic
ondrej.santolik@mff.cuni.cz

Peter H. Yoon
University of Maryland
Inst. Physical Sci. & Techn.
College Park, MD 20742, USA
yoonp@ipst.umd.edu

Philippe Louarn
CNRS/CESR
9 av. Colonel Roche
Toulouse, 31329 France
philippe.louarn@cesr.fr

Phillip A. Webb
NASA/GSFC
Code 612.3
Greenbelt, MD 20771-0000, USA
pwebb@pop600.gsfc.nasa.gov

Raymond Pottelette
CNRS/CETP
4 av. de Neptune
St. Maur des Fossés Cedex, France
raymond.pottelette@ipsl.cetp.fr

Robert F. Benson
NASA/Goddard Space Flight Center
MC 692, Bldg. 21, Room 252
Greenbelt, MD 20771-0001, USA
u2rfb@lepvax.gsfc.nasa.gov

Roger R. Anderson
University of Iowa
Dept. Phys. & Astron.
Iowa City, IA 52242-1479, USA
rra@space.phsics.uiowa.edu

Rudolf A. Treumann
Ludwig-Maximilians Universität
München, Sektion Geophysik
Theresienstr. 41, 80333 Munich
Germany
art@mpe.mpg.de

Vikas S. Sonwalkar
Univ. Alaska Fairbanks
Dept. Electr. Engng.
306 Tanana Dr. Room 229 Duckering
Fairbanks, AK 99775-0000, USA
ffvss@uaf.edu

Yuri V. Khotyaintsev
Swedish Inst. Space Physics
Box 537, Uppsala 75121, Sweden
yuri@irfu.se

Part I

High-Frequency Radiation

1

Active Wave Experiments in Space Plasmas: The Z Mode

R.F. Benson[1], P.A. Webb[1], J.L. Green[1], D.L. Carpenter[2], V.S. Sonwalkar[3], H.G. James[4], and B.W. Reinisch[5]

[1] NASA/Goddard Space Flight Center, Greenbelt, MD USA
   Robert.F.Benson@nasa.gov
[2] Stanford University, California USA
[3] University of Alaska Fairbanks, Alaska USA
[4] Communications Research Centre, Toronto, Canada
[5] University of Massachusetts Lowell, Massachusetts USA

Abstract. The term Z mode is space physics notation for the low-frequency branch of the extraordinary (X) mode. It is an internal, or trapped, mode of the plasma confined in frequency between the cutoff frequency f_z and the upper-hybrid frequency f_{uh} which is related to the electron plasma frequency f_{pe} and the electron cyclotron frequency f_{ce} by the expression $f_{uh}^2 = f_{pe}^2 + f_{ce}^2$; f_z is a function of f_{pe} and f_{ce}. These characteristic frequencies are directly related to the electron number density N_e and the magnetic field strength $|\mathbf{B}|$, i.e., $f_{pe}(\mathrm{kHz})^2 \approx 80.6 N_e (\mathrm{cm}^{-3})$ and $f_{ce}(\mathrm{kHz})^2 \approx 0.028 |\mathbf{B}|(\mathrm{nT})$. The Z mode is further classified as slow or fast depending on whether the phase velocity is lower or higher than the speed of light in vacuum. The Z mode provides a link between the short wavelength λ (large wave number $k = 2\pi/\lambda$) electrostatic (es) domain and the long λ (small k) electromagnetic (em) domain. An understanding of the generation, propagation and reception of Z-mode waves in space plasma leads to fundamental information on wave/particle interactions, N_e, and field-aligned N_e irregularities (FAI) in both active and passive wave experiments. Here we review Z-mode observations and their interpretations from both radio sounders on rockets and satellites and from plasma-wave receivers on satellites. The emphasis will be on the scattering and ducting of sounder-generated Z-mode waves by FAI and on the passive reception of Z-mode waves generated by natural processes such as Cherenkov and cyclotron emission. The diagnostic applications of the observations to understanding ionospheric and magnetospheric plasma processes and structures benefit from the complementary nature of passive and active plasma-wave experiments.

Key words: Auroral kilometric radiation, Z-mode, free space radiation, wave transformation, radiation escape, cavity modes, active experiments

R.F. Benson et al.: *Active Wave Experiments in Space Plasmas: The Z Mode*, Lect. Notes Phys.
687, 3–35 (2006)
www.springerlink.com

1.1 Introduction

According to cold plasma theory, at high frequencies there are two characteristic electromagnetic (em) waves, or modes, that can propagate in a magnetoplasma. They are often referred to as the free-space ordinary (O) and extraordinary (X) modes because waves propagating in these modes can smoothly connect to free space. The X mode has two branches. In addition to the free-space mode, it has a mode called the slow branch. This name is used because it is restricted to propagation velocities less than the vacuum speed of light c. Since this mode only exists within a plasma, there was considerable interest in explaining observations indicating that it was responsible for a unique signature on early ground-based radars designed to probe the ionosphere. In their most common application these radars, called ionosondes, operate by transmitting a radio pulse of short time duration at a particular frequency and receiving, at the same frequency, for a time interval sufficient to receive an echo from the ionosphere overhead. This process is repeated over a range of frequencies likely to produce reflections. The resulting record is called an ionogram.

Normally, there are two ionospheric reflections, one due to the O mode and one due to the X mode. "On rare occasions", as first reported by Eckersley [23], there is a third reflection with the same polarization as the O mode. This third reflection trace, corresponding to the slow X-mode branch, was dubbed the Z mode in ionospheric research; a designation commonly used in space physics. In order to explain the presence of the Z mode at ground level, i.e., far below the ionospheric plasma, and the polarization (same as the O mode), a Z-O mode coupling process involving obliquely-propagating O-mode waves was introduced by Ellis [24] as discussed in Sect. 13.5 of Ratcliffe [57]. An ionogram showing this triple splitting of the ionospheric reflection is schematically illustrated in Fig. 1.1. Here the apparent height h' (or apparent, or virtual, range) corresponds to $ct/2$, where t is the round trip echo delay time, and the frequency f is the sounding frequency. For a description of the sounding technique, and the inversion from $h'(f)$ to $N_e(h)$, where N_e is the electron number density and h is the true altitude, see Thomas [64] and Reinisch [58] and references therein.

In order to understand how the Z mode is related to the free-space O and X modes it is necessary to discuss plasma-wave dispersion. This topic will be addressed in Sect. 1.2. Since the Z mode is an internal (or trapped) mode of the plasma, the emphasis in this paper will be on the reception of the Z mode by space-borne receivers during active and passive experiments. Sections 1.3 and 1.4 will deal with sounder-stimulated Z-mode waves in the ionosphere and the magnetosphere, respectively. Particular attention will be given to the information that the sounder-stimulated Z-mode waves provide concerning magnetic-field aligned N_e irregularities (FAI). FAI are irregularities in N_e transverse to the direction of the background magnetic field **B** that are maintained for long distances along **B**. They efficiently scatter and

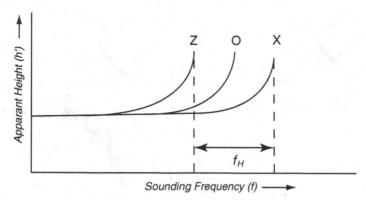

Fig. 1.1. Ground-based ionogram schematic illustrating Z-, O-, and X-mode reflection traces. Here the ionospheric notation for the electron cyclotron frequency, f_H, is used [adapted from 57]

duct sounder-stimulated Z-mode waves. Section 1.5 discusses a combined active/passive investigation of Z-mode waves generated by natural processes. A summary is presented in Sect. 1.6.

There have been many spacecraft that have generated Z-mode waves in the ionosphere and magnetosphere using radio sounders. Similarly, there have been many satellites that have detected Z-mode waves of magnetospheric origin using plasma wave receivers [LaBelle and Treumann, 43, included a review of auroral Z-mode observations and theory]. Our goal is not to review the Z-mode observations from all of these missions. Rather, it is to select specific examples that illustrate the range of Z-mode phenomena observed in active space wave-injection experiments and to demonstrate their diagnostic capability. In the case of the ionosphere, we will mainly use data from two missions, namely, (1) the ISIS (International Satellites for Ionospheric Studies) satellites [Jackson and Warren, 33] and (2) the OEDIPUS sounding rocket double payloads (Observations by Electric-field Determinations in the Ionospheric Plasma-A Unique Strategy) [see, e.g., 30, 36]. In the case of the magnetosphere, data from the Radio Plasma Imager (RPI) [Reinisch et al., 59] on the IMAGE (Imager for Magnetopause-to-Aurora Global Exploration) satellite [Burch, 15] will be used.

1.2 Plasma Wave Dispersion

Waves in a cold plasma are described by a dispersion relation, i.e., the scalar relation expressing the angular frequency $\omega = 2\pi f$ in terms of the propagation vector $\mathbf{k}$, which is related to the refractive index $\mathbf{n}$ by $\mathbf{n} = \mathbf{k}c/\omega$ where $k = |\mathbf{k}| = (2\pi/\lambda)$ and λ is the wavelength. This description has been given in a number of books and review papers [see, e.g., 1, 19, 27, 39, 57, 63]. Figure 1.2 presents dispersion curves for waves propagating in a homogeneous cold

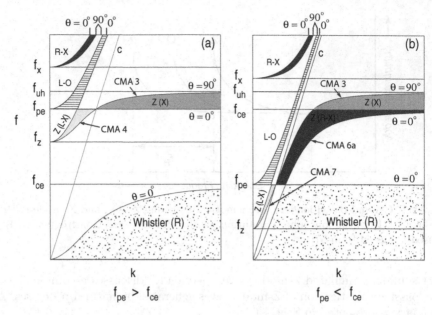

Fig. 1.2. Schematic dispersion diagrams. (a) Example when $f_{pe}/f_{ce} > 1$. (b) Example when $f_{pe}/f_{ce} < 1$ [adapted from 27, 62] (Reprinted with permission of the American Geophysical Union)

plasma, where ion motions are neglected, with $\mathbf{k}$ making an angle θ relative to $\mathbf{B}$. The figure shows the dispersion curves for $\theta = 0$ and $\theta = \pi/2$ cases for a range of frequency and wave number. The region between these two limiting cases is shown by various shades of gray – indicating various modes – where the propagation at oblique wave normal angles is permitted. The waves are labelled based on their polarization for parallel or perpendicular propagation, i.e., R or L for right- or left-hand polarization (with respect to the direction of $\mathbf{B}$) when $\theta = 0$, and X or O for extraordinary and ordinary mode polarization when $\theta = \pi/2$. In some regions, only one letter is used indicating that propagation is not possible for both $\theta = 0$ and $\theta = \pi/2$. Thus Z(X) indicates that the Z mode does not include the condition $\theta = 0$ in the region indicated based on the cold-plasma approximation. The Z-mode regions in Fig. 1.2 are also labelled with the CMA designation using the notation of Stix [63]. Thus Z(X) occurs in CMA region 3 where k, and thus $n = |\mathbf{n}|$, can become large leading to a condition ($n = \infty$) known as *resonance*; the condition $k = 0$ (or $n = 0$) is known as a *cutoff*.

The plasma resonances and cutoffs in Fig. 1.2 are given by the following expressions:

$$f_{ce}(\text{kHz}) = \frac{|e|}{2\pi m_e)}|\mathbf{B}| \approx 0.028\,|\mathbf{B}(\text{nT})| \tag{1.1}$$

$$f_{pe}(\text{kHz}) = \frac{e^2}{(4\pi^2\epsilon_0 m_e)}^{\frac{1}{2}} N_e^{\frac{1}{2}} \approx 80.6 \, N_e(\text{cm}^{-3})^{\frac{1}{2}} \tag{1.2}$$

$$f_{uh} = (f_{pe}^2 + f_{ce}^2)^{\frac{1}{2}} \tag{1.3}$$

$$f_x = \frac{f_{ce}}{2}\left[1 + \left(1 + 4\frac{f_{pe}^2}{f_{ce}^2}\right)^{\frac{1}{2}}\right] \tag{1.4}$$

$$f_z = \frac{f_{ce}}{2}\left[-1 + \left(1 + 4\frac{f_{pe}^2}{f_{ce}^2}\right)^{\frac{1}{2}}\right] \equiv f_x - f_{ce} \tag{1.5}$$

where e is the electron charge, m_e is the electron mass and ϵ_0 is the permittivity of free space. For $\theta < \pi/2$, the resonance condition that replaces (3) above, in CMA 3 of Fig. 1.2, is known as the Z-infinity and is given by

$$f_{ZI} = \frac{1}{\sqrt{2}}\left[\left(f_{uh}^2 + (f_{uh}^4 - 4f_{ce}^2 f_{pe}^2 \cos^2\theta)^{\frac{1}{2}}\right)\right]^{\frac{1}{2}} \tag{1.6}$$

The Z infinity is also referred to as the upper oblique resonance [see, e.g., Beghin et al., 4]. The above cutoffs and resonances are described using different notations in Sects. 6.4 and 6.5 of Ratcliffe [57] and Sects. 1–5 of Stix [63].

Figure 1.2 is often presented in the form of ω vs. k. In this presentation the magnitudes of the phase and group velocities,

$$|\mathbf{v}_p| = \left|\frac{\omega}{\mathbf{k}}\right| \tag{1.7}$$

and

$$|\mathbf{v}_g| = \left|\frac{\partial\omega}{\partial\mathbf{k}}\right|, \tag{1.8}$$

respectively, correspond to the slope of the line from the origin to a particular point on a dispersion curve, and to the slope of a line tangent to the dispersion curve at that point, respectively. In Fig. 1.2 there is a slanting line labelled c to indicate that it corresponds to free-space propagation. The curves to the left of this line (labelled R-X, L-O and L-X) have $v_p \geq c$ and those to the right have $v_p < c$. Accordingly, the Z-mode waves labelled L-X are called fast Z (CMA 4 in Fig. 1.2a and CMA 7 in Fig. 1.2b) and those labelled Z(X) are called slow Z (CMA 3 in Figs. 1.2a and 2b). Both fast and slow Z-mode waves can be found in the region labelled R-X (CMA 6a in Fig. 1.2b).

The ionosphere and magnetosphere, contrary to the conditions appropriate to Fig. 1.2, are neither homogeneous nor cold and the ions are not motionless. Yet the dispersion properties derived by using these assumptions, and illustrated by Fig. 1.2, have proved very successful in describing many phenomena. The standard approach is to consider the ionosphere as a horizontally-stratified medium with N_e varying only in the vertical direction. Then the wave is considered to behave as if it were in a homogeneous medium at each ionospheric level. Both diagrams in Fig. 1.2 correspond to a specific value of f_{pe}/f_{ce}. The curves change shape as f_{pe}/f_{ce} changes. For example, in Fig. 1.2a, the band of no propagation between f_{ce} and f_Z only appears when $f_{pe}/f_{ce} > \sqrt{2}$. Also, the Z(X) region, i.e., CMA-region 3 in Figs. 1.2a and 1.2b, is maximum for the condition $f_{pe} = f_{ce}$.

The progress of a radio wave through the non-homogeneous ionosphere can be modelled by considering the change in the shape of the refractive-index surface as f_{pe}/f_{ce} changes. This process is illustrated in Fig. 1.3 to illustrate how Z-mode signals at a frequency f from a high-latitude source of natural origin could propagate over great distances in the horizontal direction. The conditions correspond to a source location where $f \approx f_{pe} < f_{ce}$. The left side of the diagram shows the evolution of the refractive-index surfaces from low to high altitudes corresponding to plasma conditions changing from CMA regions 7 to 6a to 3 in Fig. 1.2b. Gurnett et al. [28] used a construction technique introduced by Poeverlein [54], based on Snell's law, to argue that a wave at frequency f originating in the region where $f < f_{ce}$ will be refracted at the $f = f_{ce}$ level and will be able to propagate long distances in the horizontal direction. In Fig. 1.3, the arrows originating at the intersections of the vertical dashed line with these refractive index surfaces indicate the direction of $\mathbf{v}_g$ in (benson-eq8). Note the change from a closed refractive index surface to an open surface as CMA region 3 is encountered.

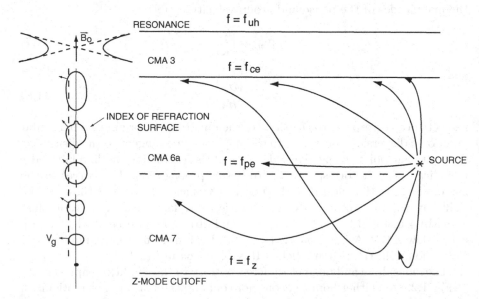

Fig. 1.3. Illustration of the large change in the shape of the refractive-index surface as CMA region 3 is encountered and the ability for long-range horizontal propagation in the Z mode in the polar regions where **B** is nearly vertical [adapted from 28] (Reprinted with permission of the American Geophysical Union)

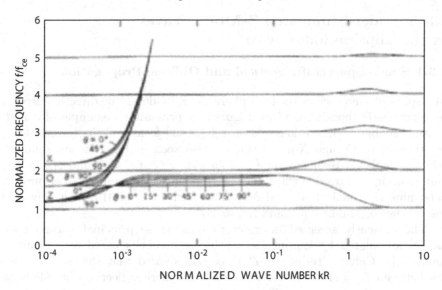

Fig. 1.4. Normalized calculated dispersion diagram for a wide range of wavelengths with electron thermal motions included where f is normalized by f_{ce} and k is normalized by $1/R$ where R is the electron cyclotron radius for the case $f_{pe}/f_{ce} = 1.6$ [adapted from Oya, 53] (Reprinted with permission of the American Geophysical Union)

Figure 1.4 shows the result of numerical solutions of the dispersion relations, for a particular case of $f_{pe}/f_{ce} > 1$, for all θ, and for θ ranging from 0 to $\pi/2$, when hot-plasma effects are included. In these solutions, electron thermal motions are included and a Maxwellian electron velocity distribution function is assumed but collisions are neglected and the ions are still considered to provide an immobile neutralizing background [Oya, 53]. The right portion of the diagram corresponds to the electrostatic (*es*) domain and the left portion corresponds to the electromagnetic (*em*) domain. New wave modes, known as the Bernstein modes [Bernstein, 13], now appear between the nf_{ce} harmonics in the *es-domain*. The Bernstein modes correspond to undamped modes with $\theta = \pi/2$. Damping rapidly increases for these modes as θ departs from $\pi/2$. These *es-modes* are coupled to the *em-domain*, with negligible damping, through the Z-mode when $\theta \approx \pi/2$, corresponding to frequencies close to f_{uh}, each, in turn, as f_{uh} increases for increasing f_{pe}/f_{ce} values. Thus the Z-mode near $\theta = \pi/2$, i.e., near f_{uh}, is of prime importance in the coupling of energy resulting from wave/particle interactions in the *es-domain* into the *em-domain* where the information can be transmitted out of the plasma.

1.3 Sounder-Stimulated Z-Mode Waves in the Topside Ionosphere

1.3.1 Single Spacecraft: Vertical and Oblique Propagation

In a space plasma such as the ionosphere the Z mode can be directly detected by ionospheric topside sounders. Figure 1.5a presents an example of a mid-latitude ionogram where ionospheric reflections form a clear Z mode trace in addition to O- and X-mode traces. Also seen in Fig. 1.5a are sounder-stimulated plasma resonances at $f_{ce} = f_H, f_{pe} = f_N, f_{uh} = f_T$ and $2f_{ce} = 2f_H$ and an oblique Z trace labelled Z'. The resonances can be used to accurately determine the ambient $|\mathbf{B}|$ and N_e values from (1.1)–(1.3) as, e.g., given in the reviews of Muldrew [50] and Benson [7].

The presently accepted interpretation for these principal plasma resonances stimulated by ionospheric topside sounders is based on the investigation by Calvert [16] of the Z' trace. He showed that this trace, which lies between f_{pe} and f_{uh}, is the result of ionospheric reflections of obliquely-propagating Z-mode waves. These oblique reflections result from the shape of the refractive index surfaces, in particular from the change in the shape to form a sphere plus a line parallel to $\mathbf{B}$ when the downward-propagating Z-mode wave encounters the level where $f = f_{pe}$ (see the left side of Fig. 1.3). The Z' trace is caused by waves reflected at this level. This condition was called a "spitze" by Poeverlein [54] and is discussed in some detail in Budden [14]. The Z' trace could be explained by using ray tracing with the cold plasma theory.

Calvert [16] did not restrict his calculations to the cold plasma approximation, however, and found ray paths that could return to the spacecraft that included propagation beyond the resonance-cone angle limit of cold plasma theory. These echoes were electrostatic in nature and had echo times much greater than the observed Z' echoes so the solutions were discarded. McAfee [46], in his investigation of the plasma resonance observed at f_{pe}, found that these hot-plasma solutions had echo delay times comparable to the delays observed on topside ionograms when frequencies very close to f_{pe} were considered. This oblique-echo model was later extended to the plasma resonances observed at f_{uh} [47] and at $n f_{ce}$ [50]. Thus, the investigation of oblique Z-mode propagation by Calvert [16] provided the fundamental first step toward our understanding of sounder-stimulated plasma resonances.

Even though the Z-mode is defined as the slow branch of the X mode, it is important to note that the Z and X modes have comparable group velocities near their respective cutoffs. This behavior is clearly illustrated in Fig. 1.5b. Though Z-mode waves cannot travel as far from the spacecraft as the free-space O- and X-mode waves, they are useful for determining the vertical N_e distribution out to a few hundred km from the spacecraft as seen in Fig. 1.5c. The good agreement between the N_e values obtained in Fig. 1.5c by inverting the Z-, O- and X-mode ionospheric-reflection traces of Fig. 1.5a provides

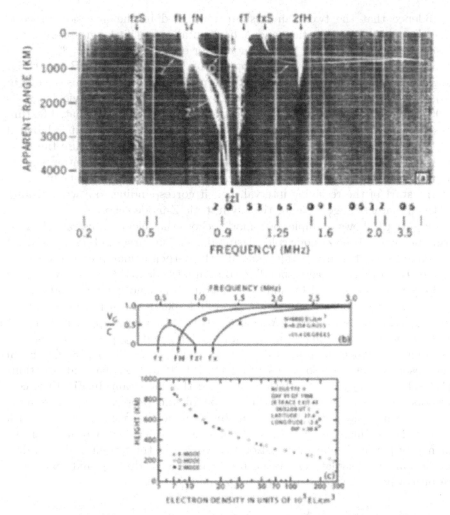

Fig. 1.5. (a) Alouette-2 mid-latitude ionogram, in negative format with signal reception in white on a black background, showing Z, O and X traces. Here the ionospheric notation of H, N, and T is used to represent the subscripts $ce, pe,$ and uh, respectively, in the present work. Also, "S" is used to designate the cutoff frequencies at the satellite. The f_{zI} label and arrow below the ionogram, and near the 0.9 MHz frequency marker, identifies the Z infinity condition given by (6). (b) Corresponding group velocities ($N_e = 6800\,\mathrm{cm}^{-3}$ corresponds to $f_{pe} = 0.74\,\mathrm{MHz}$ from (2), $B = 0.254 = 0.254 \times 10^{-4}\,\mathrm{T} = 0.254 \times 10^5\,\mathrm{nT}$ corresponds to $f_{ce} = 0.7112\,\mathrm{MHz}$ from (1) and $f_{uh} = 1.026\,\mathrm{MHz}$ from (3)). (c) Calculated N_e values from each of the traces [adapted from Jackson, 32]

confidence that the two main assumptions used in the inversion process, namely, vertical propagation and a horizontally-stratified ionosphere, were justified since this process is independent for each trace [Jackson, 32].

For vertical propagation the propagation angle relative to **B** is 90°-dip angle, or $\phi = 51.4°$ using Jackson's notation of Fig. 1.5b. Using this ϕ value in (6), with the other values from Fig. 1.5b where $f_{pe}/f_{ce} = 1.04$, yields $f_{ZI} = 0.968$ MHz in agreement with the observed narrow vertical modulated feature observed from zero to approximately 3,000 km apparent range; it is labelled f_{zI} just below the ionogram. This feature can be observed because the wide receiver bandwidth (3 dB bandwidth of 37 kHz) allows long-duration signals returning from the previous pulse (differing by only a few kHz) to be observed at the start of the receiving interval, i.e., it corresponding to a wrap-around of the apparent-range scale for the asymptotic Z-mode echo.

Figure 1.6 shows examples of Z-mode echoes that clearly illustrate the limiting behavior of the Z-mode cold-plasma dispersion curves in Figs. 1.2 and 1.4 for conditions of nearly parallel and nearly perpendicular propagation relative to **B**. In Fig. 1.6a, corresponding to high latitude and thus nearly parallel propagation, the Z- and O-mode traces touch one another near 700 km apparent range and 0.9 MHz suggesting coupling like in the dispersion diagrams in Figs. 1.2 and 1.4 for $\theta = 0$. Also, the Z trace has a large apparent range at this frequency, coinciding with the combined f_{pe} and f_{ce} plasma resonances, as would be expected from (1.6) with $\theta = 0$, i.e., $f_{ZI} = f_{pe} = f_{ce}$. From the observed plasma resonances and wave cutoffs in Fig. 1.6a, and equations (1.3)–(1.5), $f_{pe}/f_{ce} = 0.92/0.935 = 0.98$ for this ionogram. In Fig. 1.6b, corresponding to low latitude and thus perpendicular propagation, the Z-mode trace becomes asymptotic to f_{uh}, again, as expected from (1.6) (now with $\theta = \pi/2$). In this case, $f_{pe}/f_{ce} = 1.89/0.565 = 3.35$. When nearly parallel or nearly perpendicular propagation is involved in the presence of FAI, dramatic ionogram signatures can be produced due to ducting and scattering, respectively, of the sounder-generated Z-mode signals.

1.3.2 Single Spacecraft: Ducted Propagation

When ionospheric topside sounders encounter equatorial plasma bubbles, dramatic floating X-mode echoes are observed that resemble epsilons [Dyson and Benson, 22]. They are called floating because they are not tied to the zero-time baseline at the top of the figure. These traces are the result of sounder-generated signals that echo in both the local and conjugate hemispheres (relative to the location of the satellite) due to ducted propagation in FAI that are maintained from one hemisphere to the other. The bottom portion of Fig. 1.7 illustrates the top segment of such an X-mode epsilon in the frequency range above about 2.9 MHz and at apparent ranges beyond 2200 km. The X-mode echo just above this echo signature, i.e., corresponding to virtual ranges less than 2200 km in the bottom portion of Fig. 1.7, is due to ducted propagation in the local hemisphere. In the top portion of Fig. 1.7 the distances to the

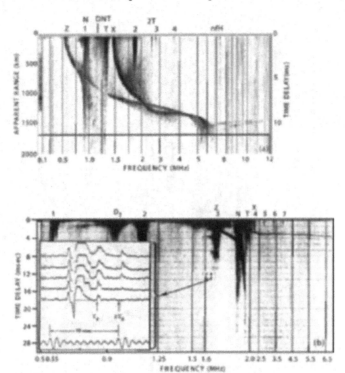

Fig. 1.6. (a) ISIS-2 high-latitude ionogram showing Z, O and X traces under conditions of nearly parallel propagation (Resolute Bay digital ionogram obtained from http://nssdc.gsfc.nasa.gov/space/isis/isis-status.html corresponding to 1500:28 UT on day 126 of 1973; 62.8° latitude, −101.8° longitude, 1394 km altitude, 83° dip). (b) Low-latitude Alouette-2 ionogram showing Z, O and X traces under conditions of nearly perpendicular propagation [adapted from Benson, 6]. As in Fig. 1.5, the ionospheric notation of N, and T is used to represent the subscripts pe, and uh, respectively, and numerals are used to identify the $nf_H = nf_{ce}$ resonances. The $3f_{ce}$ resonance contains two spurs on the low-frequency side (see Sect. 1.3.6) with delay times near 4 and 7 ms. The insert shows five selected receiver amplitude vs. time traces corresponding to these spur observations. Each trace represents a vertical scan line on the ionogram display, the amplitude modulation on the insert traces corresponding to the intensity modulation on the ionogram scan lines. The initial systematic positive and negative spikes on each of the insert traces are calibration and sync pulses [see Fig. 31 of Franklin and Mclean, 25]; the time-delay zero point was taken as the left side of the dashed line segment on the insert trace labelled (a). The nf_{ce} resonances indicated that $f_{ce} = 0.565$ MHz corresponding to $\tau_p = 3.25$ ms (Reprinted with permission of American Geophysical Union)

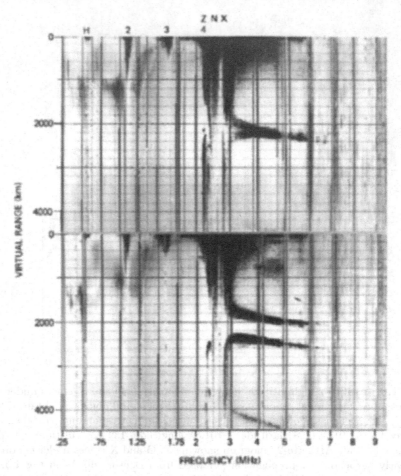

Fig. 1.7. ISIS-1 ionograms recorded 1000 km above the dip equator showing Z- and X-mode echoes from within an equatorial plasma bubble [9]. The ionospheric notation of H, and N is used to represent the subscripts ce and pe, respectively; numerals are used to identify the $nf_H = nf_{ce}$ resonances (Reprinted with permission of American Geophysical Union)

reflection levels in each hemisphere were the same, as the two traces merge and have the same virtual ranges from about 2200 to 2400 km.

Z-mode echoes from waves that are ducted along FAI can form truncated versions of these floating X-mode epsilons. In both portions of Fig. 1.7 the ducted Z-mode waves are confined to a narrow frequency range below the label Z at the top of the figure. As in the case of the ducted X-mode signals, the Z-mode traces tied to the zero virtual-range scale correspond to signals ducted within the local hemisphere and those beyond about 2000 km in virtual range correspond to signals that experience ducted propagation into both

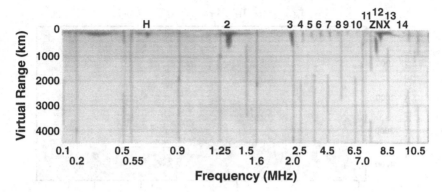

Fig. 1.8. A portion of an Alouette-2 ionogram showing Z-mode floating echoes, attributed to wave ducting in FAI, from f_Z, to a maximum frequency prior to f_{pe} labelled Z and N at the figure, respectively. The label notation is the same as in Fig. 1.7 [adapted from Benson, 9] (Reprinted with permission of the American Geophysical Union)

hemispheres. Note that the ducted Z-mode traces terminate before they reach the local f_{pe}, designated by N at the top of the figure. Similar frequency restrictions of the reception of sounder-stimulated Z-mode waves attributed to wave ducting were commonly observed in an investigation of FAI near 500 km based on Alouette-2 perigee observations [9]; an example is shown in Fig. 1.8.

In a comparison of wave ducting in different wave modes, assuming small propagation angles inside a duct produced by a small increase in refractive index, Calvert [17] showed that under conditions of high N_e, similar to the conditions of Fig. 1.8, Z-mode ducting should be stronger than X- and O-mode ducting. The strongest Z-mode ducting was found to occur in the frequency range from f_Z to midway between f_Z and f_{pe} where a transition from trough ducting to crest ducting occurs due to a curvature reversal in the refractive-index surface. Calvert [17] argues that ducting cannot be maintained across the curvature reversal in agreement with the upper-frequency truncation of the floating Z-mode signals attributed to FAI wave ducting in Fig. 1.7 and Fig. 1.8.

1.3.3 Single Spacecraft: Wave Scattering

Sounder-generated Z-mode waves that are scattered by FAI lead to strong signal returns in the frequency region between the greater of f_{ce} and f_{pe} and less than f_{uh}, i.e., in CMA region 3 in Figs. 1.2 and 1.4. Z-mode ray-tracing calculations indicate that the ray becomes horizontal in the ionosphere as the refractive index tends to infinity [Lockwood, 45] leading to a condition Muldrew [49] termed "wave trapping". The resulting signature is labelled as a noise band in Fig. 1.9. Denisenko et al. [21] used such signatures in the

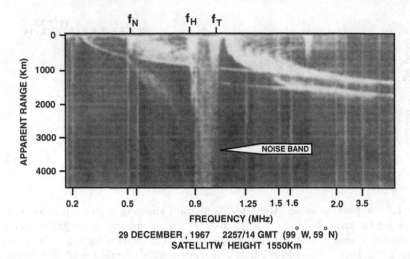

Fig. 1.9. An Alouette-2 ionogram showing sounder-generated Z-mode signal returns attributed to wave scattering in FAI (labelled "noise band") [Muldrew, 49]. The label notation is the same as in Fig. 1.5a (Copyright 1969 by IEEE; reprinted with permission)

COSMOS-1809 topside-sounder data to investigate the global distribution of small-scale ($\sim$10–100 m) FAI in the 940 to 980 km altitude range.

1.3.4 Dual Payloads: Slow Z

The above discussion pertained to observations where the receiver and transmitter were on the same spacecraft and they shared a common antenna. Thus the received Z-mode signals correspond to echo returns either from short distances, due to scattering from FAI in the vicinity of the satellite (Fig. 1.9), or from long distances ($\sim$100's km) due to vertical propagation (Z trace in Fig. 1.5a), oblique propagation (Z′ trace in Fig. 1.5a), or to nearly parallel propagation that is ducted along FAI (Figs. 7 and 8). The Z mode has also been investigated during space experiments involving wave propagation between receivers and transmitters on different payloads.

James [34, 35] performed such experiments during a high-latitude rendezvous between ISIS 1 and ISIS 2. The operating frequency was in the range $f_{pe} < f_{ce} < f < f_{uh}$, i.e., corresponding to "slow Z-mode" propagation in CMA-region 3 of Fig. 1.2b. These waves were observed to propagate over distances of several hundred km. It was found [James, 35] that the observed transmission-reception signal delay times could be explained by ray optics but that the observed distortion of the received pulse relative to the transmitted pulse indicated the importance of signal scattering by FAI.

The OEDIPUS-A dual-payload rocket was launched from the Andøya Rocket range in Norway on January 30, 1989. It was dedicated to such two-point measurements and provided an additional opportunity to investigate slow Z-mode propagation [James, 36]. In this case, $f_{ce} < f_{pe} < f < f_{uh}$, i.e., corresponding to slow Z-mode propagation in CMA-region 3 of Fig. 1.2a. The transmitting and receiving payloads were separated by nearly 1 km and the separation direction differed from the **B** direction by only a few degrees. The observed delay times for the Z-mode were too large to be explained by free-space propagation and the full *em* solution of the hot plasma dispersion equation, based on the work of Lewis and Keller [44] and Muldrew and Estabrooks [51], was used to investigate the problem. Using this hot-plasma approach, James [36] constructed refractive-index surfaces appropriate to the problem (see Fig. 1.10).

Note how different these hot-plasma Z-mode refractive-index surfaces are from the cold plasma surfaces illustrated on the left side of Fig. 1.3 in the CMA-region 3. James [36] found that direct ray paths connecting the two payloads had $|\mathbf{v}_g|$ values too small to explain the observed delays in the large n region assuming a smooth horizontally-stratified medium. Thus waves corresponding to such large-n dispersion solutions would arrive well after the OEDIPUS A ionogram display time limit (by approximately a factor of 10) and would not be detected. When the region of smaller n was investigated, labelled "electromagnetic and quasi-electrostatic domain" in Fig. 1.10, and the payloads were assumed to be within an N_e depletion duct (with cross-**B** dimension $\sim$ 100 m), ducted ray paths could be found that were consistent with the observations. James [36] suggested that such ducting may be common in the auroral ionosphere and that it should be considered when trying to interpret natural Z-mode emissions.

The OEDIPUS-C rocket dual-payload was launched from the Poker Flat rocket range in Alaska on November 7, 1995. Again, the sounder-transmitter was on one payload and the sounder-receiver was on the other, and the separation direction between the payloads, now separated by more than a 1 km, differed from the **B** direction by only a few degrees. In this case, James [37] investigated the slow Z-mode propagation corresponding to $f_{pe} < f_{ce} < f < f_{uh}$, i.e., to the CMA-region 3 of Fig. 1.2b. He found, using hot-plasma dispersion theory, that the calculated propagation times for rays directly connecting the observed payloads were typically more than a factor of three greater than the observed time delays between signal transmission and reception.

Thus waves corresponding to such solutions would arrive well beyond the observing time base and would not be detected. In all other cases, there were no solutions corresponding to the desired direction. The received signals could be explained, however, in terms of incoherent Cherenkov and cyclotron radiation from sounder-accelerated electrons (SAE). Particle detectors on both payloads detected SAE following sounder transmissions from the transmitting payload. James [37] could reproduce the observed signal delay times and could predict (within an order of magnitude) the observed signal intensities.

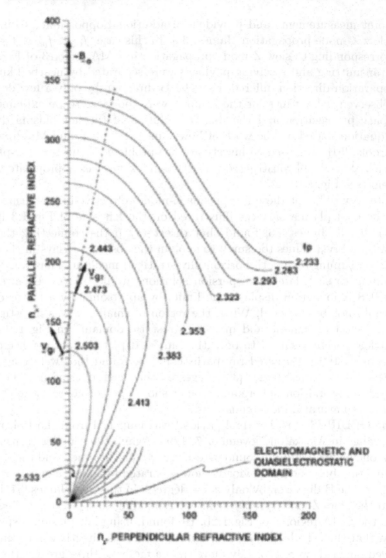

Fig. 1.10. Hot-plasma Z-mode refractive-index surfaces for $f = 2.534\,\text{MHz}$, $f_{ce} = 1.2\,\text{Hz}$, $T_e = 2000\,\text{K}$ and f_{pe}, labelled f_p in the figure, in the range $2.233 = f_{pe} = 2.533\,\text{MHz}$ [James, 36] (Reprinted with permission of the American Geophysical Union)

1.3.5 Dual Payloads: Fast Z

Horita and James [29, 30] investigated fast Z-mode propagation using OEDIPUS-C data corresponding to frequencies below f_{pe} and above the greater of f_Z or f_{ce}, i.e., to CMA-region 4 in Fig. 1.2a where there are no competing cold-plasma wave modes to complicate the interpretation. The Z mode was found to be stronger than almost all the other cold-plasma modes and to be strongest at frequencies just below f_{pe} for f_{pe}/f_{ce} near and greater than 1 but strongest just above f_Z for higher f_{pe}/f_{ce} values (between 2 and 3). Using cold-plasma dispersion theory, the Balmain [2] antenna-impedance theory and the Kuehl [40] dipole radiation theory they found that the observed and calculated signal intensities were generally in good agreement. They attribute the strength of the Z-mode signals, relative to the other free-space modes, to antenna-impedance values that permit efficient coupling between the antenna and the transmitter and receiver.

1.3.6 Possible Role of Z-Mode Waves in Sounder/Plasma Interactions

Among the plasma instability and nonlinear phenomena stimulated by ionospheric topside sounders investigated by Benson [8] was a diffuse feature observed in the frequency range above the greater of f_{ce} and f_{pe} and below f_{uh}, i.e., in the CMA 3 slow Z-mode regions of Fig. 1.2. It was designated as the DNT resonance because of its generally diffuse appearance on topside ionograms and its location between f_N and f_T (ionospheric notation for f_{pe} and f_{uh}). A weak short-duration ($< 1\,\mathrm{ms}$) example of this resonance is shown in Fig. 1.6a. It is often observed for up to about $5\,\mathrm{ms}$. It is not observed over the entire listening range, however, and thus is distinguished from the noise band in this frequency range. This noise band, attributed to wave scattering (see Sect. 1.3.3), is evident in Fig. 1.6a and, more prominently, in Fig. 1.9.

No theoretical interpretation has been offered that explains the frequency and time-duration characteristics of this resonance. It has been attributed by Pulinets [55], Pulinets et al. [56] to the scattering of sounder-generated Z-mode waves and used as a diagnostic tool for the investigation of the distribution of small-scale FAI in the topside ionosphere. James [38] attempted to explain the DNT resonance as observed by the ISIS-II sounder in terms of radiation from SAE in analogy with the successful explanation of Z-mode signals observed by the OEDIPUS-C sounder receiver as described in Sect. 1.3.4. While this explanation was not found to explain the ISIS-II observations, he concluded that SAE may still play a role because SAE pulses that persisted for milliseconds were observed when the OEDIPUS-C sounder transmitter was tuned to the DNT frequencies.

Features have been observed on topside ionograms that imply that ion motions must be considered for a proper interpretation. They appear either as prominent protrusions (called spurs) on the electron resonances (most often from the low-frequency side) or as narrow (in time delay, i.e., apparent range) emissions between the resonances. In either case, they appear with delay times that correspond to multiples of the proton gyroperiod $1/f_{cp}$. One of these phenomena, the proton spurs on the nf_{ce} resonances, appears to be strongly influenced by Z-mode transmissions [Benson, 6]. The spurs are greatly enhanced when f_Z, from (1.5), is near, but slightly less than, nf_{ce} for $n = 2, 3, 4, \ldots$ and the largest spurs are observed for large n.

Figure 1.6 illustrates the spurs observed on the $3f_{ce}$ resonance when $f_Z \approx 3f_{ce}$. It was suggested that the Z mode may be more efficient at coupling energy into the plasma under these conditions. Note that this frequency region just above f_Z corresponds to the fast Z region where Horita and James [29, 30] found the strongest Z-mode signals which they attributed to optimum antenna-impedance values (see Sect. 1.3.5). Their larger f_{pe}/f_{ce} values, which produce the strongest signals just above f_Z, would correspond to larger f_Z values which, in turn, would correspond to higher nf_{ce} resonances satisfying the condition $f_Z \approx nf_{ce}$; the largest proton spurs were observed under just such conditions [Benson, 6].

Unique Z-mode topside-ionogram signatures have been observed in low latitudes that suggest topside sounders are capable of stimulating, or enhancing, FAI when they encounter the plasma conditions $f_{pe}/f_{ce} \approx n$ where n is an integer larger than 3 [Benson, 10]. An illustration for the case of $n \approx 5$ is presented in Fig. 1.11. Note that well-defined Z, O and X traces are clearly seen for the cases $f_{pe}/f_{ce} < 5$ (top panel) and $f_{pe}/f_{ce} > 5$ (bottom panel) but that the Z trace is masked by a long-duration diffuse signal that extends from f_Z to part way to f_{pe}. It was argued that the frequent occurrence of such signatures made it unlikely that the spacecraft was just encountering FAI when the ambient conditions $f_{pe}/f_{ce} \approx n$ were satisfied. Thus the sounder-generated Z-mode waves were considered to be ducted in FAI stimulated, or enhanced, by the sounder on a short time scale ($\ll 1$ s).

The possibility that this sensitive diagnostic role of the Z-mode waves could be due to efficient scattering when $f_{pe}/f_{ce} \approx n$ was investigated by Zabotin et al. [69]. They did not find any sensitivity in the scattering of Z-mode waves by FAI near the magnetic equator to the conditions $f_{pe}/f_{ce} \approx n$ and concluded that the above examples were either due to sounder stimulation, or enhancement, as proposed or to ducting conditions sensitive to these conditions. No study of the sensitivity of Z-mode ducting by FAI to the conditions $f_{pe}/f_{ce} \approx n$ has been made. As pointed out in Sect. 1.3.2, however, Calvert [17] found that under high N_e conditions (as indicated in Fig. 1.11) Z-mode ducting in the frequency range between f_Z and midway to f_{pe} should be stronger than O- or X-mode ducting, a prediction consistent with the observations in the middle panel of Fig. 1.11. Thus the sensitivity to the $f_{pe}/f_{ce} \approx n$ condition is likely in the generation, or enhancement of existing, FAI by the sounder. Benson [10] gave other examples of sounder-stimulated plasma phenomena when $f_{pe}/f_{ce} \approx n$ and suggested that more energy is deposited into the plasma under these conditions and, particularly, when $f_{pe}/f_{ce} \geq 4$.

Osherovich et al. [52] investigated large-amplitude cylindrical electron oscillations appropriate to FAI, with initial conditions chosen so as to favor Z-mode stimulation, and found that the resulting frequency spectrum was very sensitive to the f_{pe}/f_{ce} value, with larger amplitudes, and more nonlinear frequency components, observed when $f_{pe}/f_{ce} \approx n$, and that the effect

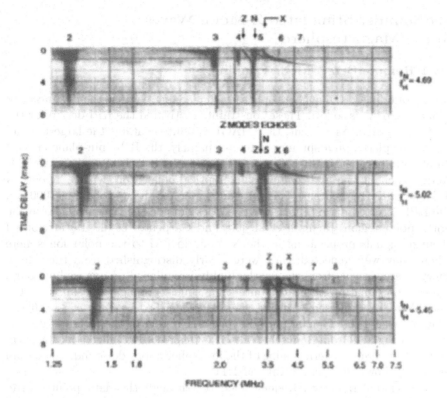

Fig. 1.11. Consecutive Alouette-2 low-latitude ionograms revealing long-duration Z-mode echoes only when $f_{pe}/f_{ce} \approx 5$ at 05:15 UT on 17 June 1969 [Benson, 10]. The label notation is the same as in Fig. 1.7 (Reprinted with permission from Elsevier)

was observed to increase with increasing n. Kuo et al. [41] investigated the creation of FAI by a Z-mode pump wave under the above resonant conditions. They added the constraint that the Z-mode pump wave frequency $f_o \approx f_{pe}$ at a short distance from the satellite where a four-wave coupling process takes place. One of the products in this coupling process corresponds to short-scale (meter size) FAI. They propose that the observed Z-mode diffuse signals, like those in the middle panel of Fig. 1.11, are caused by scattering from these FAI but stress the need for additional research to identify mechanisms that could generate (or enhance) large-scale FAI (>100 m) capable of supporting wave ducting.

1.4 Sounder-Stimulated Z-Mode Waves in the Magnetosphere

1.4.1 Remote O-Z-O-Mode Coupling

Radio sounding in the magnetosphere is challenging because the distances are so large and N_e is so low. These constraints motivated the RPI design for the IMAGE mission. As a result, the IMAGE satellite contains the largest structures ever placed on a spinning satellite, namely, the RPI spin-plane crossed dipole antennas (originally 500 m tip-to-tip length for each). Soon after the IMAGE launch on 25 March 2000 into an elliptical polar orbit, with an apogee of 8 Earth radii (R_E) geocentric distance and a perigee altitude of about 1000 km, the RPI detected discrete long-range echoes outside the plasmasphere in the north polar region. Reinisch et al. [60] and Carpenter et al. [19] attributed them to signals propagating in the X mode in FAI to the polar ionosphere where they were reflected; they were clearly distinguished from the diffuse shorter-range echoes from the nearby highly-irregular plasmapause boundary. Later, echo signatures indicating inter-hemisphere propagation, similar to the ionospheric example shown in Fig. 1.7, were identified by Fung et al. [26]; an example where the Z mode played a prominent role in the interpretation is shown in Fig. 1.12 from Reinisch et al. [61]. This record is called a plasmagram and is the magnetospheric analog of the ionospheric topside-sounder ionogram examples shown in Figs. 1.5–1.9 and 1.11.

The virtual range vs. frequency curves through the data points corresponding to X-mode reflections from the local and conjugate hemispheres (labelled SX and NX, respectively) were used to derive the hemisphere-to-hemisphere N_e distribution along the magnetic field line through the satellite [31]. This field-aligned N_e distribution was then used to calculate the reflections expected for transmitted O-mode signals that coupled to Z-mode signals, which reflected at the distances where the transmitted frequencies were equal to the f_Z cutoff frequencies given by (1.5), and then coupled back to the O-mode signals that were received at the satellite. These O-Z-O traces are labelled NZ and SZ for the echoes of this type from the northern and southern hemispheres, respectively, in Fig. 1.12. This mode-coupling interpretation of Reinisch et al. [61], involving the Z-mode to explain the weaker companion echoes to the inter-hemisphere magnetospheric RPI X-mode echoes, differs from the interpretation of Muldrew [48] of inter-hemisphere ionospheric echoes observed by Alouette 1 in that Muldrew [48] did not invoke O-Z-O mode coupling.

1.4.2 Local Z-Mode Echoes

Echoes of Z-mode signals are often directly observed by RPI in the magnetosphere. Figure 1.13 shows an example when IMAGE was near the plasmapause and a strong N_e gradient could be determined from multiple plasma

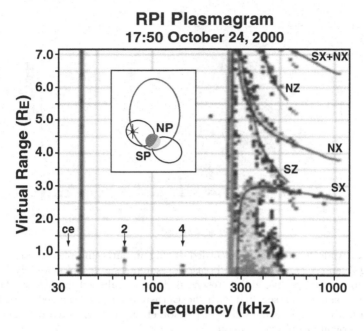

Fig. 1.12. An RPI plasmagram showing echoes from both hemispheres in the frequency region above about 300 kHz, with interpretive traces through the data points, and short-duration plasma resonances at f_{ce}, $2f_{ce}$ and $4f_{ce}$ (lower left). The labels S, N, X, and Z denote southern and northern hemispheres and X and O-Z-O traces respectively. The NX trace has been extrapolated to higher frequencies. The insert shows the orbit, location of IMAGE (x) and the $L = 4$ dipole field lines; NP and SP indicate the magnetic poles [adapted from Reinisch et al., 61] (Reprinted with permission of the American Geophysical Union)

resonances and wave cutoffs. The frequency range of this high-resolution plasmagram fortuitously included the Z- and X-mode cutoffs (f_Z and f_X, respectively) and several plasma resonances. (The linear frequency step size between transmissions of single 3.2 ms pulses was equal to the RPI bandwidth of 300 Hz.) The plasma conditions corresponded to Fig. 1.2a; the Z mode from f_Z to f_{pe} is in CMA region 4 and is the only cold-plasma wave mode. The diffuse nature is attributed to scatter returns from FAI. These scatter returns become more prominent in the CMA region 3 between f_{pe} and f_{uh} (see Fig. 1.2a). This enhanced scatter forms a noise band analogous to the noise band in the ionospheric example of Fig. 1.9 (which, however, corresponds to the CMA 3 region of Fig. 1.2b). Using the values scaled from Fig. 1.13 in (1.3)–(1.5) reveals that consistent solutions cannot be obtained for these equations with constant f_{ce} and f_{pe} values over the 41-s time interval required to record this plasmagram. Three independent f_{pe} determinations can be made, however, if the Tsy 96–1 model magnetic-field values [Tsyganenko, 65, 66, 67],

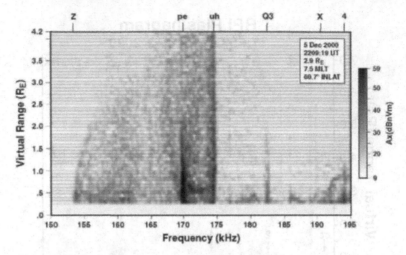

Fig. 1.13. An RPI plasmagram recorded during an outbound plasmapause crossing revealing Z- and X-mode cutoffs and plasma resonances at f_{pe}, f_{uh}, and $4f_{ce}$ labelled at the top by z, x, pe, uh, and 4, respectively. Also labelled, as $Q3$, at the top is one of the resonances associated with the Bernstein modes discussed in Sect. 1.2 in connection with Fig. 1.4 [adapted from Benson et al. 11] (Reprinted with permission of the American Geophysical Union)

with a percentage offset correction based on the observed $4f_{ce}$ plasma resonance in Fig. 1.13, are used corresponding to the spacecraft locations at the times of the recording of f_Z, f_{uh}, and f_X. The plasma resonance observed at f_{pe} in Fig. 1.13 provides a fourth measurement of f_{pe} and it is independent of the f_{ce} value. While the deduced N_e gradient is large (an order-of-magnitude decrease in a change in L value of approximately 1) [Benson et al., 11], it is only about 1/10 the gradient of a well-developed plasmapause [Carpenter and Anderson, 18].

1.4.3 Z-Mode Refractive-Index Cavities

Among the most spectacular echo signatures observed on RPI plasmagrams are those corresponding to the direct transmission and reception of Z-mode waves that are ducted along FAI within refractive-index cavities [Carpenter et al., 20]. Examples are shown in Fig. 1.14. They were obtained when RPI was operating with a linear frequency step size of 900 Hz. From the observed f_Z values in Fig. 1.14 (corresponding to the onset of the Z-mode traces), and the model values for f_{ce} (off scale to the right in both plasmagrams), it is deduced from (1.5) that f_{pe} is also off scale to the right in both plasmagrams. In both cases, $f_{pe} < f_{ce}$, i.e., propagation corresponding to CMA region 7 in Fig. 1.2b is involved.

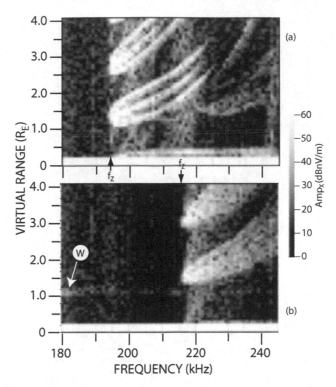

Fig. 1.14. RPI plasmagrams showing multicomponent Z-mode echoes recorded near perigee in the region of N_e gradients between the southern aurora zone and the plasmasphere. (a) $L \approx 3.2$, altitude ~ 3800 km, $f_Z = 194$ kHz, f_{ce}(model) $= 382$ kHz implies $f_{pe} \approx 334$ kHz or $f_{pe}/f_{ce} \approx 0.9$; 0824 UT on 26 July 2001. (b) $L \approx 2.9$, altitude ~ 2700 km, $f_Z = 216$ kHz, f_{ce}(model) $= 469$ kHz implies $f_{pe} \approx 384$ kHz or $f_{pe}/f_{ce} \approx 0.8$; 0245 UT on 12 July 2001 (whistler-mode echoes, due to reflections from the bottom side of the ionosphere [62], are marked "W") [adapted from Carpenter et al. 20] (Reprinted with permission of the American Geophysical Union)

The virtual ranges of the observed echoes starting near $1.5\,R_E$, which appear as upward slanting epsilons in Figs. 1.14a and 1.14b, are too short to be explained in the same manner as used for the echoes shown in Fig. 1.12, i.e., they are too short to be attributed to echoes from the conjugate hemisphere. They can be explained, however, in terms of ducted echoes returned from within a refractive-index cavity in the hemisphere containing the IMAGE satellite. The interpretation presented by Carpenter et al. [20] is illustrated in Fig. 1.15 for the case when the IMAGE satellite is assumed to be located below a relative minimum in the profile of f_Z along **B**, as deduced from (1.5) with f_{ce} and f_{pe} values based on models and RPI observations.

The most prominent features in Fig. 1.14a are reproduced in Fig. 1.15a with labels corresponding to the ray paths defined in Fig. 1.15b which displays

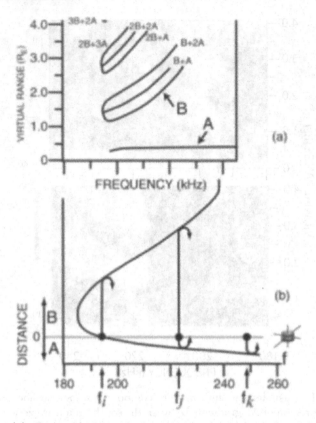

Fig. 1.15. (a) Reproduction of the most prominent echoes in Fig. 1.14a. (b) Schematic f_Z profile limiting ray paths A and B for representative frequencies f_i, f_j, and f_k, relative to the location of the IMAGE satellite. These ray paths correspond to the labels used for the echo traces in (a) [Carpenter et al., 20] (Reprinted with permission of American Geophysical Union)

an idealized f_Z profile along **B**. When the sounder frequency reaches the frequency f_i, corresponding to the condition $f_i = f_Z$ at the satellite level as illustrated in Fig. 1.15b, a wave can propagate upward along path B into the region where $f_i > f_Z$ and be reflected at a higher altitude where the condition $f_i = f_Z$ is again satisfied. This returning wave is responsible for the echo with a virtual range of $\approx 1.5\,R_E$, i.e., the nose of the first upward slanting epsilon signature in Fig. 1.15a. This wave is reflected again at the $f_i = f_Z$ condition at the satellite and the process is repeated. Two such repetitions are evident in the data of Fig. 1.14a and the reproduction in Fig. 1.15a. Higher sounder frequencies, such as f_j in Fig. 1.15b, correspond to the condition $f_j > f_Z$ at the satellite level, and waves can now propagate both upward along path B and downward along path A, within the region where $f_j > f_Z$, and be reflected at both higher and lower altitudes where the condition $f_j = f_Z$

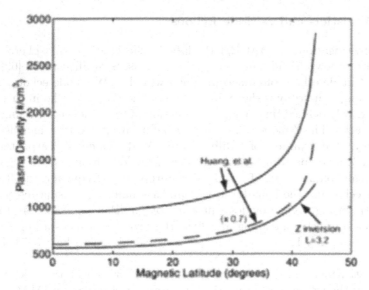

Fig. 1.16. N_e along the $L = 3.2$ magnetic-field line above the IMAGE satellite derived from inverting the "B" Z trace of Fig. 1.15a compared with an RPI-derived empirical model of Huang et al. [31] for a different day (8 June 2001) and a scaling of that model by a factor of 0.7. The portion of the lowest curve corresponding to magnetic latitude values less than 13° is an extrapolation [Carpenter et al. 20] (Reprinted with permission of the American Geophysical Union)

is satisfied. Multiple combinations of these echoes produced the elements of the epsilon signatures seen in Figs. 1.14a and 1.15a. Carpenter et al. [20] also presented examples of a special form of echo signature on RPI plasmagrams that corresponds to ducted Z-mode propagation along FAI within Z-mode refractive-index cavities when IMAGE is assumed to be located above the relative minimum in the profile of f_Z along **B**.

Carpenter et al. [20] introduced an inversion method to determine the N_e distribution along **B** from the upward propagating signals within Z-mode refractive-index cavities of the type discussed above. The results of applying this method to the trace corresponding to B in Fig. 1.15a are presented in Fig. 1.16. Also presented in Fig. 1.16 are predicted values from an empirical N_e model based on the inversion of RPI X-mode echoes from signals that propagated on multiple field-aligned paths on a different day, namely, 8 June 2001 [Huang et al., 31]. The lower N_e values derived from the Z-mode data were attributed to the movement of IMAGE through a region of plasmapause N_e gradients at the time of the measurements. Since these Z-mode echoes from waves trapped in Z-mode refractive-index cavities are often the only prominent echoes observed on a single plasmagram, this inversion method provides a valuable diagnostic tool for determining the N_e distribution along **B**.

1.4.4 Whistler- and Z-Mode Echoes

In an investigation of IMAGE/RPI data in the inner plasmasphere and at moderate to low altitudes over the polar regions, Sonwalkar et al. [62] found diffuse Z-mode echoes often accompanied whistler (W)-mode echoes. An example from their study is shown in Fig. 1.17. The W-mode echo in this figure, with narrowly defined time delay as a function of frequency, is an example of a discrete echo. The Z-mode echo, with a time-delay spread that increases with frequency, is an example of a diffuse echo. As discussed by Carpenter et al. [20], this Z-mode pattern is characteristic of the low altitude polar region and the plasma condition $f_{pe}/f_{ce} < 1$. The abrupt high-frequency cutoff of this Z-mode echo, and the long-duration sounder-stimulated plasma resonance at $\sim$787 kHz in Fig. 1.17, provides a measure of f_{uh} [Benson et al., 14]; the gap, or decrease in echo spreading at $\sim$685 kHz, provides a measure of f_{ce} [Carpenter et al., 20]. From these values f_{pe} is calculated to be $\sim$387 kHz from (1.3).

Whistler-mode echoes with a much broader range of time delays with frequency than those shown in Fig. 1.17 are also observed on IMAGE. They are called diffuse W-mode echoes [Sonwalkar et al., 62]. In regions poleward of the plasmasphere, diffuse Z-mode echoes of the kind illustrated in Fig. 1.17 were found to accompany both discrete and diffuse W-mode echoes 90% of the time, and were also present during 90% of the soundings when no W-mode echoes were detected.

Based on comparisons of ray tracing simulations with the observed dispersion of W- and Z-mode echoes, Sonwalkar et al. [62] proposed that: (1) the observed discrete W-mode echoes are due to RPI signal reflections from the

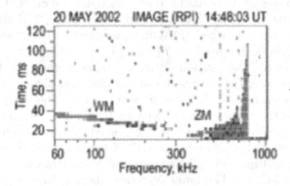

Fig. 1.17. RPI plasmagram displaying both W-mode echoes (frequencies below $\sim$300 kHz) and diffuse Z-mode echoes (frequencies above $\sim$300 kHz) labelled WM and ZM, respectively. The minimum observable time delay is 13 ms due to the 3.2-ms minimum transmitted pulse length and additional time needed for the receiver to recover from the high voltage generated during the transmitter pulses. The amplitude scale is coded from 10 to 50 dB nV/m [adapted from Sonwalkar et al., 62] (Reprinted with permission of the American Geophysical Union)

lower boundary of the ionosphere, (2) the diffuse W-mode echoes are due to scattering of RPI signals by FAI located within 2000 km earthward of IMAGE and in directions close to that of the field line passing through IMAGE, and (3) the diffuse Z-mode echoes are due to scattering of RPI signals from FAI within 3000 km of IMAGE, particularly to signals propagating in the generally cross-**B** direction.

This interpretation suggests that Z-mode echoes occur most frequently ($\sim 90\%$), both in the presence and absence of whistler-mode echoes, because the Z-mode waves capable of returning to the sounder can propagate long distances in all directions, i.e., not only close to the field lines as in the case of the whistler mode waves that are capable of returning to the sounder. Thus there is a much larger probability of encountering plasma irregularities which may lead to Z-mode echoes. These RPI results are consistent with previous investigations in that they indicate that the high-latitude magnetosphere is highly structured with FAI that exist over cross-**B** scales ranging from 10 m to 100 km and that these FAI profoundly effect W- and Z-mode propagation.

1.5 Active/Passive Investigation of Z-Mode Waves of Magnetospheric Origin

The first observation of enhanced Z-mode radio emissions of natural origin, corresponding to CMA-region 3 in Fig. 1.2, were made during the radio-astronomy rocket experiment of Walsh et al. [68]. They ruled out a thermal source due to the large signal intensities and suggested Cherenkov radiation as a likely source mechanism because of the large refractive index (and hence low wave phase velocities) in this frequency domain. Bauer and Stone [3], using satellite observations, were the first to show that the observed frequency limits of this CMA region 3 Z-mode radiation could be used to determine the magnetospheric N_e. Several later experiments have investigated these emissions attributed to CMA region 3 Z-mode radiation by comparing the observed frequencies with f_{pe} values determined by active techniques. Beghin et al. [4] used the AUREOL/ARCADE-3 mutual impedance probe in the 400–2000 km altitude region, Kurth et al. [42] used a sounder during a brief (5 min) period of the single pass through the terrestrial magnetosphere by Cassini, and Benson et al. [12] used active soundings by the RPI on four passes of IMAGE in the vicinity of the plasmapause region.

In each of these investigations, it was concluded that the upper and lower frequency boundaries of an observed intense upper-hybrid band corresponded to f_{uh} and f_{pe}, respectively, in the region where $f_{pe} > f_{ce}$. In the study by Benson et al. [12] these frequency identifications were found to hold to an accuracy of a few per cent in f_{pe} by interpolating between active soundings to the intervening passive dynamic spectra. Figure 1.18 shows the results of superimposing the plasmagram-determined f_{ce}, f_{pe} and f_{uh} values from active

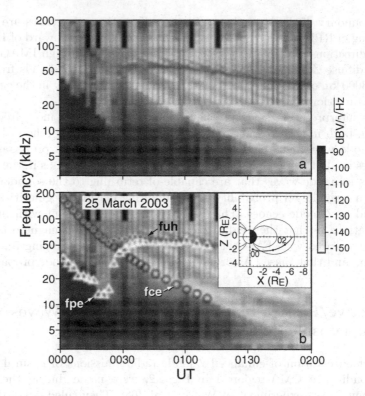

Fig. 1.18. (a) Passive RPI dynamic spectrum. (b) Same as (a) except with superimposed f_{ce}, f_{pe} and f_{uh} values determined from active RPI plasmagrams [adapted from Benson et al., 12] (Reprinted with permission of the American Geophysical Union)

sounding on the passive RPI dynamic spectrum corresponding to the same time interval.

Comparing Figs. 1.18a and 1.18b illustrates the benefit of having active soundings to confidently determine f_{pe}, particularly when $f_{pe} < f_{ce}$. The sounder-derived f_{pe} values (white triangles) follow the upper edge of an intense, presumably whistler mode, emission. They deviate from this upper edge, however, at a location that would be difficult to determine without the active soundings. Also, in this $f_{pe} < f_{ce}$ frequency domain, the upper-hybrid band enhancement, in this case between f_{ce} and f_{uh}, is often not very well defined as f_{uh} approaches f_{ce}. Beghin et al. [4] never observed an enhancement when $f_{pe} < f_{ce}$; they attribute this finding to a lack of instability growth of Z-mode waves in the upper-hybrid band under these conditions. The upper-hybrid band is relatively broad (extending from ~50 to 60 kHz) near 00:30 UT where $f_{pe} \sim f_{ce}$, and it narrows in bandwidth as time progresses. The scaled f_{pe} and f_{uh} frequencies in the region beyond about 00:30 UT, i.e., in the region where $f_{pe} > f_{ce}$, allow the boundaries of the upper-hybrid band to be identified and distinguished from the slanting finger-like higher-frequency emissions which are attributed to Bernstein-mode emissions discussed in connection with Fig. 1.4.

Figure 1.18 suggests that there are two different sources of the observed W-mode emissions, one more intense than the other. The most intense one extends out to slightly beyond 01:00 UT and is limited by the minimum in f_{pe} near 10 kHz as determined from the active sounding. The weaker one extends out to about 01:45 UT and is limited by f_{ce} from this point backward in time to 00:30 where it is limited by f_{pe}. At earlier times, the weak emissions in the frequency domain from f_{pe} to f_{uh} could be either L-O or CMA region 6 a Z-mode emissions (see Fig. 1.2b).

Confirming identifications of passive dynamic-spectral features by nearly-simultaneous active determinations of f_{pe}, such as illustrated in Fig. 1.18, provides confidence in the interpretation of the passive dynamic spectra when supporting active measurements are not available. It also provides confidence in the use of the passive dynamic spectra to help interpret plasmagrams when the spectrum of sounder-stimulated resonances is complex [Benson et al., 11].

1.6 Summary

Even though the Z mode is an internal, or trapped, mode of the plasma it has valuable diagnostic applications in space plasmas in both active and passive wave experiments. In active experiments discrete Z-mode echo traces can be inverted to provide N_e profiles, diffuse traces provide information about FAI, and two-point propagation studies provide information concerning wave propagation, wave ducting and wave/particle interactions. In passive experiments, intense Z-mode signals of magnetospheric origin provide valuable ambient N_e information.

Acknowledgements

We are grateful to the reviewer for many helpful comments on the manuscript. The work at the University of Alaska Fairbanks was supported by NASA under contract NNG04GI67G. Support for B. W. R. was provided by NASA under subcontract 83822 from SwRI.

References

[1] Allis, W.P., S.J. Buchsbaum, and A. Bers: Waves in anisotropic plasmas, MIT Press, Cambridge, 1963.

[2] Balmain, K.G.: Dipole admittance for magnetoplasma diagnostics, IEEE Trans. Antennas Propag. 17, 389–392, 1969.

[3] Bauer, S.J. and R.G. Stone: Satellite observations of radio noise in the magnetosphere, Nature 218, 1145–1147, 1968.

[4] Beghin, C., J.L. Rauch, and J.M. Bosqued: Electrostatic plasma waves and HF auroral hiss generated at low altitude, J. Geophys. Res., 94, 1359–1379, 1989.

[5] Beghin, C., J.L. Rauch, and J.M. Bosqued: Electrostatic plasma waves and HF auroral hiss generated at low altitude, J. Geophys. Res. 94, 1359–1379, 1989.

[6] Benson, R.F.: Ion effects on ionospheric electron resonance phenomena, Radio Sci. 10, 173–185, 1975.

[7] Benson, R.F.: Stimulated plasma waves in the ionosphere, Radio Sci. 12, 861–878, 1977.

[8] Benson, R.F.: Stimulated plasma instability and nonlinear phenomena in the ionosphere, Radio Sci. 17, 1637–1659, 1982.

[9] Benson, R.F.: Field-aligned electron density irregularities near 500 km – Equator to polar cap topside sounder Z mode observations, Radio Sci. 20, 477–485, 1985.

[10] Benson, R.F.: Evidence for the stimulation of field-aligned electron density irregularities on a short time scale by ionospheric topside sounders, J. Atm. and Solar-Terr. Phys. 59, 2281–2293, 1997.

[11] Benson, R.F., V.A. Osherovich, J. Fainberg, and B.W. Reinisch: Classification of IMAGE/RPI-stimulated plasma resonances for the accurate determination of magnetospheric electron-density and magnetic field values, J. Geophys. Res. 108, 1207, doi:10.1029/2002JA009589, 2003.

[12] Benson, R.F., P.A. Webb, J.L. Green, L. Garcia, and B.W. Reinisch: Magnetospheric electron densities inferred from upper-hybrid band emissions, Geophys. Res. Lett. 31, L20803,doi:1029/2004GL020847, 2004.

[13] Bernstein, I.B.: Waves in a plasma in a magnetic field, Phys. Rev. 109, 10–21, 1958.

[14] Budden, K.G.: The propagation of radio waves, the theory of radio waves of low power in the ionosphere and magnetosphere, 669 pp., Cambridge University Press, New York, 1985.

[15] Burch, J.L.: The first two years of IMAGE, Space Sci. Rev. 109, 1–24, 2003.

[16] Calvert, W., Oblique z-mode echoes in the topside ionosphere, J. Geophys. Res. 71, 5579–5583, 1966.

[17] Calvert, W.: Wave ducting in different wave modes, J. Geophys. Res., 100 (A9), 17,491–17,497, 1995.

[18] Carpenter, D.L. and R.R. Anderson: An ISEE/whistler model of equatorial electron density in the magnetosphere, J. Geophys. Res. 97, 1097–1108, 1992.

[19] Carpenter, D.L., M.A. Spasojevic, T.F. Bell, U.S. Inan, B.W. Reinisch, I.A. Galkin, R.F. Benson, J.L. Green, S.F. Fung, and S.A. Boardsen: Small-scale field-aligned plasmaspheric density structures inferred from RPI on IMAGE, J. Geophys. Res., 107(A9), 1258, doi:10.1029/2001JA009199, 2002.

[20] Carpenter, D.L., T.F. Bell, U.S. Inan, R.F. Benson, V.S. Sonwalkar, B.W. Reinisch, and D.L. Gallagher: Z-mode sounding within propagation "cavities" and other inner magnetospheric regions by the RPI instrument on the IMAGE satellite, J. Geophys. Res. 108, 1421, doi:10.1029/2003JA010025, 2003.

[21] Denisenko, P.F., N.A. Zabotin, D.S. Bratsun, and S.A. Pulinets: Detection and mapping of small-scale irregularities by topside sounder data, Ann. Geophysicae 11, 595–600, 1993.

[22] Dyson, P.L. and R.F. Benson: Topside sounder observations of equatorial bubbles, Geophys. Res. Lett. 5, 795–798, 1978.

[23] Eckersley, T.L.: Discussion of the ionosphere, Proc. Roy. Soc. A141, 708–715, 1933.

[24] Ellis, G.R.: The Z propagation hole in the ionosphere, J. Atmosph. Terr. Phys. 8, 43–54, 1956.

[25] Franklin, C.A. and M.A. Maclean: The design of swept-frequency topside sounders, Proc. IEEE 57, 897–929, 1969.

[26] Fung, S.F., R.F. Benson, D.L. Carpenter, J.L. Green, V. Jayanti, I.A. Galkan, and B.W. Reinisch: Guided Echoes in the Magnetosphere: Observations by Radio Plasma Imager on IMAGE, Geophys. Res. Lett. 3, 1589, doi:10.1029/2002GL016531, 2003.

[27] Goertz, C.K. and R.J. Strangeway: Plasma waves, in: Introduction to Space Physics, edited by M.G. Kivelson, and C.T. Russell, pp. 356–399, Cambridge University Press, New York, 1995.

[28] Gurnett, D.A., S.D. Shawhan, and R.R. Shaw: Auroral hiss, Z mode radiation, and auroral kilometric radiation in the polar magnetosphere: DE 1 observations, J. Geophys. Res. 88, 329–340, 1983.

[29] Horita, R.E. and H.G. James: Enhanced Z-mode radiation from a dipole, Adv. Space Res. 29, 1375–1378, 2002.

[30] Horita, R.E. and H.G. James: Two-point studies of fast Z-mode waves with dipoles in the ionosphere, Radio Sci., 39, RS4001, doi:10.1029/2003RS002994, 2004.

[31] Huang, X., B.W. Reinisch, P. Song, P. Nsumei, J.L. Green, and D.L. Gallagher: Developing an empirical density model of the plasmasphere using IMAGE/RPI observations, Adv. Space Res. 33, 829–832, 2004.

[32] Jackson, J.E.: The reduction of topside ionograms to electron-density profiles, Proc. IEEE 57, 960–976, 1969.

[33] Jackson, J.E. and E.S. Warren: Objectives, history, and principal achievments of the topside sounder and ISIS programs, Proc. IEEE 57, 861–865, 1969.

[34] James, H.G.: Wave propagation experiments at medium frequencies between two ionospheric satellites, 1, General results, Radio Sci. 13, 531–542, 1978.

[35] James, H.G.: Wave propagation experiments at medium frequencies between two ionospheric satellites 3. Z mode pulses, J. Geophys. Res. 84, 499–506, 1979.

[36] James, H.G.: Guided Z mode propagation observed in the OEDIPUS A tethered rocket experiment, J. Geophys. Res. 96, 17,865–17,878, 1991.

[37] James, H.G.: Slow Z-mode radiation from sounder-accelerated electrons, J. Atmos. Solar-Terr. Phys. 66, 1755–1765, 2004.

[38] James, H.G., Radiation from sounder-accelerated electrons, Adv. Space Res., in press, 2005.

[39] Kelso, J.M.: Radio ray propagation in the ionosphere, 408 pp., McGraw Hill, New York, 1964.

[40] Kuehl, H.H.: Electromagnetic radiation from an electric dipole in a cold anisotropic plasma, Phys. Fluids 5, 1095–1103, 1962.

34   R.F. Benson et al.

[41] Kuo, S.P., M.C. Lee, and P. Kossey: Excitation of short-scale field-aligned electron density irregularities by ionospheric topside sounders, J. Geophys. Res. 104, 19,889–19,894, 1999.

[42] Kurth, W.S., G.B. Hospodarsky, D.A. Gurnett, M.L. Kaiser, J.-E. Wahlund, A. Roux, P. Canu, P. Zarka, and Y. Tokarev: An overview of observations by the Cassini radio and plasma wave investigation at Earth, J. Geophys. Res. 106, 30239–30252, 2001.

[43] LaBelle, J. and R.A. Treumann: Auroral radio emissions, 1. hisses, roars, and bursts, Space Sci. Rev. 101, 295–440, 2002.

[44] Lewis, R.M. and J.B. Keller: Conductivity tensor and dispersion equation for a plasma, Phys. Fluids, 5, 1248–1264, 1962.

[45] Lockwood, G.E.K.: A ray-tracing investigation of ionospheric Z-mode propagation, Can. J. Phys. 40, 1840–1843, 1962.

[46] McAfee, J.R.: Ray trajectories in an anisotropic plasma near plasma resonance, J. Geophys. Res. 73, 5577–5583, 1968.

[47] McAfee, J.R.: Topside ray trajectories near the upper hybrid resonance, J. Geophys. Res. 74, 6403–6408, 1969.

[48] Muldrew, D.B.: Radio propagation along magnetic field-aligned sheets of ionization observed by the Alouette topside sounder, J. Geophys. Res. 68, 5355–5370, 1963.

[49] Muldrew, D.B.: Nonvertical propagation and delayed-echo generation observed by the topside sounders, Proc. IEEE 57, 1097–1107, 1969.

[50] Muldrew, D.B.: Electron resonances observed with topside sounders, Radio Sci. 7, 779–789, 1972.

[51] Muldrew, D.B. and M.F. Estabrooks: Computation of dispersion curves for a hot magnetoplasma with application to the upper-hybrid and cyclotron frequencies, Radio Sci. 7, 579–586, 1972.

[52] Osherovich, V.A., J. Fainberg, R.F. Benson, and R.G. Stone: Theoretical analysis of resonance conditions in magnetized plasmas when the plasma/gyro frequency ratio is close to an integer, J. Atm. and Solar-Terr. Phys. 59, 2361–2366, 1997.

[53] Oya, H.: Conversion of electrostatic plasma waves into electromagnetic waves: numerical calculation of the dispersion relation for all wavelengths, Radio Sci. 6, 1131–1141, 1971.

[54] Poeverlein, H.: Strahlwege von Radiowellen in der Ionosphäre, Z. Angew. Phys. 1, 517, 1949.

[55] Pulinets, S.A.: Prospects of topside sounding, in WITS handbook N2, edited by C.H. Liu, pp. 99–127, SCOSTEP Publishing, Urbana, Illinois, 1989.

[56] Pulinets, S.A., P.F. Denisenko, N.A. Zabotin, and T.A. Klimanova: New method for small-scale irregularities diagnostics from topside sounder data, in: SUNDIAL Workshop, McLean, Virginia, 1989.

[57] Ratcliffe, J.A.: The Magneto-Ionic Theory and its Applications to the Ionosphere, 206 pp., Cambridge University Press, New York, 1959.

[58] Reinisch, B.W.: Modern Ionosondes, in: Modern Ionospheric Science, edited by H. Kohl, R. Ruster, and K. Schlegel, pp. 440–458, European Geophysical Society, Katlenburg-Lindau, Germany, 1996.

[59] Reinisch, B.W., D.M. Haines, K. Bibl, G. Cheney, I.A. Gulkin, X. Huang, S.H. Myers, G.S. Sales, R.F. Benson, S.F. Fung, J.L. Green, W.W.L. Taylor, J.-L. Bougeret, R. Manning, N. Meyer-Vernet, M. Moncuquet, D.L. Carpenter, D.L.

Gallagher, and P. Reiff: The radio plasma imager investigation on the IMAGE spacecraft, Space Sci. Rev. 91, 319–359, 2000.

[60] Reinisch, B.W., X. Huang, D.M. Haines, I.A. Galkin, J.L. Green, R.F. Benson, S.F. Fung, W.W.L. Taylor, P.H. Reiff, D.L. Gallagher, J.-L. Bougeret, R. Manning, D.L. Carpenter, and S.A. Boardsen: First results from the radio plasma imager on IMAGE, Geophys. Res. Lett. 28, 1167–1170, 2001a.

[61] Reinisch, B.W., X. Huang, P. Song, G.S. Sales, S.F. Fung, J.L. Green, D.L. Gallagher, and V.M. Vasyliunas: Plasma density distribution along the magnetospheric field: RPI observations from IMAGE, Geophys. Res. Lett. 28, 4521–4524, 2001b.

[62] Sonwalkar, V.S., D.L. Carpenter, T.F. Bell, M.A. Spasojevic, U.S. Inan, J. Li, X. Chen, A. Venkatasubramanian, J. Harikumar, R.F. Benson, W.W.L. Taylor, and B.W. Reinisch: Diagnostics of magnetospheric electron density and irregularities at altitudes <5000 km using whistler and Z mode echoes from radio sounding on the IMAGE satellite, J. Geophys. Res. 109, A11212, doi:10.1029/2004JA010471, 2004.

[63] Stix, T.H.: The Theory of Plasma Waves, 283 pp., McGraw-Hill, New York, 1962.

[64] Thomas, J.O.: The distribution of electrons in the ionosphere, Proc. IRE 47, 162–175, 1959.

[65] Tsyganenko, N.A.: Modeling the Earth's magnetospheric magnetic field confined within a realistic magnetopause, J. Geophys. Res. 100, 5599–5612, 1995.

[66] Tsyganenko, N.A.: Effects of the solar wind conditions on the global magnetospheric configuration as deduced from data-based field models, in 3rd International Conference on Substorms (ICS-3), pp. 181–185, ESA SP-389, Versailles, France, 1996.

[67] Tsyganenko, N.A. and D.P. Stern: Modeling the global magnetic field of the large-scale Birkeland current systems, J. Geophys. Res. 101, 27,187–27,198, 1996.

[68] Walsh, D., F.T. Haddock, and H.F. Schulte: Cosmic radio intensities at 1.225 and 2.0 Mc measured up to an altitude of 1700 km, in: Space Research, edited by P. Muller, pp. 935–959, North Holland Publishing Company, Amsterdam, 1964.

[69] Zabotin, N.A., D.S. Bratsun, S.A. Pulinets, and R.F. Benson: Response of topside radio sounding signals to small-scale field-aligned ionospheric irregularities, J. Atmosph. Solar-Terr. Phys. 59, 2231–2246, 1997.

2

Review of Kilometric Continuum

K. Hashimoto[1], J.L. Green[2], R.R. Anderson[3], and H. Matsumoto[1]

[1] Research Institute for Sustainable Humanosphere, Kyoto University, Uji, Kyoto
611-0011, Japan
{kozo,matsumot}@rish.kyoto-u.ac.jp
[2] Code 605, NASA/GSFC, Greenbelt, MD 20771, USA
James.L.Green@nasa.gov
[3] The University of Iowa, Department of Physics and Astronomy, Iowa City, IA
52242-1479, USA.
rra@space.physics.uiowa.edu

Abstract. Kilometric continuum radiation is a non-thermal magnetospheric radio
emission. It is one of the fundamental electromagnetic emissions in all planetary
magnetospheres [cf. the review by Kaiser, 27]. We review its observational properties in view of their agreement with theoretical models. Although this emission
has been observed and studied for more than 35 years, there are still several unverified theories on how this emission is generated. It is by now quite certain that
it is emitted from the plasmasphere, in particular from the plasmapause and from
density notches and cavity gradients. Mode conversion at density gradients plays an
important role. Observations show that the radiation consists of a magnetospherically trapped and an escaping component. It exhibits a narrow-band fine structure
that is barely understood, but beaming models can be safely excluded based on
the observations of the frequency-time structure of the radiation. We investigate its
relation to geomagnetic activity and solar activity.

Key words: Magnetosphere, continuum radiation, non-thermal radiation,
plasmapause source

2.1 Introduction

A magnetospheric electromagnetic emission that is associated with intense
narrow-band electrostatic emissions in the vicinity of the plasmapause at the
geomagnetic equator is the non-thermal continuum (NTC) radiation (see for
example: [17, 29]). NTC is observed over a very broad frequency range from
as low as 5 kHz [Gurnett, 15]. Its highest frequency was known to be 200 kHz
[Kurth et al., 30] before the identification of "kilometric continuum" by Geotail [Hashimoto et al., 19]. The conventional lower frequency component of
continuum has been called the "normal continuum" by a number of authors

[eg., 19, 28] and as terrestrial myriametric radiation [Jones, 23] based on its wavelength range ($\lambda \sim 10\,\mathrm{km}$ at $f \sim 30\,\mathrm{kHz}$). On the other hand, a new component, kilometric continuum, is its high-frequency extension with frequencies up to as high as $800\,\mathrm{kHz}$.

The NTC is generated in the free space L-O mode above f_p, where f_p is the local electron plasma frequency, from sources at or very near the plasmapause. The strong electrostatic bands occur at frequencies where the frequency of the electrostatic upper hybrid resonance (f_uhr) is equal to the frequency of the electrostatic $(n + \frac{1}{2})f_\mathrm{g}$ resonance, where f_g is the local electron cyclotron frequency [Kurth, 31]. The kilometric continuum is not merely a high frequency extension. It triggered new investigations since this frequency range is above the maximum plasma frequency of a few hundred kHz observed at the plasmapause. This is believed to be generated in events separate from the lower frequency non-thermal continuum. Recent NTC research has focused on improving our understanding of the source location, emission cone characteristics, propagation characteristics, and detailed spectral measurements primarily in the kilometric frequency range.

Much of what has emerged from these studies in terms of source location is summarized in Fig. 2.1 adapted from Green et al. [12]. The lower frequency trapped and escaping continuum is typically generated in the pre-noon sector and has been called the "normal continuum", the continuum enhancement is generated in the morning sector [6, 13, 28], and the kilometric continuum

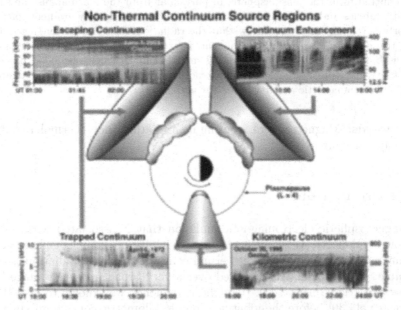

Fig. 2.1. The observed source locations of the escaping after [5], trapped [16], kilometric [19] and continuum enhancement [28] emissions [12]

is generated in deep plasmaspheric notch[1] structures that corotate with the plasmasphere [10] and other density irregularities [20]. The purpose of this paper is to introduce NTC briefly and to review our current understanding of the kilometric continuum radiation.

2.2 Trapped and Escaping NTC

From its unique polar orbiting vantage point, the IMAGE/RPI instrument [Reinisch et al., 44] has observed NTC over its entire frequency range at many local times. Figure 2.2 shows a frequency time spectrogram from the RPI instrument during a pass of IMAGE through the magnetic equator. The NTC extends from about 29 kHz to about 500 kHz forming a Christmas tree pattern in the spectrogram nearly symmetric about the magnetic equator which is clearly delineated by the increased intensity of the $(n + \frac{1}{2})f_g$ electrostatic emission bands. The orbital position of the IMAGE during these observations is shown in the upper left panel of the spectrogram in Fig. 2.2. Due to the relatively weak nature of the emission, the NTC observations in

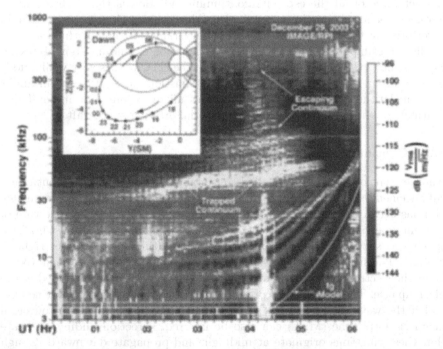

Fig. 2.2. An RPI frequency-time spectrogram taken during a passage through the magnetic equator on the dawn side (see orbit insert). The Christmas tree pattern of nonthermal continuum is clearly shown nearly centered about the magnetic equator delineated by the intense electrostatic emissions

---

[1]Notches were also called bite-outs [9].

Fig. 2.2 were made during quiet geomagnetic times on the dawn side of the magnetosphere. At frequencies less than the magnetopause plasma frequency (∼50 kHz in this example), the continuum radiation has been referred to as the "trapped" component by Gurnett and Shaw [14] since it is observed primarily in the magnetospheric density cavity between the plasmapause and the magnetopause. The trapped continuum spectrum is observed as a broadband emission with very little frequency structure. The broadband structure of the trapped continuum spectrum is believed to be produced from a series of narrow band emissions at slightly different frequencies from an extended source region at the plasmapause whose emission then mixes due to multiple reflections (with some Doppler broadening) in the magnetospheric density cavity. As shown in Fig. 2.2, the trapped continuum is observed first (the widest part of the Christmas tree pattern) since multiple reflections from the magnetopause broaden its angular distribution [9].

NTC at frequencies above the magnetopause plasma frequency has been referred to as the "escaping" component by Kurth et al. [30] since it propagates from the Earth's plasmapause to well outside the magnetosphere. As shown in Fig. 2.2 the escaping component extends from above 50 kHz. A common characteristic of all the escaping continuum radiation is that it has narrow frequency bands of emissions showing that the name continuum is not entirely descriptive of the radiation in this frequency range. Although there are few published examples of normal continuum radiation extending above 100 kHz, Fig. 2.2 shows the high frequency portion of the NTC that is very weak and is beamed around the magnetic equator into emission cones of less than about ±15° in latitude. These two characteristics might account for it not being routinely observed and reported by equatorial orbiting spacecraft.

2.2.1 Continuum Enhancement

Another type of continuum is called continuum enhancement. The continuum enhancement is characterized by a strong variation in intensity and frequency and has been observed to last for only 1-2 hours [6, 13, 28]. The spectrum of continuum enhancement is distinct, with discrete emissions at almost uniform spacing as shown in the frequencies less than 100 kHz in the upper right of Fig. 2.1. The spacings of the discrete continuum enhancement are believed to be at the electron cyclotron frequency with the source being embedded in the plasmapause. This has allowed prediction of the radial location of the source, and if the wave direction can be determined, the location of the source in magnetic local time (MLT) can also be inferred. Direction findings indicate that these emissions originate at midnight and propagate dawnward. Gough [13] interpreted this to be due to the inward motion of the plasmapause due to enhanced convection, while Filbert and Kellogg [6] associated it with the dusk-to-dawn motions of injected electrons. Kasaba et al. [28] used the continuum enhancement to analyze the displacement of the plasmapause during substorms.

2.3 Generation Theories

There are three lines of theoretical models [see, 2, 32, 36, 46, for reviews] that try to explain continuum radiation in general:

- synchrotron radiation [7, 48],
- linear mode conversion models [21, 22, 23], and
- nonlinear mode conversion models [8, 38, 45].

The synchrotron radiation model is believed to be a factor of ten too weak at the nominal plasmapause location. The linear and nonlinear models assume that the sources of the freely propagating electromagnetic waves are electrostatic waves generated near the upper hybrid-resonance frequency. The sources of the electrostatic waves are most likely energetic electrons in the tens of keV energy range with highly anisotropic phase-space density distributions which become unstable when the resonant wave-particle interaction conditions are satisfied. Sharp boundaries like the plasmapause are good regions for instability to occur because the resonance conditions vary greatly when moving across that boundary.

In the linear model there is a narrow radio window through which the electrostatic waves can propagate and be converted into electromagnetic waves at the same frequency [Jones, 21, 22, 23]. Okuda et al. [42] performed calculation on O mode radiation from electric fluctuations. The electrostatic waves are in the Bernstein modes which are connected to the Z mode at smaller wavenumbers. The Z mode wave is mode-converted to the O mode where the local plasma frequency is equal to the wave frequency. A ray path to the mode conversion point at the local plasma frequency from the Bernstein mode was verified through ray tracing studies in a hot plasma [18, 49]. The linear model predicts that the angle between the equator and the wave normal direction of the O mode, the beaming angle, is defined as $\alpha = \tan^{-1}(f_{\mathrm{g}}/f_{\mathrm{p}})$ [22, 23]. Almost no conversion occurs in the equatorial direction according to full-wave, warm plasma computations [26].

In the nonlinear model of Melrose [38], density irregularities formed by low frequency waves coalesce with upper hybrid waves generating this radiation. In the nonlinear model of [8], an electrostatic wave propagating into a density gradient can nonlinearly interact with its reflected wave to generate an electromagnetic wave at twice the electrostatic wave frequency. Rönnmark [45] showed coalescence of upper hybrid waves with high-harmonic ion Bernstein waves to produce electromagnetic radiation. Both linear and nonlinear models predict that the electromagnetic waves will be beamed in magnetic latitude with the beaming becoming more perpendicular to the magnetic field as the ratio of the electron plasma to cyclotron frequency increases. At a sharp plasmapause, since the cyclotron frequency is almost constant, this means that the higher the wave frequency, the closer it is beamed to the magnetic equator.

2.4 Kilometric Continuum

Kilometric continuum (KC) is a major component of the escaping continuum radiation in the 100–800 kHz frequency range first identified by Hashimoto et al. [19] from the Sweep Frequency Analyzer (SFA) data of the Geotail Plasma Wave Instrument (PWI) [Matsumoto et al., 37]. Although the escaping continuum is not "continuum" as shown in the previous section, the new extension is named "kilometric continuum" to distinguish it from auroral kilometric radiation or AKR (which is in the same frequency range) and yet indicate that it is a high frequency extension of NTC.

KC intensities are similar to those of normal continuum and much weaker than those of AKR. A typical example observed by Geotail/PWI above 100 kHz is shown in Fig. 2.3. This emission is a collection of narrowband emissions and often lasts more than six hours. The spacings of the discrete narrowband emissions are often irregular. It is important to note that KC is typically observed without the accompanying lower frequency component.

KC has sparked considerable interest in further understanding of various aspects of this radiation that make it different from its lower frequency trapped and escaping (< 100 kHz) counterparts generated in the pre-noon sector as shown in Fig. 2.1. The frequency range for the kilometric continuum is approximately that of AKR but, as shown in the lower right hand panel of Fig. 2.1, there are significant differences that can be used to easily distinguish

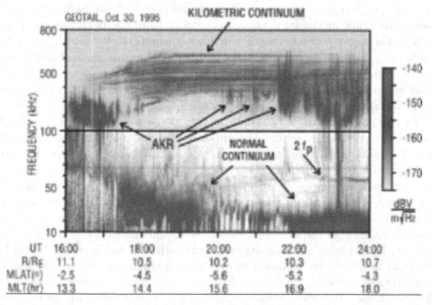

Fig. 2.3. Kilometric Continuum Radiation [19] (Reprinted with permission of the American Geophysical Union)

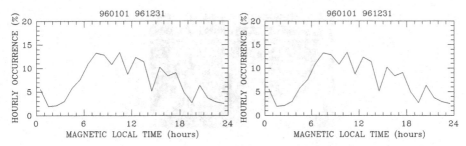

Fig. 2.4. (a) (*left*) Hourly occurrence of KC as a function of geomagnetic latitude and (b) (*right*) hourly occurrence as function of geomagnetic local time. The data are adopted for 1996 from Hashimoto et al. [19] (Reprinted with permission of the American Geophysical Union)

between these two emissions. KC has a narrow band structure over a number of discrete frequencies while AKR is observed at a distance to be a much stronger broadband and sporadic emission and can be seen from 16:00-17:10 and from 21:30 to 24:00 UT in that spectrogram.

The kilometric continuum has been observed at all local times, although it has been difficult to make a positive identification of the emission during the times when Geotail was in the late evening or early morning local time sector when AKR was active [19]. From Geotail and IMAGE observations, Hashimoto et al. [19] and Green et al. [11] have found that the kilometric continuum is confined to a narrow latitude range of approximately ±15° about the magnetic equator. The hourly occurrences as a function of geomagnetic latitude and local time as observed near solar minimum are shown in Fig. 2.4. The almost equatorial orbit of Geotail was advantageous to detect this emission.

Although these characteristics make it different from the lower frequency continuum discussed in the previous section, the similar spectral characteristics of the emission and its relationship to the plasmapause support the conclusion of Menietti et al. [39] from Polar high-resolution electric and magnetic field observations of KC that the radiation is generated by the same mechanism. Through the high-resolution Polar and Cluster observations of the normal continuum, Menietti et al. [40] confirmed that they are analogous to those of KC.

At lower frequencies, beaming of continuum radiation around the magnetic equator to latitudes as high as 50° has also been observed by Jones et al. [25] (from 80 to 100 kHz), Morgan and Gurnett [41] (from 45 to 154 kHz), and by Green and Boardsen [9] (from 24 to 56 kHz). The narrow beaming of the kilometric continuum in magnetic latitude has made this emission difficult to observe routinely or observable for only short periods of time except for equatorial orbiting spacecraft with the proper instrumentation, such as Geotail.

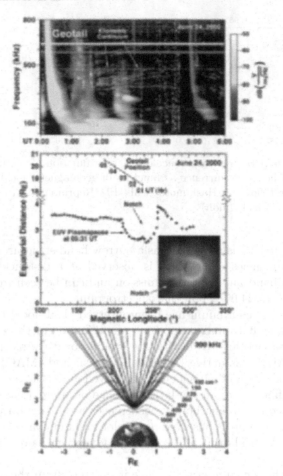

Fig. 2.5. KC wave observations from Geotail/PWI (*top panel*) map to a plasmaspheric notch structure as observed by IMAGE/EUV (*middle panel*) where the resulting emission cone pattern (*bottom panel*) is modelled with ray tracing calculations [10, 11] (Reprinted with permission of the American Geophysical Union)

The source region for KC was originally identified by Carpenter et al. [4] as coming from emissions trapped in plasmaspheric cavities from CRRES observations [Anderson et al., 1]. More recently, Green et al. [10] and Green et al. [11] clearly identified KC as being generated at the plasmapause, deep within notch structures that corotate with the Earth. Figure 2.5 has been adapted from Fig. 8 of Green et al. [10] and Fig. 1 of Green et al. [11] and illustrates that the location of the KC source region and resulting emission cone pattern of the radiation is consistent with the observations. The top panel of Fig. 2.5 is a frequency-time spectrogram from the PWI instrument on Geotail showing the banded structure of KC. The middle panel shows the magnetic longitude versus the equatorial radial distance of the plasmapause (derived from the inserted EUV image of the plasmasphere) and the position of Geotail during the KC observations of the top panel. The bottom panel is a ray tracing analysis which shows that the structure of the plasmaspheric notch has a significant effect on the shape of the resulting emission cone through refraction. The process by which the notch structure is produced in the plasmasphere is not completely understood at this time.

Figures 2.6 and 2.7 show other source regions observed by CRRES [20]. The plasma frequency at the plasmapause is 200 to 300 kHz in Fig. 2.6. KC is radiated from the plasmapause and is a simple extension of the normal continuum. Emissions trapped in a notch are seen at about 300 kHz near 16 UT. If a satellite moves across a notch structure, the emissions look trapped inside the structure as stated by Carpenter et al. [4]. This phenomenon is also called "donkey ear" in EXOS-D (Akebono) observations [43]. Based on IMAGE observations, the structure is open to the magnetosphere, and waves

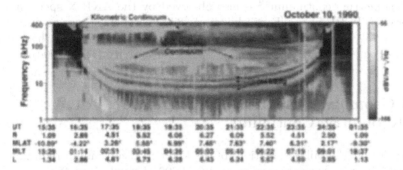

Fig. 2.6. CRRES observations on October 10, 1990. Notch-like structure is seen around 1600 UT and the emissions above 200 kHz starting from 1640 UT are kilometric continuum [20] (Reprinted with permission of the American Geophysical Union)

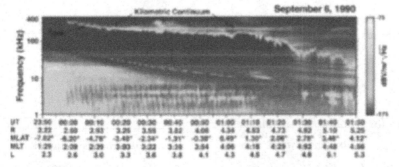

Fig. 2.7. CRRES observations on September 6 and 7, 1990. The electromagnetic radiation starting from 0010 at about 200 kHz is kilometric continuum [20] (Reprinted with permission of the American Geophysical Union)

are not really trapped. On the other hand, kilometric continuum is generated at density irregularities deep inside the plasmapause in the case of Fig. 2.7.

The Christmas-tree emission pattern of KC (see Fig. 2.2) provides important information on the emission mechanism. The characteristic pattern observed by IMAGE in a polar pass across the equator shows that KC at higher frequencies tends to have smaller beaming angles (see Fig. 2.2). This suggests that the higher frequency waves are generated deeper in a plasmapause notch and thus are more confined, while the lower frequency waves may be produced along the sides of the notch. Similar patterns have been analyzed by Hashimoto et al. [20] where it has been clarified that KC is observed at the equator contrary to the beaming predicted by the linear mode conversion of Jones [23, 26].

While IMAGE RPI can observe frequencies higher than 800 kHz and Geotail PWI cannot, IMAGE RPI never observed KC at such high frequencies. Therefore, we can conclude that the maximum frequency of KC is around 800 kHz.

Kilometric "continuum"[2] is also observed by the AKR-X spectrum analyzer of INTERBALL-1 [Kuril'chik et al., 34, 35]. Their observing frequencies were at 252 and 500 kHz, and they claim that the occurrence of the emission was extremely rare during a time of high solar activity (1999–2000).

2.5 Geomagnetic Activity Dependence of Kilometric Continuum

Statistics on Geotail observations of the kilometric continuum radiation were derived for a one year period from July 2000 through June 2001 near solar maximum. The results for the dependence of the hourly occurrence on the geomagnetic latitude are shown in Fig. 2.8a and on the magnetic local time in Fig. 2.8b. Although these results are quite similar to those of Fig. 2.4, there are a few differences which may be attributed to the solar cycle. In the data

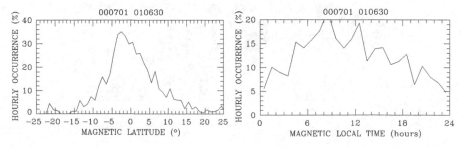

Fig. 2.8. (a) (*left*) Hourly occurrence as a function of geomagnetic latitude and (b) (*right*) hourly occurrence as a function of geomagnetic local time for 2000-2001

---

[2]They quoted the term "continuum" throughout their papers.

in Fig. 2.8 at solar maximum there is no dip near the equator on the latitude dependence, the occurrence probabilities are higher, kilometric continuum radiation was observed at more than 20° in the northern and southern latitudes, and more observations were made at the morning hours. These results are in direct conflict with Kuril'chik et al. [35].

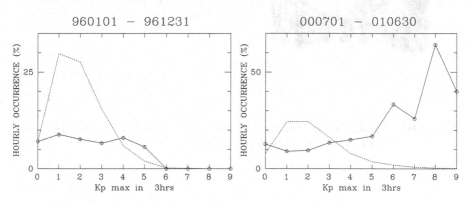

Fig. 2.9. Dependence on the maximum Kp in the preceding 3 hours. (**a**) (*left*) for 1996 and (**b**) (*right*) for 2000–2001

The Kp dependencies of the kilometric continuum radiation are shown in Fig. 2.9a for 1996 and Fig. 2.9b for 2000–2001. The solid and dashed lines indicate the hourly occurrence probability in percent and one tenth of the number of observed cases in hours, respectively. The maximum Kp for the three hours preceding the observation is used. Although Fig. 2.9a shows almost no Kp dependence, Fig. 2.9b shows a modest Kp dependence that increases dramatically for Kp > 5. There were almost no Geotail kilometric continuum radiation observations when Kp > 5 in 1996. Figure 2.9 provides dramatic evidence for the enhancement of KC at solar maximum (right panel) over solar minimum (left panel). The cause of this solar cycle difference is unknown.

Both studies show that kilometric continuum radiation is observed even if Kp = 0. In order to check the Kp dependence, an example when kilometric continuum radiation was observed and Kp = 0 is shown in Fig. 2.10. The kilometric continuum is observed from 06–18 UT on Geotail on December 20, 1996, as shown in the top dynamic spectra. The magnetic latitudes of the satellite are shown for one day in the center of the upper right panel next to the spectrogram and the circles indicate the observations of kilometric continuum. Kp indices are shown above the panel for the day in the second line and for the previous day in the first line. Kp had been less or equal to 1 for more than 24 hours. This demonstrates that kilometric continuum occurs even in such a very quiet time. Dst had been almost zero as seen from the lower right panel. These observations pose a problem for the energy source of the emissions.

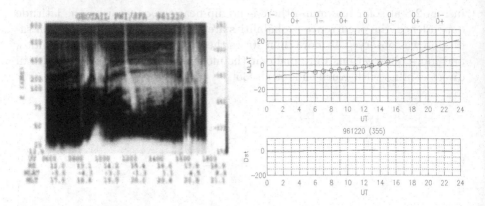

Fig. 2.10. Geotail observations for December 20, 1996, when Kp was always less than 1. Kilometric continuum radiation is observed from 06 UT to 18 UT

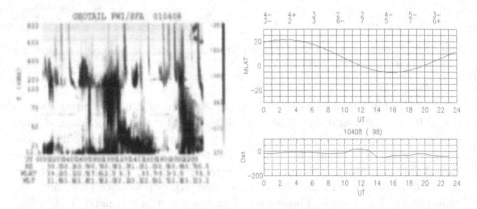

Fig. 2.11. Geotail observations April 8, 2001, when Kp ranged from 2 to 7. Kilometric continuum radiation was not observed until after 22 UT even though Geotail was within 10 degrees of the equator since 09 UT

Figure 2.11 is an example of occurrence of large Kp without kilometric continuum. The continuum is observed only after 22 UT although the satellite was within 10 degrees of the equator since 09 UT, Kp was large for a long time, and AKR was often observed. It should be noted that the kilometric continuum radiation from about 600 kHz to about 800 kHz was quite strong from 22 UT to 05 UT the following day.

2.6 Simultaneous Wave Observations by Geotail and IMAGE

Data from the extreme ultraviolet (EUV) imager [47] of the IMAGE satellite demonstrated the existence of notches, where the electron densities are low near the equatorial plasmapause, and the notches are the source of kilometric continuum radiation. The radio plasma imager (RPI) of the satellite indicated that the electromagnetic waves were not only trapped in the low density region but also escape radially out or are generated outside as kilometric continuum. The IMAGE satellite observed the notch near 4 UT in the dayside as shown in Fig. 2.12 [10]. The orbit is the black curve and the plasmapause is shown as the light curve. Near 190° in the magnetic longitude, the density is decreased and the plasmapause position was inside the normal one. Kilometric continuum is observed inside the notch region.

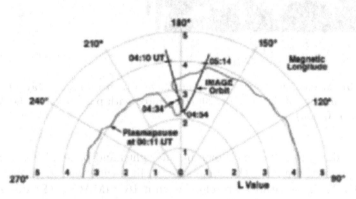

Fig. 2.12. Notch observation by IMAGE on April 8, 2001 [10] (Reprinted with permission of the American Geophysical Union)

On the same day, Geotail did not observe kilometric continuum under very high Kp as shown in Fig. 2.11 until 18 hours later. The notch was observed around local noon, but Geotail was in the nightside. This fact indicates that kilometric continuum is radiated during high Kp, but Geotail was not able to observe it. The satellite position could be the cause of the low occurrence probability at high Kp.

Simultaneous observations of kilometric continuum with IMAGE RPI and Geotail PWI are displayed in a frequency range of $300 - 800$ kHz in the top and the bottom of Fig. 2.13, respectively [Hashimoto et al., 20]. Intense kilometric continuum was received during the disturbed time, especially for Kp > 7 from 20 UT to 03 UT. It should be noted that both kilometric continuum spectra show quite good similarity including the fine structures from 21 UT to 06 UT. IMAGE moved from the southern hemisphere to 30°N. On the other hand, Geotail moved in the equatorial region from 4.4°N to 12.3°N at 01 UT and

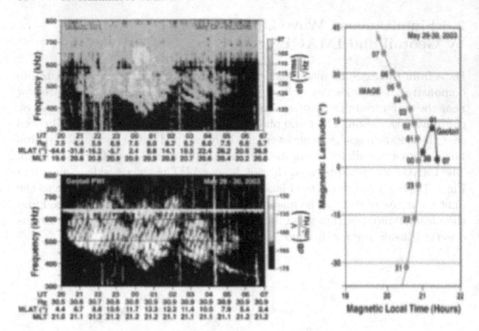

Fig. 2.13. IMAGE and Geotail observations and their orbits on May 29–30, 2003. Note the similarity of the spectra [20] (Reprinted with permission of the American Geophysical Union)

then back down to 2.4° N as shown in the right hand panel of Fig. 2.13. Both satellites observed almost the same spectra in a wide latitude range of more than 30°. Their longitudes are close within 10°. IMAGE RPI observed the emission in a wide latitudinal range different from general trends reported by Hashimoto et al. [19] and Green et al. [10].

The vertical line at 0420 UT is a type III burst. The intensity observed by IMAGE is weaker around 400 kHz after 0430 UT when the satellite is at latitudes higher than 25°. It would be difficult to explain these quite similar spectra by multiple narrow beam sources. Rather, this can be explained if the sources radiate uniformly in wide emission cones in latitude and both satellites receive the emissions from the same sources, which is contrary to the beaming theory.

2.7 Summary and Conclusions

Non-thermal continuum radiation is one of the fundamental electromagnetic emissions in planetary magnetospheres [cf. the review by Kaiser, 27]. It has been observed in every planetary magnetosphere visited by spacecraft armed with wave instruments and has even been found to be generated in the magnetosphere of the Galilean moon Ganymede [Kurth et al., 33]. Although this emission has been observed and studied for more than 35 years, there are still several unverified theories on how this emission is generated. There is also much more which we do not know about this emission and its relationship to the dynamics of the plasmasphere.

Many of the characteristics of the lower frequency portion of the non-thermal continuum (trapped component) have been difficult to determine due to the multiple reflections of the emission from the magnetopause and plasmapause. Recently there is renewed interest in studying the high frequency extension of this emission (the escaping component), especially the extension into the kilometric frequency range. Kilometric continuum has been reported to be observed by Polar and Cluster [Menietti et al., 39, 40] and INTERBALL-1 [Kuril'chik et al., 34, 35] in addition to Geotail, IMAGE, and CRRES [10, 11, 19, 20, and the present paper].

It has been confirmed that the kilometric continuum is generated at steep density gradients at density irregularities in the equatorial region. These irregularities do not only exist at the plasmapause, but also inside the plasmapause and in notches. Although the observations are consistent with the mode conversion mechanism at the plasma frequency, they are not consistent with the beaming model of Jones [22, 23]. The simultaneous observations given in Fig. 2.13 provide striking evidence against the latter although this is discussed in more detail by Hashimoto et al. [20]. The relations to solar and geomagnetic activities are also interesting topics.

Several new features of the high frequency escaping kilometric continuum, such as the narrow latitudinal beam structure and relationship to plasmaspheric notch or notch structures, provide a new opportunity to observe the triggering of this emission and its relationship to plasmaspheric dynamics. The insight gained in performing multi-spacecraft correlative measurements should provide key measurements on separating spatial from temporal effects that are essential to verifying existing theories. Observing the radiation while the instability has been initiated and grows, and examining the dynamics of the large-scale plasmasphere should lead to significant advances in delineating the best theory for the generation of this emission.

References

[1] Anderson, R.R., D.A. Gurnett, and D.L. Odem: CRRES plasma wave experiment, J. Spacecraft Rockets 29, 570, 1992.

[2] Barbosa, D.D.: Low-level VLF and LR radio emissions observed at earth and Jupiter, Rev. Geophys. Space Phys. 20, 316, 1982.

[3] Carpenter, D.L. and R.R. Anderson: An ISEE/whistler model of equatorial electron density in the magnetosphere, J. Geophys. Res. 97, 1097, 1992.

[4] Carpenter, D.L., R.R. Anderson, W. Calvert, and M.B. Moldwin: CRRES observations of density cavities inside the plasmasphere, J. Geophys. Res. 105, 23323, 2000.

[5] Décréau, P.M.E. et al.: Observation of continuum radiations from the Cluster fleet: First results from direction finding, Ann. Geophys. 22, 2607, 2004.

[6] Filbert, P.C. and P.J. Kellogg: Observations of low-frequency radio emissions in the earth's magnetosphere, J. Geophys. Res. 94, 8867, 1989.

[7] Frankel, M.S.: LF radio noise form the Earth's magnetosphere, Radio Sci. 8, 991, 1973.

[8] Fung, S.F. and K. Papadopoulos: The emission of narrow-band Jovian kilometric radiation, J. Geophys. Res. 92, 8579, 1987.

[9] Green, J.L. and S.A. Boardsen: Confinement of nonthermal continuum radiation to low latitudes, J. Geophys. Res. 104, 10307, 1999.

[10] Green, J.L., B.R. Sandel, S.F. Fung, D.L. Gallagher, and B.W. Reinisch: On the origin of kilometric continuum, J. Geophys. Res. 107, doi: 10.1029/2001JA000193, 2002.

[11] Green, J.L. et al.: Association of kilometric continuum radiation with plasmaspheric structures, J. Geophys. Res. 109, A03203, doi: 10.1029/2003JA010093, 2004.

[12] Green, J.L. and S.F. Fung: *Advances in Inner Magnetospheric Passive and Active Wave Research*, In: AGU Monograph on Physics and Modeling of the Inner Magnetosphere, AGU, Washington D.C., in press, 2005.

[13] Gough, M.P.: Nonthermal continuum emissions associated with electron injections: Remote plasmapause sounding, Planet. Space Sci. 30, 657, 1982.

[14] Gurnett, D.A. and R.R. Shaw: Electromagnetic radiation trapped in the magnetosphere above the plasma frequency, J. Geophys. Res. 78, 8136, 1973.

[15] Gurnett, D.A.: The Earth as a radio source: The nonthermal continuum, J. Geophys. Res. 80, 2751, 1975.

[16] Gurnett, D.A. and L.A. Frank: Continuum radiation associated with low-energy electrons in the outer radiation zone, J. Geophys. Res. 81, 3875, 1976.

[17] Gurnett, D.A., W. Calvert, R.L. Huff, D. Jones, and M. Sugiura: The polarization of escaping terrestrial continuum radiation, J. Geophys. Res. 93, 12817, 1988.

[18] Hashimoto, K., K. Yamaashi, and I. Kimura: Three-dimensional ray tracing of electrostatic cyclotron harmonic waves and Z mode electromagnetic waves in the magnetosphere, Radio Science 22, 579, 1987.

[19] Hashimoto, K., W. Calvert, and H. Matsumoto: Kilometric continuum detected by Geotail, J. Geophys. Res. 104, 28645, 1999.

[20] Hashimoto, K., R.R. Anderson, J.L. Green, and H. Matsumoto: Source and Propagation Characteristics of kilometric continuum observed with multiple satellite, J. Geophys. Res., 110, A09229, doi: 10.1029/2004JA010729, 2005.

[21] Jones, D.: Source of terrestrial nonthermal radiation, Nature 260, 686, 1976.

[22] Jones, D.: Latitudinal beaming of planetary radio emissions, Nature 288, 225, 1980.

[23] Jones, D.: Beaming of terrestrial myriametric radiation, Adv. Space Res. 1, 373, 1981.

[24] Jones, D.: Terrestrial myriametric radiation from the Earth's plasmapause, Planet. Space Sci. 30, 399, 1982.

[25] Jones, D., W. Calvert, D. A. Gurnett, and R. L. Huff: Observed beaming of terrestrial myriametric radiation, Nature 328, 391, 1987.

[26] Jones, D.: Planetary radio emissions from low magnetic latitudes – Observations and theories, In: *Planetary Radio Emissions II* (H.O. Rucker, S.J. Bauer, and B.M. Pedersen. Eds., Austrian Acad. Sci., Vienna) p. 245, 1988.

[27] Kaiser, M.L.: Observations of non-thermal radiation from planets, In: *Plasma Waves and Instabilities at Comets and in Magnetospheres*, (B. Tsurutani and H. Oya, Eds., AGU, Washington) p. 221, 1989.

[28] Kasaba, Y., H. Matsumoto, K. Hashimoto, R.R. Anderson, J.-L. Bougeret, M.L. Kaiser, X.Y. Wu, and I. Nagano: Remote sensing of the plasmapause during substorms: GEOTAIL observation of nonthermal continuum enhancement, J. Geophys. Res. 107, 20389, 1998.

[29] Kurth, W.S., J.D. Craven, L.A. Frank, and D.A. Gurnett: Intense electrostatic waves near the upper hybrid resonance frequency, J. Geophys. Res. 84, 4145, 1979.

[30] Kurth, W.S., D.A. Gurnett, and R.R. Anderson: Escaping nonthermal continuum radiation, J. Geophys. Res. 86, 5519, 1981.

[31] Kurth, W.S.: Detailed observations of the source of terrestrial narrowband electromagnetic radiation, Geophys. Res. Lett. 9, 1341, 1982.

[32] Kurth, W.S.: Continuum radiation in planetary magnetospheres, In: *Planetary Radio Emissions III* (H. O. Rucker, S. J. Bauer, and M. L. Kaiser, Eds., Austrian Acad. Sci., Vienna) p. 329, 1992.

[33] Kurth, W.S., D.A. Gurnett, A. Roux, and S.J. Bolton: Ganymede: A new radio source, Geophys. Res. Lett. 24, 2167, 1997.

[34] Kuril'chik, V.N., M.Y. Boudjada, and H.O. Rucker: Interball-1 observations of the plasmaspheric emissions related to terrestrial "continuum" radio emissions, In: *Planetary Radio Emissions V* (H.O. Rucker, M.L. Kaiser and Y. Leblanc, Eds., Austrian Acad. Sci., Vienna) p. 325, 2001.

[35] Kuril'chik, V.N., I.F. Kopaeva, and S.V. Mironov: INTERBALL-1 observations of the kilometric "continuum" of the Earth's magnetosphere, Cosmic Research 42, 1, 2004.

[36] Lee, L.C.: Theories of non-thermal radiations from planets, In: *Plasma Waves and Instabilities at Comets and in Magnetospheres* (B. Tsurutani and H. Oya, Eds., AGU, Washington) p. 239, 1989.

[37] Matsumoto, H., I. Nagano, R.R. Anderson, H. Kojima, K. Hashimoto, M. Tsutsui, T. Okada, I. Kimura, Y. Omura, and M. Okada: Plasma wave observations with GEOTAIL spacecraft, J. Geomagn. Geoelectr. 46, 59, 1994.

[38] Melrose, D.B.: A theory for the nonthermal radio continuum in the terrestrial and Jovian magnetospheres, J. Geophys. Res. 86, 30, 1981.

[39] Menietti, J.D., R.R. Anderson, J.S. Pickett, D.A. Gurnett, and H. Matsumoto: Near-source and remote observations of kilometric continuum radiation from multi-spacecraft observations, J. Geophys. Res. 108, 1393, doi: 10.1029/2003JA009826, 2003.

[40] Menietti, J.D., O. Santolik, J.S. Pickett, and D. A. Gurnett: High resolution observations of continuum radiation, Planet. Space Sci. 53, 283, 2005.

[41] Morgan, D.D. and D.A. Gurnett: The source location and beaming of terrestrial continuum radiation, J. Geophys. Res. 96, 9595, 1991.

[42] Okuda, H., M. Ashour-Abdalla, M.S. Chance, and W.S. Kurth: Generation of nonthermal continuum radiation in the magnetosphere, J. Geophys. Res. 87, 10457, 1982.

54   K. Hashimoto et al.

[43] Oya, H., M. Iizima, and A. Morioka: Plasma turbulence disc circulating the equatorial region of the plasmasphere identified by the plasma wave detector (PWS) onboard the Akebono (EXOS-D) satellite, Geophys. Res. Lett. 18, 329, 1991.)

[44] Reinisch, B.W. et al.: The Radio Plasma Imager investigation on the IMAGE spacecraft, Space Sci. Rev. 91, 319, 2000.

[45] Rönnmark, K.: Emission of myriametric radiation by coalescence of upper hybrid waves with low frequency waves, Ann. Geophys. 1, 187, 1983.

[46] Rönnmark, K.: Conversion of upper hybrid waves into magnetospheric radiation, In: *Planetary Radio Emissions III* (H.O. Rucker, S.J. Bauer, and M.L. Kaiser, Eds., Austrian Acad. Sci., Vienna) p. 405, 1992.

[47] Sandel, B.R., R.A. King, W.A. King, W.T. Forrester, D.L. Gallagher, A.L. Broadfoot, and C.C. Curtis: Initial results from the IMAGE Extreme Ultraviolet Imager, Geophys. Res. Letts. 28, 1439, 2001.

[48] Vesecky, J.F. and M.S. Frankel: Observations of a low-frequency cutoff in magnetospheric radio noise received on IMP 6, J. Geophys. Res. 80, 2771, 1975.

[49] Yamaashi, K., K. Hashimoto, and I. Kimura: 3-D electrostatic and electromagnetic ray tracing in the magnetosphere, Mem. Natl. Inst. Polar Res., Spec. Issue 47, 192, 1987.

3

Generation of Auroral Kilometric Radiation in Bounded Source Regions

P. Louarn

Centre d'Etudes Spatiales du Rayonnement, Toulouse, France
philippe.louarn@cesr.fr

Abstract. Owing to the complete and precise wave/particle measurements performed by the Viking and FAST spacecraft in the sources of the terrestrial Auroral Kilometric Radiation, the conditions of generation of this coherent radio emission are now well documented. It has been demonstrated that the production of radio waves occurs in thin laminar regions that exactly correspond to the regions of auroral particle acceleration. We present and discuss observations, mainly performed by Viking, that have led to the conception of a model of wave generation by finite geometry sources. The corresponding theoretical analysis is also presented. This model has certainly a very broad domain of applications, including radio emissions from other planets, from the Sun and more distant astrophysical objects.

Key words: Auroral Kilometric Radiation, free space radiation, magneto-ionic modes, effect of inhomogeneity, cyclotron maser mechanism, loss cone distribution, horseshoe, ring distribution

3.1 Introduction

As seen from deep space, the Earth is a powerful natural radio source emitting 10^7 to 10^8 W with a maximum spectral density at $\sim$250 kHz. Its dominant emission – the Auroral Kilometric Radiation (AKR hereafter) – is generated in the night sector of the auroral zone, at magnetic latitudes larger than 65° and altitudes ranging from 5000 km to 15000 km. Its power increases with magnetospheric activity, especially when substorms develop. This has been realized first by Benediktov et al. [6] and confirmed by others [9, 20, 21, 23]. This radiation is characterized by a brightness temperature ($>10^{15}$ K) much larger than any plasma temperature in planetary magnetospheres. It cannot be explained by a thermal generation mechanism and coherent emission processes such as plasma instabilities must be invoked for its generation.

The AKR is not a singular phenomenon. In its non-thermal origin, high polarization, temporal variability and spectral fine structure AKR is similar

P. Louarn: *Generation of Auroral Kilometric Radiation in Bounded Source Regions*, Lect. Notes Phys. **687**, 55–86 (2006)
www.springerlink.com

to the radio emissions that emanate from other magnetospheres, at Jupiter, Saturn, Uranus and Neptune [see, e.g., Zarka, 60]. It can be compared to solar radio bursts (microwave spike bursts) [1, 15] and stellar radio emissions [32, 39]. The AKR generation mechanism has thus to be considered as a general process able to efficiently convert some forms of free energy present in magnetized plasmas into radiating electromagnetic waves.

Due to the proximity of its sources, the study of the AKR offers a splendid opportunity to test all the details of this radiation process. It was actually possible to determine some important characteristics of AKR from early space measurements [6, 21, 22]. The evidence for a dominant X polarization came from Voyager measurements [Kaiser et al., 26], as the probes began their odyssey. This was confirmed later from DE-1 measurements published by Shawhan and Gurnett [55] and Mellott et al. [38], although the existence of a fainter O mode was also detected. The good correlation between inverted V precipitation and AKR [Green et al., 20] has given support to the idea that the AKR is generated by electrons accelerated above the auroral regions. Important progress has been made by the crossing of the sources regions themselves, first reported by Benson and Calvert [7] using ISIS 1 spacecraft. These direct in-situ measurements have revealed that the AKR is generated:

- in strongly magnetized regions ($f_p/f_c \ll 1$ where f_p is the plasma frequency and f_c the electron gyrofrequency),
- at frequencies close to the local cut-off of the X mode: $f_x = \frac{1}{2}[f_c + (f_c^2 + 4f_p^2)^{1/2}]$,
- with waves propagating at large angles with respect to the geomagnetic field B_0 [7, 8].

Numerous theories have been proposed for the interpretation of this coherent radiation: nonlinear mode coupling, soliton radiation and linear instabilities. In view of the observational constraints summarized above, a linear instability – the cyclotron maser – first proposed by Wu and Lee [59] and Lee and Wu [27] and further developed by several authors [28, 29, 30, 40, 41, 42], has been acknowledged by a majority as being the most promising generation mechanism. Though this relativistic instability was known before [see Trubnikov, 57], the main progress realized by Wu and Lee [59] was to show that the relativistic correction in the electron motion is crucial even at the modest keV energies of auroral electrons.

In the tenuous auroral plasmas, the X mode cut-off frequency is indeed just a few percent above f_c and the relativistic correction must be considered in the wave/particle resonant condition: $\omega - k_\parallel v_\parallel - \omega_c/\Gamma = 0$ where $\Gamma = (1 - v^2/c^2)^{-\frac{1}{2}}$ is the Lorentz factor and $k_\parallel$ the wave vector component parallel to the static magnetic field. The relativistic effects permit an exchange of energy between X mode waves and electrons. It can be shown that the plasma may act as a coherent radiating source if the electron distribution presents an inversion of population in the form of positive gradients $\partial f/\partial v_\perp > 0$. Since the most common distribution functions observed in the auroral zone – the loss-cone distributions – precisely present this type of free energy, an efficient X mode amplification may take place at frequencies close to f_c. This good agreement between theoretical predictions and observations has explained the success of this direct amplification mechanism in the community. The cyclotron maser model was thus soon recognized as the most plausible generation mechanism of non-thermal radio emission radiated by highly magnetized planetary, solar and even stellar plasmas.

Our understanding of the AKR generation was greatly improved by the measurements made by the Swedish satellite Viking in the heart of the AKR sources. It was possible to understand the connection between the particle acceleration and the production of the radiations and to get a precise knowledge of the plasma conditions inside the sources and in their near vicinity. As it is reviewed here, these observations have motivated new theoretical analyses that were severely tested. These studies have contributed to establish a sophisticated model of the AKR generation and, by extension, of the planetary radio emissions. Ten years after Viking, the superb FAST measurements have confirmed and updated this picture of wave amplification in thin laminar sources. In Sect. 3.2, we present some of the Viking observations performed in the AKR sources. The observational effects linked to the finite geometry of the sources are discussed in Sect. 3.3. The theoretical analysis in made in Sect. 3.4 before the discussion and the conclusion proposed in Sect. 3.5.

3.2 Spacecraft Observations in the Sources of the Auroral Kilometric Radiation

3.2.1 Structure of Sources and Wave Properties

The Viking spacecraft was the first to explore, with a complete set of experiments, the central part of the AKR sources, at altitudes of the order of one Earth's radius (R_E) [see the review by Roux et al., 53]. Frequency-time spectrograms corresponding to AKR source crossings by Viking have been presented by Bahnsen et al. [2, 3], Pottelette et al. [44], Louarn et al. [33], and Ungstrup et al. [58]. Figure 3.1 shows such a source crossing. The source is defined in the data by an intensification of the signal at frequencies very close to f_c. For the example shown in Fig. 3.1 (orbit 165), this occurs from ∼2033:00 to ∼2033:30 UT. During this 30 s time period, the spectral density of the AKR is maximum just at the frequency channel that contains f_c. Two additional sources are also crossed during short time periods, at ∼2032:30 and 2034:00 UT.

For the purpose of a statistical study, Hilgers et al. [24] computed the integrated electric energy E_{AKR} in the vicinity of the electron gyrofrequency (from $f_c - 5\,kHz$ to $f_c + 10\,kHz$) for the whole set of Viking data. They defined the sources as the regions along the orbit where this quantity is maximum.

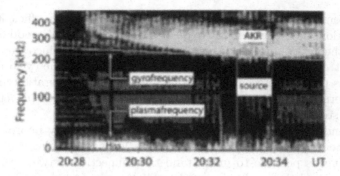

Fig. 3.1. An example of source crossed by Viking. The AKR source is indicated and is characterized by an increase of the AKR wave power at and, in some cases, even below the gyrofrequency

Almost 40 source crossings have been identified using this method. The statistical wave properties inside or near the sources were then determined with the following results:

- The lowest frequency peak f_{peak} of the AKR is on average a few percent above f_c, but it may also be up to 3% below it.
- To determine the lowest frequency of the AKR, f_{lc} is defined as the first frequency below f_{peak} where the spectral energy has decreased by 10%. This frequency is significantly below f_{peak}, by 2% in the average. A significant power is thus produced down to and often below f_c.
- The refractive index is close to unity everywhere but close to the cutoff where values smaller than unity are measured [Bahnsen et al., 3].
- Concerning the polarization, Hilgers et al. [25] showed that the orientation of the electric fields is not the same within the source at $f \sim f_c$ and outside the sources at $f \gg f_c$. Inside the sources, the modulation pattern due to the spin of the spacecraft is consistent with a wave electric field being confined within $10°$ in the plane perpendicular to B_0, as expected for X mode waves propagating close to the direction perpendicular to B_0. Outside the sources, the modulation pattern is less pronounced. The waves being transverse, this suggests that they propagate at a more oblique angle with respect to B_0.

Concerning the source extension and its lifetime, the fact that well-identified AKR patterns are seen in the dynamic spectrograms over several tens of minutes shows that the source exists for at least that long, essentially at the same location and, even on the same field lines. In most situations, the spectrum covers a broad frequency range. Since the emission is generated at or very close to f_c, it is possible to translate the frequency bandwidth into an altitude range for the source. The 100–200 kHz bandwidth of the AKR then corresponds to an extension of the order of 2500 km below the spacecraft (at a typical altitude of 6000 km). As discussed by De Feraudy et al. [13], the observation of broad spectra also implies that propagation takes place in a filled cone or, more likely, that the source is extended in east-west direction along the auroral oval. The north-south extent is always small, often less than 100 km, and is directly measured since the satellite orbit is nearly in the meridian plane. Altogether, this suggests that the AKR sources are relatively long lived regions, at a mean altitude of 6000 km, with a dimension along the Earth magnetic field of $\sim$2000 km and more. They would thus be ribbons or laminar structures, limited to a width of 20–200 km north-south and much more extending east-west.

3.2.2 AKR Sources as Regions of Particle Accelerations

An example of an AKR source was discussed in detail by Louarn et al. [33]. It occurred during the orbit 849 and has lasted a sufficiently long time to permit the complete measurements of the distribution functions. The corresponding wave and particle measurements are displayed in Fig. 3.2.

Using the criteria presented in Sect. 3.2.1, the source crossing occurred in between the two vertical red lines. This time period corresponds to important modifications in the electron and ions energy spectrograms. The energy of the electrons is maximum just outside the source where it reaches 10 keV. This energy decreases to 5 keV inside the source as upward propagating ion beams are simultaneously detected with energies of $\sim$5 keV. As discussed in Louarn et al. [33], this can be explained by assuming that the AKR source is an acceleration region characterized by V-shaped iso-potential surfaces. These observations indeed suggest that Viking was successively below, inside, and again below a zone of nonzero parallel electric field, this acceleration region coinciding with the AKR source (see the lower panels in Figs. 3.2 and 3.3). As discussed below, the fact that the sources are regions of parallel particle acceleration has several important consequences regarding the wave generation mechanism.

3.2.3 AKR Sources as Plasma Cavities

That regions of nonzero parallel electric field are regions where the plasma density is perturbed is not a surprise. This has, nevertheless, a special importance here. The plasma density is a crucial parameter for the cyclotron maser instability: it contributes to determine the different regimes of the process and the properties of the most unstable waves.

Different methods can be used to obtain the density. For example, the low frequency emission seen below 50 kHz in Figs. 3.1 and 3.2 (the auroral "hiss") are whistler waves with a upper-cutoff frequency at the local plasma frequency [see the comparison with relaxation sounder measurements by Perraut et al., 43]. This cut-off is larger than 20 kHz outside the source. This corresponds to densities above 6 cm^{-3}. It clearly decreases in the source (see the line in Fig. 3.2). A careful analysis shows that it at that point approaches or even

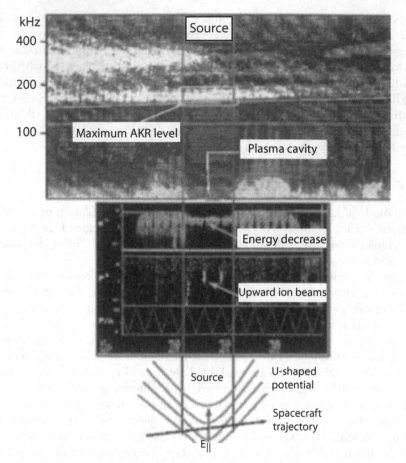

Fig. 3.2. Wave and particle measurements performed by Viking inside a source of AKR. The particle measurements are fully consistent with the crossing of a region of particle acceleration. The white line in the upper panel is the gyro frequency

goes below 10 kHz, the lower- frequency threshold of the receiver. The density of the plasma is then below 1.1 cm^{-3} in the central part of the source. This observation can be generalized to other source crossings. Hilgers et al. [24] have even shown that the auroral small-scale cavities with densities typically of the order of or less than 1.5 cm^{-3} detected by Viking coincide almost systematically with AKR sources.

To get more indications on the plasma parameters, Louarn et al. [33] have estimated the electron density within and in the vicinity of the source from the electron detector. While the density of high-energy (1–40 keV) electrons remains almost constant across the source, the density of low-energy electrons (below 1 keV) decreases by a factor larger than 10. The density variation is thus associated with a large decrease of the density of cold plasma. A

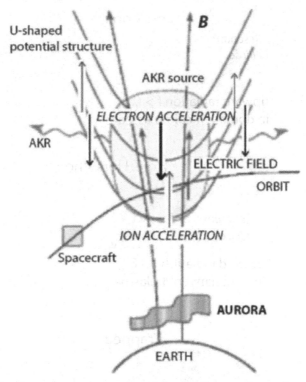

Fig. 3.3. A schematic representation of the AKR source/auroral acceleration region

simple interpretation would be that the low energy particles coming from the ionosphere have been evacuated from the source by the accelerating electric field.

3.2.4 Free Energy for the Maser Process

Let us determine the free energy that may drive the maser. At first glance, the problem could be reduced to the search for positive $\partial f / \partial v_\perp$ slopes in the distribution functions. Nevertheless, the determination of the non-thermal feature that drives the maser is ambiguous:

- several $\partial f / \partial v_\perp$ features are able to feed the instability, and
- the diffusion associated with the waves can level out, at least partly, the "operating" $\partial f / \partial v_\perp$ slopes.

Theoretical arguments are thus useful here for the identification of the most efficient form of free energy.

   In Fig. 3.4 examples of non-thermal auroral electron distribution functions are given, together with the properties of the waves that can be amplified via

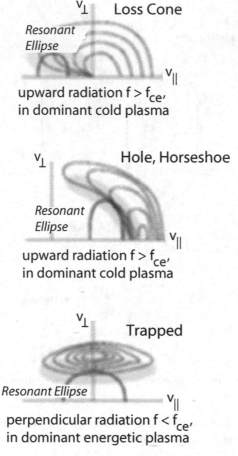

Fig. 3.4. Forms of free energy that may drive the maser process and expected properties of the unstable waves. In gray: regions of positive $\partial f/\partial v_\perp$

the cyclotron maser instability by such free energy sources. A crucial parameter for determining the dominant free energy source is the proportion of hot electrons in the plasma. In a plasma dominated by cold electrons, X mode waves propagate above the electron gyrofrequency and the resonance condition, $\omega - k_\| v_\| - \omega_c/\Gamma = 0$, can only be fulfilled if $k_\|$ is not null. Therefore, the resonant curve – an ellipse in the $v_\|, v_\perp$ phase space – is not centered at $v_\| = 0, v_\perp = 0$ and the growth rate which is proportional to the integral of $v_\perp \partial f/\partial v_\perp$ along the resonant curve is large only if free energy exists for $v_\| \neq 0$. This is the case of the "loss cone" and the "hole" distribution functions, as indicated in Fig. 3.4. Conversely, in an energetic plasma with no or a very tenuous cold component, X mode propagation is possible below the electron gyro-frequency. The Doppler shift term ($k_\| v_\|$) is no longer required to fulfill the resonant condition and free energy centered at $v_{\parallel} = 0$ can be used. The detailed theoretical analysis even shows that, for a given energy of the particles, inversions of populations centered at $v_{\parallel} = 0$ are the most efficient to drive the maser.

The three-dimensional plots displayed in Fig. 3.5 show the electron distribution functions just outside and inside an AKR source. The distribution function measured just outside the source is characterized by a dense thermal core and an enhanced population of energetic particles up to 10 keV. Two kinds of $\partial f / \partial v_{\perp}$ non-thermal features are clearly apparent: a loss cone, for electrons moving upward and a "hole" (Fig. 3.5b) for electrons moving downward. Note that the "hole" distribution functions are similar to the "horse-shoe" distributions described by Delory et al. [14], using FAST data. The source of the AKR, as identified from the criteria discussed in Sect. 3.2.1, does not coincide with these non-thermal features. This suggests that neither a loss cone nor a hole alone suffices to generate the AKR. Inside the potential structure (distributions D, E, F), the thermal core is highly depleted. The loss cone and the hole are still present but less pronounced than outside the source. The most prominent feature is a broad plateau in $v_{\perp}$ (with sometimes a faint indication of positive $\partial f / \partial v_{\perp}$ slopes), which indicates that electrons tend to accumulate in this region of the phase space. This characteristic feature of the distribution functions inside the source leads Louarn et al. [33] to propose that the trapped electrons are the free energy source of the AKR. As shown by Eliasson et al. [16] and by Louarn et al. [33, 34], this trapped electron component can be produced by a time-varying or by a spatially-varying parallel electric field.

3.2.5 FAST Observations

To date, the FAST satellite, launched in 1996, was the most recent spacecraft to explore the high altitude auroral zones. It crossed the source regions of the AKR with experiments characterized by a remarkable time resolution, the electron distributions being measured in less than a few 10 ms for example [see, 12, 19]. FAST has largely confirmed the results of Viking with even more precision on several crucial points [Ergun et al., 17, 18]:

- the correspondence between the acceleration regions and the sources of AKR,
- the fact that only accelerated electrons are present in the source, the low energy component being completely evacuated,
- the presence of the trapped distribution,
- the relatively small dimension of the sources,
- the fine temporal/spectral structuring of the radiations, in bursts of several tenths of seconds and frequency bandwidth thinner than 1 kHz. This point was not accessible to Viking observations.

These observations have reinforced the status of the cyclotron maser process as the most plausible wave generation mechanism. It was nevertheless

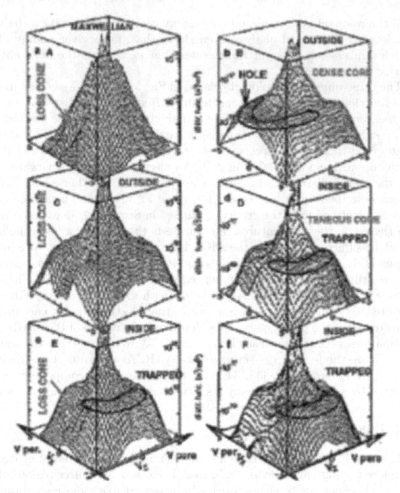

Fig. 3.5. Distribution functions measured outside and inside the AKR source. Different forms of free energy are indicated. The trapped population in only found in the source

advocated that horseshoe distribution functions, thus different from the trapped distribution, would be used by the maser [see Delory et al., 14]. Horseshoe distributions result from the same processes as trapped distributions: a combination of electric field acceleration and the motion in phase space due to the conservation of the first adiabatic invariant. This cannot be considered as a fundamental difference between the Viking and the FAST observations.

3.2.6 Summary

To conclude, an important result from Viking was to reveal that the AKR sources correspond to acceleration regions. They are laminar structures with a small north-south extension (a few 10 km, typically). The plasma in these regions is tenuous ($n \sim 1$ cm^{-3}) and essentially constituted by energetic particles ($\langle E \rangle = 3 - 10$ keV). The sources are separated from the denser and colder external plasma ($n \sim 5 - 10$ cm^{-3}, $\langle E \rangle \sim$ a few 10 eV) by sharp density gradients with typical scale lengths smaller than 1 km. An important feature of the electron distribution functions observed inside the sources is an electron accumulation at low parallel velocities and keV energies. These electrons are trapped between their magnetic mirror point and an electrostatic reflection point, leading to distributions presenting positive $\partial f/\partial v_\perp$ slopes at low $v_\parallel$ and large $v_\perp$. The theory shows that this constitutes a very efficient source of free energy for the maser.

In the tenuous plasma that fills the sources and even at moderate energy ($\langle E \rangle < 1$ keV), it can be shown that relativistic effects must be taken into account in the full X mode propagation and not only in the perturbation effect that leads to wave amplification [47, 48, 49, 56]. The relativistic dispersion is then the "zero order" theory for any realistic model of the generation of the AKR. The cyclotron maser instability is just the relativistic negative-absorption effect linked to the presence of positive $\partial f/\partial v_\perp$ slopes in the electron distributions. However, the limited extension of the sources modifies the properties of the generation process, with observable consequences regarding the radiated waves. An ensemble of papers [35, 36, 50, 51, 52] have considered this specific problem of AKR generation in finite geometry sources. The observation of finite geometry effects, the model of laminar sources, and its mathematical analysis are discussed in the next sections.

3.3 Cyclotron Maser in Finite Geometry Sources

The first detailed mentioning of finite geometry effects on AKR generation was due to Calvert [10, 11]. He proposed that the finest structures in the AKR observed by ISIS 1 could be explained by a feedback lasing effect due to wave reflection in the sources. However, this model is considerably different from the one discussed here. Before Viking, the AKR source regions were indeed assimilated to the auroral "cavity", a generic term that designates a wide region of low plasma density corresponding, as a whole, to the region of auroral precipitations. The lasing effect studied by Calvert was supposed to take place due to reflections on the edges of this large scale region. The sources explored by Viking are actually much smaller and are embedded in this large scale auroral cavity.

3.3.1 A Simple Model of AKR Sources

We have chosen the simplest model of the sources (see Fig. 3.6). They are supposed to be laminar structures, limited in one direction perpendicular to the static **B** field (x direction, the source width being $2l$) and unlimited in the y-z directions (z being the direction of $\mathbf{B}_0$). Inside the source, the electron

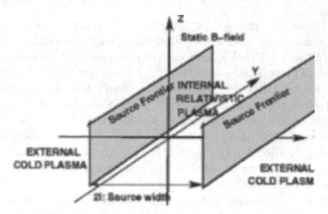

Fig. 3.6. A simple slab model for the sources of AKR

population is constituted by energetic particles with an idealized ring-like distribution: $f(v_\parallel, v_\perp) = (2\pi v_0)^{-1}\delta(v_\parallel)(v_\perp - v_0)$. The external plasma is cold and the source boundaries are sharp density gradients.

The idealized distribution neglects the thermal effects and could appear oversimplified. However, as shown by Pritchett [47], Strangeway [56], and Le Quéau and Louarn [31], it leads to dispersion equations that take into account the relativistic effects without mathematical complications. The use of more realistic distributions does not significantly improve the model. For a first approach of finite geometry effects, this simple description can be considered as precise enough. Only four parameters must be taken into account:

- the density inside the source: n_i (plasma frequency: ω_{pi}),
- the energy of the internal electron population: $\langle E \rangle = \frac{1}{2}mv_0^2$,
- the density of the external plasma: n_0 (plasma frequency: ω_{p0}) and,
- the width (l) of the source.

The ionic component is supposed to be a motionless neutralizing background. We will use the following normalized parameters: $\epsilon_i = (\omega_{pi}/\omega_c)^2, \epsilon_o = (\omega_{po}/\omega_c)^2, \delta = (v_0/c)^2$ and $L = \omega_c l/c$. For a typical source, $n_i \sim 1$ cm^{-3}, $n_o \sim 10$ cm^{-3}, energy of the electron $\sim$5 keV, $f_c \sim 200$ kHz and $l \sim 15 - 30$ km; these parameters are: $\epsilon_i = 0.2 \times 10^{-2}, \epsilon_o = 2 \times 10^{-2}, \delta = 10^{-2}$ and $L = 60$. The non-homogeneity of the geomagnetic field is another important parameter. For a dipolar field: one has $B(z) \sim B_0(l - z/H)$; with $H = R/3$ where R is the distance to the center of the planet.

3.3.2 Generation in Plasma Cavities: A Simple Approach

Some of the physical effects linked to finite geometry can be discussed in a simple way by comparing the relativistic dispersion curve and the "cold" one. As it is calculated in Sect. 3.4.1, the relativistic equation that describes the wave propagation at frequencies close to f_c has two solutions [Le Quéau and Louarn, 31]:

$$N_\perp^2 = \left(\frac{k_\perp c}{\omega_c}\right)^2 = 1 - \chi_z \qquad \text{and} \qquad N_\perp^2 = \frac{1 - 2\chi - N_\parallel^2}{1 - \chi} \qquad (3.1)$$

where $\chi_z = (\omega_p/\omega_c)^2$, and $\chi = \frac{1}{2}[\epsilon(\Delta\omega - \delta N_\parallel^2)]/(\Delta\omega + \delta)^2$. The first equation is the usual dispersion equation of the ordinary mode (O mode), here developed at the zero order in $\Delta\omega = (\omega - \omega_c)/\omega_c$. The second equation corresponds to the extraordinary (X or Z) mode. In Fig. 3.7, the corresponding dispersion curves are displayed, both for an energetic plasma (left panel) and a cold plasma (right panel). At leading order in $\Delta\omega$ and $\epsilon_{i,o}$ (which are small quantities of the order of 10^{-2} throughout this study), the cut-off and the resonance frequencies assume simple expressions: (1) in cold plasma, the X mode cut-off and Z mode resonance are $\Delta\omega_X = \epsilon_o/2$ and $\Delta\omega_Z = \epsilon_o/4$, (2) in relativistic plasma, they become: $\Delta\omega_X = \epsilon_o/2 - \delta$ and $\Delta\omega_Z = \epsilon_o/4 - \delta$.

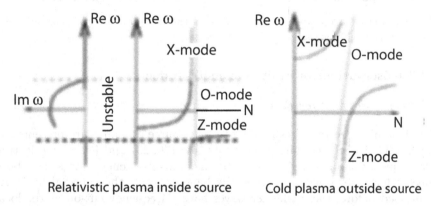

Fig. 3.7. Dispersion curves in relativistic and cold plasmas, for quasi-perpendicular propagation. Only a narrow frequency band centered at the gyrofrequency is considered here. The width of the unstable domain is δ

At first order in δ, the O-mode dispersion is not modified by relativistic effects and is stable. Conversely, due to the resonant denominator $(\Delta\omega + \delta)^2$, an instability develops on the respective frequency domains $[\Delta\omega_X, \delta]$ for the X mode and $[-\delta, \Delta\omega_Z]$, for the Z mode [Le Quéau and Louarn, 31]. In the complex frequency plane, the unstable frequencies lie on the circle $\|\Delta\omega\| = \delta$. The maximum growth rate (Im $[\Delta\omega] = \delta$) is obtained for Re $[\Delta\omega] = 0$. Let us note the existence of an energy threshold: for $\delta < 3\epsilon_i/8$ the maximum growth rate is on the Z mode, when for $\delta > 3\epsilon_i/8$ it is on the X mode. Taking $\epsilon_i = 2.5 \times 10^{-2}$ (corresponding to $f_p = 10\,\text{kHz}$ and $f_c = 200\,\text{kHz}$), this critical electron energy is $\delta = 0.094 \times 10^{-3}$ corresponding to 500 eV. This is below the typical electron energy measured in the majority of sources. The most frequent regime is rather $\delta > \epsilon_i$ which is consistent with the observation of a dominant X mode inside the source.

The examination of the dispersion curves suggests that the internally unstable waves are not easily connected to the external X mode, when the connections to the external Z and O modes are systematically possible. The direct internal/external X connection would imply: $\delta > \epsilon_o/2$. For an external plasma frequency of $\sim$30 kHz, this corresponds to $\delta = 1.12 \times 10^{-2}$ or, to energies of $\sim$5.5 keV. In most cases, the energy measured by Viking is smaller, with $\delta < \epsilon_o/2$. Let us also note that the maximum instability takes place at $\mathrm{Re}(\Delta\omega) = 0$ thus always below the external X mode cut-off. One concludes that the most unstable internal waves are rarely directly connected to the external X mode. This leads to a few questions:

- If the most unstable waves are directly connected to the external Z and O modes only, why is the level of Z and O modes as low as it is observed?
- Can the electromagnetic energy created at $\mathrm{Re}(\Delta\omega) = 0$ be transmitted to the external X mode, and by what mechanism?
- How important is the condition $\delta > \epsilon_o/2$ in the generation of AKR?

These questions are addressed by analyzing the dynamic spectra measured by Viking during the crossing of AKR sources.

3.3.3 Observations of Finite Geometry Effects

Selection of Source Crossings

The dynamic spectra corresponding to four source crossings are presented in Fig. 3.8. They occurred during the orbits 165, 176, 237 and 1260. The electron gyrofrequency is indicated by a white line. Two frequency ranges have been selected for each orbit. The 10–60 kHz band contains the "auroral hiss". As mentioned before, these whistler waves have a frequency cut-off at the local plasma frequency providing the plasma density. The radiation in the upper band is the AKR (typically 200–400 kHz). During small periods of time (a few 10 s at most, in between the vertical bars), the AKR intensifies, and its low-frequency cut-off shifts down to f_c. These are the AKR sources. They always correspond to a decrease of the power of the hiss, its cut-off being less apparent and often shifted below the band of analysis (10 kHz). This directly illustrates that the sources are plasma cavities.

For each source, the conditions of mode connection can be studied at the two interfaces corresponding to the crossings of the source frontiers. The ratio f_p/f_c at each interface, just outside the sources, is deduced from the position of the hiss cut-off. The lower value of this ratio is observed for orbit 1260 (0.06) and the higher for orbit 176 (0.208). This corresponds to frequency-gaps between f_c and the cut-off of the external X mode (f_{xo}) that vary from 1.2 kHz to more than 7 kHz. This is also the gap between the domain of wave amplification and the frequency above which the connection with the external X mode

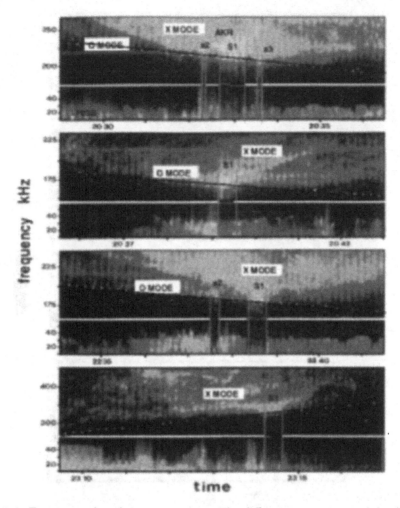

Fig. 3.8. Four examples of source crossings. The different components of the AKR are indicated. The black line is the gyro frequency

becomes possible. The density inside the source is not precisely known. Since the hiss cut-off is below 10 kHz, it is expected to be smaller than $\sim 1\,cm^{-3}$. The duration of the crossing gives an upper limit for the source width. The thinnest sources would be 20–40 km wide, the widest 150–200 km. Note that oblique crossings being possible, these numbers must thus be considered as upper limits.

Mode Identifications, Internal/External Connections

Two components in the AKR can be distinguished in Fig. 3.8. The most intense has a bowl shaped cut-off at a few frequency channels above f_c (see

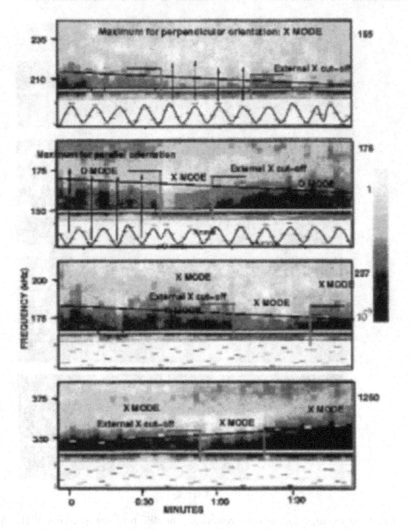

Fig. 3.9. Detailed views of the sources. The modulation due to the spin is well-apparent and is used in the mode identification

orbits 165 and 176). The analysis of the polarization shows that it is the X mode component. Between this component and f_c, a fainter O mode radiation is often detected. It can be relatively powerful as for orbit 176.

In Fig. 3.9, detailed views of the same source crossings are presented. For each orbit, three panels are shown: (1) a dynamic spectra near f_c, (2) a compressed "hiss" dynamic spectra and (3) the angle between the antenna and the geomagnetic field. This angle will be used for the study of the polarization. The temporal resolution is 2.4 s and the spectral one is 2 kHz (4 kHz above

256 kHz). The source frontiers are indicated by tick marks. We also indicate the gyrofrequency and the X mode cut-off.

Between f_c and the X cut-off in the external plasma (f_{xo}) the level of the AKR is more than 20 dB greater inside the sources than outside. This feature of the dynamic spectra shows that the electromagnetic energy is confined inside the sources, at least close to f_c. This is clearly apparent for obits 165 and 176. This difference between the internal and the external wave level decreases as the frequency increases. At frequencies a few percent above f_c, the energy propagates outward more easily or even freely. Outside the sources, the most intense external waves are then observed above f_{xo}.

The modes of propagation can be determined from the spin modulation. Inside the sources and whatever the frequency, the maxima of the AKR E-field are observed for a perpendicular orientation of the antenna which is consistent with the extraordinary mode polarization. Outside the sources and for $f > f_{xo}$, the maxima of the E-field are also observed when the antenna is perpendicular (see the example shown in the panels of orbit 165). The internal and the external fields have thus a similar polarization which explains the easy energy escape in this frequency domain. Still outside the sources but for $f_c < f < f_{xo}$, the maxima are now observed for a parallel orientation of the antenna. This is the expected polarization of O mode waves. In this frequency domain, the internal/external polarization then differs which could explain the attenuation observed at the crossing of the source frontier. It is interesting to note that the relative importance of the X and O components varies with the ratio f_c/f_p at the interfaces. For high values of this ratio, the O component is relatively intense. The O mode level is even larger than the X one for the densest case (left part of orbit 176). For low values of this ratio, the O component is comparatively less powerful and can even be undetectable (as in the case of orbit 1260). The relative proportion of O mode thus increases for sources embedded in dense plasma.

The angle of propagation of the different AKR components can also be estimated from the dynamic spectra. Given the geometry of the Viking orbit, radiation with a largely opened bowl-shaped dynamic spectra propagates at larger angles with respect to B_0 than radiation with close bowl shape. One sees that the O mode essentially propagates perpendicular to the geomagnetic field when the X mode propagates upward at oblique angles. This angle of propagation has been evaluated at the interfaces of the different sources. It clearly decreases, i.e. the propagation is more parallel, when the ratio f_p/f_c increases. For large f_p/f_c (case of the orbit 176), very oblique angles of propagation (20°) are observed [see Louarn and Le Quéau, 35].

Structure of the Electromagnetic Field and Radiating Diagram

The ratio $E_{\parallel}/E_{\perp}$ (ratio between the electric field measured when the antenna is respectively parallel and perpendicular to B_0) can be used to follow the wave refraction in the sources. This ratio increases as the frequency increases. On average for the sources presented here, one gets $E_{\parallel}/E_{\perp} \sim 0.1$ for $(f-f_c)/f_c = 0.01$ and $E_{\parallel}/E_{\perp} \sim 0.8$ for $(f-f_c)/f_c = 0.1$. This increase can be related to the refraction of the waves inside the source. Since the waves are generated very close to f_c, the analysis of waves with increasing frequencies is equivalent to the analysis of waves generated farther below. It is then possible to describe the evolution of the wave vector due to the upward propagation from the evolution of the spin modulation with the frequency. This nevertheless requires some theoretical considerations.

The three components of the wave electric field are related by the following relations:

$$E_x = -i\frac{1-\chi-N^2}{\chi}E_y \qquad (3.2)$$

$$E_z = i\frac{N_{\parallel}N_{\perp}(1-\chi-N^2)}{\chi(1-N_{\perp}^2)}E_y , \qquad (3.3)$$

where z is the direction of B_0, and the wave vector is in the x,z-plane. It can be shown that $E_{\parallel}/E_{\perp}[E_z/(E_x^2 + E - y^2)^{\frac{1}{2}}]$ depends more strongly on the variations on $N_{\parallel}$ than of the frequency. The evolution of this ratio provides thus actually an indication of the wave refraction in the source. As discussed in Louarn and Le Quéau [35], the increase of the ratio $E_{\parallel}/E_{\perp}$ from 0.1 to 0.8 would indicate that the wave vector rotates from a nearly perpendicular direction to an oblique one (~45–$60°$), over an altitude range of 100–200 km.

The measurements of $E_{\parallel}/E_{\perp}$ offer another indication. In the case of an isotropic radiation ($\langle E_x \rangle \sim \langle E_y \rangle$), it can be verified that the theoretical $E_{\parallel}/E_{\perp}$ value hardly exceeds 0.4 even for very oblique propagation. This is significantly smaller than the observed values. This can be solved by noting that the polarization can be very elliptic in the x-y plane. For example, for $N_{\parallel} \sim 0.1$ and $ky \sim 0$, one gets $E_x/E_y \sim 0.05$. If the radiating diagram is anisotropic in the x-y plane, the measurements of E may vary significantly with the orientation of the antenna in the x-y plane. Much larger $E_{\parallel}/E_{\perp}$ could be measured by antenna almost perpendicular to the dominant x/y polarization or, equivalently, if the plane of the antenna is close to the main direction of the radiating diagram (see Fig. 3.10). Given the geometry of the orbit and the cartwheel spin mode of Viking, this means that the radiation coming from the sources studied here mainly propagates in the meridian plane (north-south direction). It could appear surprising that the four orbits examined here present this particular orientation. The bias certainly comes from the choice of long-duration source crossings. This indeed leads to the selection of tangential-to-the-source crossings, thus presenting a similar meridian orientation. This cannot be considered as a generality. Despite this particularity of these source crossings, the study shows that the waves are rather emitted tangentially to the sources which suggests that they optimize their amplification path in the sources.

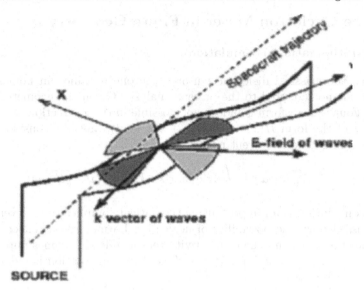

Fig. 3.10. Radiating diagram of AKR sources. The waves are preferentially emitted tangentially to the source

Summary

The main conclusions of this observational study are the following:

- The energy is systematically produced on the X mode. It is confined inside the source from f_c (the frequency of wave amplification) to f_{x0} (the frequency of possible connection with the external X mode). In this frequency range, the connection is possible with the external O and Z modes. The efficiency of transmission to the O mode is small (attenuation by 25 dB). Nevertheless, it increases as f_p/f_c increases in the external plasma. Sources embedded in a particularly dense plasma can thus even emit a dominant O mode. The Z mode transmission coefficient can be considered as null.
- The energy generated in the source propagates upward before escaping. This corresponds to a progressive refraction of the waves that is actually measured. As discussed later, the range of upward propagation is linked to the frequency gap that separates f_c and f_{xo}.
- The angle of propagation in the external plasma varies from 80°–70° to less than 20°. This angle decreases as the density gap at the source frontiers increases. A quite parallel propagation is observed for large f_p/f_c.
- The radiation diagram of a source is not isotropic: the preferred direction of emission is the tangential-to-the-source direction.

As shown by these observations, the finite geometry of the sources deeply influences the properties of the radiations. We will examine below how these effects are explained by the theory of the cyclotron maser in finite geometry.

3.4 The Cyclotron Maser in Finite Geometry

3.4.1 Mathematical Formulation

A priori, the classical dielectric tensor – the one obtained in homogeneous plasma – is not adapted to the present analysis. Given the geometry chosen here, a Fourier transform is indeed not possible in the x direction and general solutions of the form $H(x) \cdot \exp[i(k_y y + k_\| z - \omega t)]$ must be considered. The corresponding general current perturbation is:

$$\overline{J(x, k_y, k_\|, \omega)} = \int \mathrm{d}x' \sigma(x, x' k_y, k_\|, \omega) \overline{E(x', k_y, k_\|, \omega)} \qquad (3.4)$$

The x convolution is an important mathematical complication. Nevertheless, as discussed now, the possibility of neglecting Larmor radius effects greatly simplifies the problem. The conductivity tensor indeed becomes purely local: $\sigma(x, x', k_y, k_\|, \omega) = \boldsymbol{\sigma}(k_y, k_\|, \omega)\delta(x - x')$ and the convolution is reduced to a simple multiplication.

The conductivity tensor can also be directly computed starting from the Vlasov equation

$$\frac{\partial f}{\partial t} + \mathbf{v} \cdot \frac{\partial f}{\partial \mathbf{x}} - e\left(\mathbf{E} + \mathbf{v} \times \mathbf{B}\right) \cdot \frac{\partial f}{\partial \mathbf{p}} = 0 \qquad (3.5)$$

where $\mathbf{p}$ is the relativistic momentum. Using cylindrical coordinates, $\mathbf{p} = (p \cos \theta, p \sin \theta, p_\|)$, the zero and first order linear equations are:

$$\frac{\partial f_0}{\partial t} \mathbf{v} \cdot \frac{\partial f}{\partial \mathbf{x}} - e\left(\mathbf{v} \times \mathbf{B}_0\right) \cdot \frac{\partial f}{\partial \mathbf{p}} = 0 \qquad (3.6)$$

$$\left\{-i\omega + ik_\| v_\| + v_x \frac{\partial}{\partial x} - v_y \frac{\partial}{\partial y}\right.$$

$$\left. + \frac{\omega_c}{\Gamma} \frac{\partial}{\partial \theta}\right\} \delta f = e\left[\delta \mathbf{E} + \mathbf{v} \times \delta \mathbf{B}\right] \cdot \frac{\partial f_0}{\partial \mathbf{p}} \qquad (3.7)$$

The zero order equation reads

$$\frac{\Gamma v_\perp}{\omega_c} \cos \theta \frac{\partial f_0}{\partial x} + \frac{\partial f_0}{\partial \theta} = 0 \, . \qquad (3.8)$$

If the Larmor radius (ρ) is neglected, one notes that the idealized ring-like distribution (see Sect. 3.2.1) is a solution of (3.8). Concerning the first order (3.7), supposing that the transverse scale of the perturbation ($L = 1/k$ for a wave) is large compared to ρ, one can neglect the terms $v_x \partial / \partial x$ and $v_y \partial / \partial y$. They are indeed of the order $\omega_c \rho / L$ and thus very small compared to ω and ω_c. The first-order equation then reduces to

$$\left\{-i\omega + ik_\| v_\| + \frac{\omega_c}{\Gamma} \frac{\partial}{\partial \theta}\right\} \delta f = e\left[\delta \mathbf{E} + \mathbf{v} \times \delta \mathbf{B}\right] \cdot \frac{\partial f_0}{\partial \mathbf{p}} \, . \qquad (3.9)$$

This is simply the classical equation obtained in an homogeneous plasma supposing $k_\perp = 0$. The conductivity tensor can be deduced from (3.9). Equivalently, one can make the assumption $k_\perp = 0$ in the general expression of this tensor that writes [see, e.g., Bekefi, 5, p. 229]

$$\sigma = i2\pi\epsilon_0 \frac{\omega_p^2}{\omega} \int_0^\infty p_\perp \, dp_\perp \int_{-\infty}^\infty \frac{dp_\parallel}{\Gamma} \sum_{m=-\infty}^\infty \frac{\mathbf{M}_m}{\omega - k_\parallel v_\parallel - m\omega_c/\Gamma} \tag{3.10}$$

with:

$$\mathbf{M}_m = \left\{ \begin{array}{ccc} v_\perp U \left(\frac{m J_m}{b}\right)^2 & -iv_\perp U \left(\frac{m J_m J'_m}{b}\right) & v_\perp W_m \left(\frac{m J_m^2}{2}\right) \\ iv_\perp U \left(\frac{m J_m J'_m}{b}\right) & v_\perp U \left(J'_m\right)^2 & -iv_\perp W_m J_m J'_m \\ v_\parallel U \left(\frac{m J_m^2}{2}\right) & -iv_\parallel U J_m J'_m & v_\parallel W_m J_m^2 \end{array} \right\}$$

where

$$U = m_e \Gamma \omega \frac{\partial f}{\partial p_\perp} + k_\parallel \left(p_\perp \frac{\partial f}{\partial p_\parallel} - p_\parallel \frac{\partial f}{\partial p_\perp} \right)$$

$$W_m = m_e \Gamma \left[\omega \frac{\partial f}{\partial p_\parallel} - \frac{m\omega_c}{\Gamma p_\perp} \left(p_\perp \frac{\partial f}{\partial p_\parallel} - p_\parallel \frac{\partial f}{\partial p_\perp} \right) \right]$$

where the J_m are the Bessel functions of order m, with argument $b = \Gamma k_\perp v_\perp / \omega_c$, and m_e is the rest mass of the electrons. For the problem under consideration here, i.e. the generation of supra-luminous mode in a moderately energetic plasma, the argument b is small. Retaining only the leading term in the development of the Bessel functions, one gets:

$$\sigma = -i2\pi\epsilon_0 \frac{\omega_p^2}{\omega} \int_0^\infty v_\perp \, dv_\perp \int_{-\infty}^\infty dv_\parallel \left\{ \frac{\omega v_\parallel}{\omega - k_\parallel v_\parallel} \frac{\partial f}{\partial v_\parallel} \mathbf{M}_z \right.$$

$$+ \frac{v_\perp}{4} \left[\omega \frac{\partial f}{\partial v_\perp} + k_\parallel \left(v_\perp \frac{\partial f}{\partial v_\parallel} - v_\parallel \right) \right] \tag{3.11}$$

$$\left. \times \left[\frac{\mathbf{M}_-}{\omega - k_\parallel v_\parallel - \frac{\omega_c}{\Gamma}} + \frac{\mathbf{M}_+}{\omega - k_\parallel v_\parallel + \frac{\omega_c}{\Gamma}} \right] \right\}$$

where

$$\mathbf{M}_z = \left\{ \begin{array}{ccc} 0 & 0 & 0 \\ 0 & 0 & 0 \\ 0 & 0 & 1 \end{array} \right\} \qquad \mathbf{M}_- = \left\{ \begin{array}{ccc} 1 & -i & 0 \\ i & 1 & 0 \\ 0 & 0 & 0 \end{array} \right\} \qquad \mathbf{M}_+ = \left\{ \begin{array}{ccc} 1 & i & 0 \\ -i & 1 & 0 \\ 0 & 0 & 0 \end{array} \right\}$$

Assuming that the distribution function is an idealized ring-like function: $f(v_\parallel, v_\perp) = (2\pi v_0)^{-1} \delta(v_\parallel) \delta(v_\perp - v_0)$, this becomes:

$$\epsilon = \left\{ \begin{matrix} 1 - \chi_1 & i\chi_2 & 0 \\ -i\chi_2 & 1 - \chi_1 & 0 \\ 0 & 0 & 1 - \chi_z \end{matrix} \right\} \qquad (3.12)$$

with

$$\chi_{1,2} = \pm \frac{\omega_p^2}{2\omega^2} \left\{ \frac{1}{1 + \omega_c/\omega} + \frac{1}{1 - (1 - \delta)\omega_c/\omega} \pm \frac{\delta(N_\parallel^2 - 1)}{[1 - (1 - \delta)\omega_c/\omega]} \right\},$$

$$\chi_z = -(\omega_p/\omega)^2$$

With the additional assumption that the generation takes place close to the gyrofrequency or $\Delta\omega = (\omega - \omega_c)/\omega_c \ll 1$, one obtains the simplified dielectric tensor that will be systematically used here:

$$\epsilon_{i,o} = \left\{ \begin{matrix} 1 - \chi_{i,o} & i\chi_{i,o} & 0 \\ -i\chi_{i,o} & 1 - \chi_{i,o} & 0 \\ 0 & 0 & 1 - \chi_{i,o}^z \end{matrix} \right\} \qquad (3.13)$$

where the abbreviations are

$$\chi_i = \frac{\epsilon_i}{2} \frac{\Delta\omega + \delta N_\parallel^2}{(\Delta\omega + \delta)^2}, \qquad \chi_o = \frac{\epsilon_o}{2} \frac{1}{\Delta\omega}, \qquad \chi_{o,i}^z = \epsilon_{o,i}$$

and the subscripts i, o refer to the internal or the external plasma. Let us now take into account the finite geometry. In each of the three homogeneous regions that constitute the slab geometry, the Maxwell equations read:

$$\nabla \times (\nabla \times \mathbf{E}) - \frac{\omega^2}{c^2} \epsilon \cdot \mathbf{E} = 0$$

$$\nabla \times (\nabla \times \mathbf{H}) + i\omega\epsilon_0 \nabla \times (\epsilon \cdot \mathbf{E}) = 0 \qquad (3.14)$$

It can be shown that the components of the electromagnetic field can be deduced from the parallel components only ($E_\parallel$ and $H_\parallel$, with $H = \mu_0 B$):

$$E_x = -\frac{i}{\Delta} \left\{ k_\parallel \left[(N_\parallel^2 - 1 + \chi)\frac{\partial E_\parallel}{\partial x} + i\chi\frac{\partial E_\parallel}{\partial y} \right] \right.$$
$$\left. + \omega\mu_0 \left[-i\chi\frac{\partial H_\parallel}{\partial x} + (N_\parallel^2 - 1 + \chi)\frac{\partial H_\parallel}{\partial y} \right] \right\} \qquad (3.15)$$

$$E_y = -\frac{i}{\Delta} \left\{ k_\parallel \left[-i\chi\frac{\partial E_\parallel}{\partial x} + (N_\parallel^2 - 1 + \chi)\frac{\partial E_\parallel}{\partial y} \right] \right.$$
$$\left. - \omega\mu_0 \left[(N_\parallel^2 - 1 + \chi)\frac{\partial H_\parallel}{\partial x} + i\chi\frac{\partial H_\parallel}{\partial y} \right] \right\} \qquad (3.16)$$

and

$$H_x = -\frac{1}{\omega\mu_0} \left(k_\parallel E_y + i\frac{\partial E_\parallel}{\partial y} \right), \qquad H_y = \frac{1}{\omega\mu_0} \left(k_\parallel E_x + i\frac{\partial E_\parallel}{\partial x} \right) \qquad (3.17)$$

Here $\Delta = (\omega^2/c^2)[(1 - \chi - N_\parallel^2)^2 - \chi^2]$.

The parallel components of the electromagnetic field are related by the equations

$$\left[\nabla_\perp^2 + \frac{\omega^2}{c^2}\frac{1-\chi^z}{1-chi}(1-\chi+N_\parallel^2)\right]E_\parallel = i\frac{\mu_0}{c}\omega^2 N_\parallel\frac{\chi}{1-\chi}H_\parallel \qquad (3.18)$$

$$\left[\nabla_\perp^2 + \frac{\omega^2}{c^2}\left(\frac{1-2\chi}{1-\chi}-N_\parallel^2\right)\right]H_\parallel = -i\frac{\epsilon_0}{c}\omega^2 N_\parallel\frac{\chi(1-\chi^2)}{1-\chi}E_\parallel \qquad (3.19)$$

where $\nabla_\perp^2$ is the "perpendicular" Laplacean. For perpendicular propagation ($N_\parallel = 0$), these equations respectively correspond to the dispersion of the O and the extraordinary (X and Z) mode. They can be combined to obtain a fourth-order equation for either the parallel electric or magnetic components. This equation must be solved in the different plasma regions. Using the continuity conditions for $E_y, H_y, E_\parallel$ and $H_\parallel$, one gets the compatibility condition that determines the set of discrete acceptable solutions. At the leading order in $\Delta\omega$, one finds:

$$\left[\frac{\partial}{\partial\tilde{x}} + \frac{\partial}{\partial\tilde{y}} + \left(1-\frac{N_\parallel^2}{1-\chi}\right)\right]E_\parallel = ic\mu_0 N_\parallel\frac{\chi}{1-\chi}H_\parallel \qquad (3.20)$$

$$\left[\frac{\partial}{\partial\tilde{x}} + \frac{\partial}{\partial\tilde{y}} + \frac{\omega^2}{c^2}\left(\frac{1-2\chi}{1-\chi}-N_\parallel^2\right)\right]H_\parallel = -ic\epsilon_0 N_\parallel\chi 1-\chi E_\parallel \qquad (3.21)$$

where $(\tilde{x},\tilde{y}) = (\omega_c/c)(x,y)$. We assume that $\chi^z = 0$ which is justified by the low value of $(\omega_p/\omega_c)^2$ inside and near the sources of AKR. With this simplification, the fourth-order differential operator can be combined into two second-order wave operators corresponding respectively to the extraordinary and O modes:

$$\left[\frac{\partial}{\partial\tilde{x}} + \frac{\partial}{\partial\tilde{y}} + \left(\frac{1-2\chi-N_\parallel^2}{1-\chi}\right)\right]\left[\frac{\partial}{\partial\tilde{x}} + \frac{\partial}{\partial\tilde{y}} + \left(1-N_\parallel^2\right)\right]\frac{H_\parallel}{E_\parallel} = 0. \qquad (3.22)$$

At this order of simplification, the "O mode" operator simply becomes the "vacuum" operator. The O mode then "sees" no difference between the source and the external plasma. In this limiting case, it is stable, it freely propagates, and it is not coupled with the X mode. Let us consider the extraordinary mode:

$$\left[\frac{\partial}{\partial\tilde{x}} + \frac{\partial}{\partial\tilde{y}} + \left(\frac{1-2\chi-N_\parallel^2}{1-\chi}\right)\right]H_\parallel = 0. \qquad (3.23)$$

Let us choose symmetric solutions in each portion of the slab geometry. Inside the source

$$H_\parallel^i(x,y,z,t) = \cos\left(s_i\tilde{x}\exp\left[i(N_y\tilde{y} + N_\parallel\tilde{z} - \omega t)\right]\right). \qquad (3.24)$$

Outside the source (+ on the right side, − on the left side)

$$H_\|^o(x,y,z,t) = A \exp(\pm i s_o \tilde{x}) \exp\left[i(N_y \tilde{y} + N_\| \tilde{z} - \omega t)\right] . \tag{3.25}$$

A is a normalization factor. $s_{i,o}$ are the transverse wave numbers outside and inside the source, respectively. They satisfy the equation

$$s_{i,o}^2 + N_y^2 = \frac{1 - 2\chi_{i,o} - N_\|^2}{1 - \chi_{i,o}} . \tag{3.26}$$

Now, using the condition of continuity of $E_y, H_y, E_\|$ and $H_\|$, one gets a relation between the amplitude inside and outside the sources and a compatibility condition:

$$A = \cos(s_i L) \exp[-i s_{i,o} L] \tag{3.27}$$

$$(N_\|^2 - 1 - \chi_i)\frac{s_i}{\Delta_i} \sin(s_i L) + i(N_\|^2 - 1 + \chi_o)\frac{s_o}{\Delta_o} \cos(s_i L)$$

$$= N_y \left(\frac{\chi_o}{\Delta_o} - \frac{\chi_i}{\Delta_i}\right) \cos(s_i L) . \tag{3.28}$$

Equation (3.28) can be considered as the dispersion equation that contains the physical effects due to the finite geometry of the source region.

3.4.2 Solutions of the Dispersion Relation

The main results concerning the solutions of (3.28) are summed up below:

- *The solutions are discrete.* They are localized both in the complex frequency and $\mathrm{Re}(\omega)/\mathrm{Re}(k)$ dispersive plane along curves presenting strong analogies with those obtained in the homogeneous case. In particular, the frequency of the X mode cut-off and of the Z mode resonance are not modified. This is illustrated in Fig. 3.11, where the solutions (for $N_\| = 0$ and $N_y = 0$) are plotted. Two sets of discrete solutions are obtained (i) supraluminous solutions ($s < 1$) with the same cut-off as the "homogeneous" X

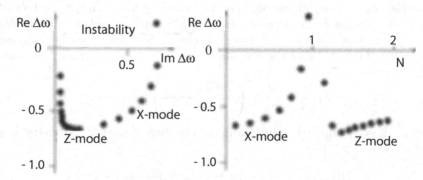

Fig. 3.11. Repartition of the discrete solutions in the complex frequency and dispersion plane. Normalized frequencies are used (see text)

mode and (ii) infra-luminous solutions ($s > 1$) that accumulate near the
Z mode resonance. The cut-off and the resonance do not depend on the
width of the source. The wave vectors of two successive solutions differ by
$2\pi/L$, the quantization rule being $s_i(n) \sim 2n\pi/L$. For $s \sim 1$, the frequency
gap between two successive solutions is thus of the order of $2c\pi/L$ which
corresponds to 200 Hz for a source of 30 km width.

- *The cyclotron maser instability presents the same regimes as in the homogeneous case.* They correspond to parameter ranges independent on the
  source width: for $\delta > \epsilon_i/2$, the most unstable solutions are on the X mode
  near f_c and $N \sim 1$. For $3\epsilon_i/8 < \delta < \epsilon_i/2$, the most unstable solutions
  are on the X mode near the cut-off and $N \sim 0$, for $3\epsilon_i/8 > \delta$, the most
  unstable solutions are on the Z mode near the resonance and $N \gg 1$.
- *Nevertheless, compared to the homogeneous case, both the maximum growth
  rate and the domain of instability are slightly reduced.*
- *Whatever the value of the electron energy, there is no possibility of a direct
  connection between the internal unstable waves and the external X mode
  waves.* From the simple approach (see Sect. 3.3.2), one could conclude
  that for very energetic sources, such that $\delta > \epsilon_o/2$, some internal unstable
  waves might directly escape the source on the X mode. The complete
  analysis shows that this is not the case. When δ is increased, the instability
  domain is indeed shifted towards lower frequencies which further increases
  the frequency gap with the external X mode cut-off. In fact, the direct
  connection is only possible for overdense source regions ($\epsilon_i > \epsilon_o$).
- *Polarization of the discrete mode.* Inside the source, for $s \sim 0$, the polar-
  ization is circular ($|E_x| \sim |E_y|$). For $s \sim 1$ it is elliptic ($|E_x| < |E_y|$) and,
  for $s \gg 1$ it becomes quasi-longitudinal ($|E_x| > |E_y|$). The external part of
  the solutions corresponds to Z mode wave. The amplitudes of the internal
  and of the external parts of the solutions are of the same order, the level
  of the external Z mode is similar to that of the internal X mode. This
  result was also obtained in the simulation studies of Pritchett [50, 52]. It
  is in contradiction with the observations since no Z mode escapes from the
  sources.

3.4.3 The Problem of Energy Escape

The problem of a too large theoretical production of the Z mode can be
solved by considering an oblique propagation ($N_y \neq 0$). The component of the
Poynting flux that corresponds to the energy escape is proportional to $S_x = E_y H_\parallel$. As already mentioned, the polarization is strongly elliptic for $N \sim 1$,
and E_y can be decreased by rotating the wave vector towards y direction.
The X/Z transmission coefficient can then be reduced by a factor larger than
20, from $S_x \sim 0.9$ if $N_y = 0$ to $S_x < 0.05$ for $N_y \sim 0.95$. Wave generation
at $N_y \sim 1$ would thus explain the very low level of Z mode production.
This implies that the radiation would be preferentially generated tangentially

rather than perpendicularly to the source. This is precisely what was deduced from the analysis of the polarization.

As discussed in Louarn and Le Quéau [36], the fact that waves escape from the sources on the X mode can be explained by considering the upward propagation inside the source. When it propagates upward, the electromagnetic energy generated at a given frequency f indeed "sees" a decreasing f_c and thus f_{xo}. At some altitude above the region of generation, f becomes larger than f_{xo} and the connection with the external X mode is possible. For this mechanism to be efficient, the connections between the internal X mode and the external O and Z modes must be as inefficient as possible. Otherwise, the energy would be converted into O and Z modes prior to the possible escape on the X mode.

From both the observations and theoretical considerations, we thus get a new scenario of generation of the different components of the AKR. During the upward propagation inside the source, the internal X mode waves are connected to the external O and Z modes and a part of the initial energy is converted into Z and O modes. Nevertheless, the transmission coefficients (internal X mode → external O or Z modes) being small, the electromagnetic energy remains confined inside the source until the connection with the external X mode becomes possible (see Fig. 3.12). This indirect production of the observable radio emission by a mode conversion at the source frontiers is one of the specific properties of the laminar source model. Let us quantify this effect. If W is the density of electromagnetic energy inside the source, L the width of the source, T the coefficient of energy transmission across the interfaces, $v_{g\perp}$ and $v_{g\parallel}$ the perpendicular and parallel group velocities, one

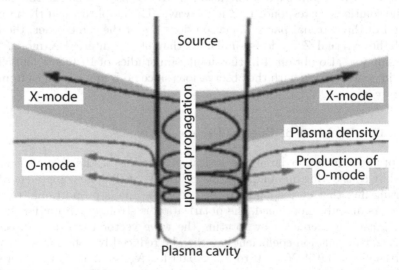

Fig. 3.12. Schematic of wave refraction and propagation inside and outside the sources. Sources corresponding to deep plasma cavities produce more O mode

has:

$$L\frac{\partial W}{\partial h} = -2T\frac{v_{g\perp}}{v_{g\|}}W \qquad (3.29)$$

where h is the altitude. The energy radiated from the source, assuming an upward propagation from h_o to h is:

$$W(h) = W_0 \exp\left[-2\int_{h_o}^{h} T\frac{v_{g\perp}}{v_{g\|}}\frac{dz}{L}\right] \qquad (3.30)$$

T can be split into three terms: T_{xx} the X/X transmission, T_{xo} the X/O and, T_{xz}, the X/Z ones. No Z mode escapes from the source which, as already presented, could be explained if the radiation is produced tangentially to the source. Thus, with this polarization, one has $T_{xz} \sim 0$. T_{xx} is zero from h_o to the altitude of possible connection with the X mode (h_x). This altitude is such that:

$$f_{xo}(h_x) = f_c(h_x)(1 + \epsilon_0/2) \le f_c(h_o) , \qquad (3.31)$$

where $f_{xo}(h)$ is the cut-off of the external X mode at altitude h. Assuming a linear variation of the magnetic field, $h = h_x - h_o$ is of the order of $R\epsilon_0/6$. This altitude range is $\sim100\,$km if $R = 15000\,$km and $\epsilon_0 = 4\times 10^{-2}$. T_{xo} can be calculated from the continuity of the electromagnetic field components. Using the dispersion relation in Sect. 4.1, it is possible to perform a parametric study of T_{xo}. One approximately gets:

$$T_{xo} = 10^{-5}\epsilon_0\delta^3 \qquad (3.32)$$

Using the equation of transfer, one obtains the relative proportion of X and O mode produced by the source, assuming that the energy not converted into O mode escapes as X mode whenever this becomes possible:

$$O/X = (1 - A)/A \quad \text{with} \quad A \sim \exp[-10^{-4}\epsilon_0\delta^3(\epsilon_0 + \delta/2)R/L] \quad (3.33)$$

This expression shows that the proportion of O mode increases if:

- the energy of the emitting particles increases (role of δ),
- the density of the external plasma increases (role of ϵ_0),
- the distance to the planet increases (R),
- or the width of the source decreases (L).

In conclusion, narrow energetic sources, embedded in a dense plasma, produce a more powerful O mode radiation. Again, as explained in Sect. 3.3, this is an observational fact.

3.5 Discussion and Conclusion

3.5.1 O and X Mode Production in Narrow Sources

An important aspect of the laminar source model is to separate the initial generation of the electromagnetic energy (the maser inside the source) and the production of the observable radio emission (by conversion across the source frontiers). One may consider that the maser operates inside the sources with maximum efficiency: this produces X mode waves at the gyrofrequency in a fully relativistic plasma. The produced electromagnetic energy is confined in the source, and mode conversion must be taken into account for explaining the energy escape. During this stage, radiation in the O mode is produced. To illustrate this scenario, we have performed a parametric study corresponding to terrestrial kilometric and Jovian decametric radiation respectively [Louarn, 37]. These two cases differ by the distances from the center of the planet ($R = 15000$ km for the Earth and $R = 70000$ km for Jupiter) and the value of the gyrofrequency, a factor of 100 larger at Jupiter than at Earth. Since f_c is used for normalization, one obtains normalized parameters ($\epsilon_{i,o}, L \ldots$) much smaller in the Jovian than the terrestrial cases. For the terrestrial case, one shows that only relatively extreme conditions (width smaller than 10 km, strong density: $\epsilon_o \sim 0.3$ and high energy: 10 keV) lead to the generation of a large fraction of O mode, as is actually observed with Viking. For the Jovian case, the expected very low values of ϵ_o due to the strong magnetization of the plasma make O mode production even more difficult. Even for very narrow sources and energetic particles, the X mode is thus expected to dominate. It is nevertheless possible that energetic very low altitude sources (so that ϵ_o could be larger) emit in the O mode.

This discussion could be extended to other types of radio emission. For example, in the solar corona, the laminar source model is consistent with the idea of fragmented energy release. During a flare, the global energy release could well take place over small scale regions that could also be the sources of non-thermal emission. The polarization of the radiation coming from such a strongly structured fibrous-like corona would simply result from the conversion processes at the frontiers of the sources, not from specific generation processes [see Sharma et al., 54].

3.5.2 Fine Structure of Radiation

What could be the radiating diagram of a laminar source? Could it explain some of the fine structures observed in the radio spectra by Baumback and Calvert [4], Ergun et al. [17], and Pottelette et al. [45]?

As already discussed, the filamentary geometry creates an asymmetry around the magnetic field. The x and y directions are indeed not equivalent: k_y is a free parameter when k_x is quantized ($k_x = 2\pi n/L$). For a given k_y, the unstable waves are thus emitted at discrete frequencies, along preferential directions of propagation, the radiation diagram being a succession of narrow beams in the plane perpendicular to $\mathbf{B}_0$. At a different k_y, a slightly modified diagram will be obtained with different quantized frequencies. Due to the possibility of continuous variation of k_y, the narrow beams likely overlap and, at some distance from the source, a continuous dynamic spectrum would be observed. This point is discussed by Pritchett et al. [52]. Without other ingredients as, for example, a mechanism that would explain wave generation at preferred altitudes or an interaction with non-linear structures [see Pottelette and Treumann, 46, this issue], it seems unlikely that the laminar source model alone may explain the fine structures.

Nevertheless, the possibility of the formation of conics has still not been analyzed. Due to the quantization of k_x, it is indeed not excluded that the energy radiated by the source concentrates along defined ray-paths. A spacecraft crossing these ray-paths would see a more powerful emission at defined frequencies. Another possibility would be that both k_y and k_x are quantized as it would be the case for sources limited in both the x and the y directions. This 3-D model also remains to be studied.

3.5.3 Conclusion and Pending Questions

To conclude, the laminar source model has been precisely tested, both observationally and theoretically, and it is certainly not exaggerated to affirm that the AKR generation is today one of the best-known magnetospheric phenomena. Given the similarity of the terrestrial auroral phenomena with those at Jupiter and Saturn, the laminar source model can also certainly be generalized.

One fundamental point is still lacking: the prediction of the power of the emission, in relation with the energy and the number of the accelerated particles. An important difficulty here is the identification of the process that saturates the maser. Is this just the convection of the energy out of the source? Is it some non-linear effect, for example the relaxation of the free energy by non-linear wave/particle interactions? The full quantitative understanding of the coherent radio wave production remains an exciting goal for the future, with many potential applications regarding the interpretation of radio emission from distant astrophysical objects.

References

[1] Aschwanden, M.J. and A. Benz: On the electron cyclotron instability: II. Pulsation in the quasi-stationary state, Astrophys. J. 332, 466, 1988.

[2] Bahnsen, A., M. Jespersen, E. Ungstrup, and I.B. Iversen: Auroral hiss and kilometric radiation measured from the Viking satellite, Geophys. Res. Lett. 14, 471, 1987.

[3] Bahnsen A., B.M. Pedersen, M. Jespersen, E. Ungstrup, L. Eliasson, J.S. Murphree, D. Elphinstone, L. Blomberg, G. Hohmgren, and L.J. Zanetti: Viking observations at the source of the AKR, J. Geophys. Res. 94, 6643, 1989.

[4] Baumback, M.M. and W. Calvert: The minimum bandwidths of auroral kilometric radiation, Geophys. Res. Lett. 14, 119, 1987.

[5] Bekefi, G.: *Emission processes in plasma*, p. 229, Gordon & Breach, New York, 1963.

[6] Benediktov, E.A., G.G. Getmantsev, Y.A. Sazonov, and A.F. Tarojov: Preliminary results of measurements of the intensity of distributed extraterrestrial radio frequency emission at 725 and 1525 kHz (in Russian), *Kosm. Issled.* 1, 614, 1965.

[7] Benson, R.F. and W. Calvert: ISIS-1 observations at the source of auroral kilometric radiation, Geophys. Res. Lett. 6, 479, 1979.

[8] Benson, R.F., W. Calvert, and D.M. Klumpar: Simultaneous wave and particle observations in the auroral kilometric radiation source region, Geophys. Res. Lett. 7, 959, 1980.

[9] Benson, R.F. and S.I. Akasofu: Auroral kilometric radiation/aurora correlation, Radio Sci. 19, 527, 1984.

[10] Calvert, W.: The auroral plasma cavity, Geophys. Res. Lett. 8, 919, 1981.

[11] Calvert, W.: A feedback model for the source of auroral kilometric radiation, J. Geophys. Res. 87, 8199, 1982.

[12] Carlson, C.W., R.F. Pfaff, and J.G. Watzin: The fast auroral snapshot (FAST) mission, Geophys. Res. Lett. 25, 2017, 1998.

[13] De Feraudy, H., B.M. Pedersen, A. Bahnsen, and M. Jespersen: Viking observations of AKR. from plasmasphere to night auroral oval source region, Geophys. Res. Lett. 14, 511, 1987.

[14] Delory, G.T., R.E. Ergun, C.W. Carlson, L. Muschietti, C.C. Chaston, W. Peria, J.P. McFadden, and R.J. Strangeway: FAST observations of electron distributions within AKR source regions, Geophys. Res. Lett. 25, 2069, 1998.

[15] Dulk, G.A.: Radio emission From the sun and the stars, Ann. Rev. Astron. Astrophys. 23, 169, 1985.

[16] Eliasson, L., G.A. Holmgren, and K. Rönnmark: Pitch-angle and energy distributions of auroral electrons measured by the ESRO-4 satellite, Planet. Space Sci. 27. 87, 1979.

[17] Ergun, R.E., et al.: FAST satellite wave observations in the AKR source region, Geophys. Res. Lett. 25, 2061, 1998.

[18] Ergun, R.E., C.W. Carlson, J.P. McFadden, G.T. Delory, R.J. Strangeway, and P.L. Pritchett: Electron-cyclotron maser driven by charged particle acceleration from magnetic field-aligned electric fields, Astrophys. J. 538, 456, 2000.

[19] Ergun, R.E., et al.: The FAST satellite fields instrument, Space Sci. Rev. 98, 67, 2001.

[20] Green, J.L., D.A. Gurnett, and R.A. Hoffman: Correlation between auroral kilometric radiation and inverted-V electron precipitations, J. Geophys. Res. 84, 5216, 1979.

[21] Gurnett, D.A.: The Earth as a radio source: terrestrial kilometric radiation, J. Geophys. Res. 79, 4227, 1974.

[22] Gurnett, D.A. and J.L. Green: On the polarization and origin of auroral kilometric radiation, J. Geophys. Res. 83, 689, 1978.

[23] Gurnett, D.A. and R.R. Anderson: The kilometric radio emission spectrum; Relation to auroral acceleration processes, in Physics of Auroral Arc Formation, Geophys. Monogr. Ser., vol. 25, edited by S.-l. Akasofu, and J. R. Kan, p. 341, AGU, Washington, D. C., 1981.

[24] Hilgers, A., A. Roux, and R. Lundin: Characteristics of AKR sources: a statistical description, Geophys. Res. Lett. 18, 1493, 1991.

[25] Hilgers, A., H. de Feraudy, and D. Le Quéau: Measurement of the direction of the auroral kilometric radiation electric field inside the source with the Viking satellite, J. Geophvs. Res. 79, 8381, 1992.

[26] Kaiser, M.L., J.K. Alexander, A.C. Riddle, J.B. Pierce, and J.W. Warwick: Direct measurements by Voyager 1 and 2 of the polarization of terrestrial kilometric radiation, Geophys. Res. Lett. 5., 857, 1978.

[27] Lee, L.C. and C.S. Wu: Amplification of radiation near cyclotron frequency due to electron population inversion, Phys. Fluids 22, 1348, 1980.

[28] Le Quéau, D., R. Pellat, and A. Roux: Direct generation of the auroral kilometric radiation by the maser synchrotron instability: An analytical approach, Phys. Fluids 27, 247, 1984a.

[29] Le Quéau, D., R. Pellat, and A. Roux: Direct generation of the auroral kilometric radiation by the maser synchrotron instability: Physical discussion of the mechanism and parametric study, J. Geophys. Res. 89, 2841, 1984b.

[30] Le Quéau, D., R. Pellat, and A. Roux: The maser synchrotron instability in an inhomogeneous medium: Application to the generation of the auroral kilometric radiation, Ann. Geophys. 3, 273, 1985.

[31] Le Quéau, D. and P. Louarn: Analytical study of the relativistic dispersion: application to the generation of the AKR, J. Geophys. Res. 94, 2605, 1989.

[32] Louarn, P., D. Le Quéau, and A. Roux: A new mechanism for stellar radio bursts: the fully relativistic electron maser, Astron. Astrophys. 165, 211, 1986.

[33] Louarn, P., A. Roux, H. de Feraudy, D. Le Quéau, M. André, and L. Matson: Trapped electrons as a free energy source for the auroral kilometric radiation, J. Geophys. Res. 95, 5983, 1990.

[34] Louarn, P., D. Le Quéau, and A. Roux: Formation of electron trapped population and conics inside and near auroral acceleration region, Ann. Geophys. 9, 553, 1991.

[35] Louarn, P. and D. Le Quéau: Generation of the auroral kilometric radiation in plasma cavities, I, Experimental study, Planet. Space Sci. 44, 199, 1996a.

[36] Louarn, P. and D. Le Quéau: Generation of the auroral kilometric radiation in plasma cavities, II, The cyclotron maser instability in small size sources, Planet. Space Sci. 44, 211, 1996b.

[37] Louarn, P.: Radio emissions from filamentary sources: a simple approach, Planetary radio emissions IV, Ed. Rucker H.O., Bauer S.J. and Lecacheux A., Verlag der Österreichischen Akademie der Wissenschaften, 1997.

[38] Mellott, M.M., W. Calvert, R.L. Huff, D.A. Gurnett, and S.D. Shawhan: DE 1 observation of ordinary mode and extraordinary mode auroral kilometric radiation, Geophys. Res. Lett. 11, 1188, 1984.

[39] Melrose, D.B. and G.A. Dulk: Electron-cyclotron masers as the source of certain solar and stellar radio bursts, Astrophys. J., 259, 844, 1982.

[40] Melrose, D.B., K.G. Rönnmark, and R.G. Hewitt: Terrestrial kilometric radiation: The cyclotron theory, J. Geophys. Res. 87, 5140, 1982.

[41] Omidi, N. and D.A. Gurnett: Growth rate calculations of auroral kilometric radiation using the relativistic resonance condition, J. Geophys. Res. 87, 2377, 1982.

[42] Omidi, N. and D.A. Gurnett: Path-integrated growth of auroral kilometric radiation, J. Geophys. Res. 89, 10,801, 1984.

[43] Perraut, S., H. de Feraudy, A. Roux, P.M.E. Décréau, J. Paris, and L. Matson: Density measurements in key regions of the Earth's magnetosphere: Cusp and auroral region, J. Geophys. Res. 95, 5997, 1990.

[44] Pottelette, R., M. Malingre, A. Bahnsen, L. Eliasson, and K. Stasiewicz: Viking observations of bursts of intense broadband noise in the source region of auroral kilometric radiation, Ann. Geophys. 6, 573, 1988.

[45] Pottelette, R., R.A. Treumann, and M. Berthomier: Auroral plasma turbulence and the cause of auroral kilometric radiation fine structure, J. Geophys. Res. 106, 8465, 2001.

[46] Pottelette, R. and R.A. Treumann: Auroral Acceleration and Radiation, in: part 1 of this book, 2005.

[47] Pritchett, P.L.: Relativistic dispersion, the cyclotron maser instability, and auroral kilometric radiation, J. Geophys. Res. 89, 8957, 1984.

[48] Pritchett, P.L. and R.J. Strangeway: A simulation study of kilometric radiation generation along an auroral field line, J. Geophys. Res. 90, 9650, 1985.

[49] Pritchett, P.L.: Electron-cyclotron maser instability in relativistic plasmas, Phys. Fluids 29, 2919, 1986a.

[50] Pritchett, P.L.: Cyclotron maser radiation from a source structure localized perpen dicular to the ambient magnetic field, J. Geophys. Res. 91, 13,569, 1986b.

[51] Pritchett, P.L. and R.M. Winglee: Generation and propagation of kilometric radiation in the auroral plasma cavity, J. Geophys. Res. 94, 129, 1989.

[52] Pritchett, P.L., R.J. Strangeway, R.E. Ergun, and C.W. Carlson: Generation and propagation of cyclotron maser emissions in the finite auroral kilometric ra-diation source cavity, J. Geophys. Res. 107, 12, doi:10.1029/2002JA009403, 2002.

[53] Roux, A., A. Hilgers, H. de Feraudy, D. Le Quéau, P. Louarn, S. Perraut, A. Bahnsen, M. Jespersen, E. Ungstrup, and M. André: Auroral kilometric radiation sources: in situ and remote sensing observations from Viking, J. Geophys. Res. 98, 11657, 1993.

[54] Sharma, R.R., L. Vlahos, and K. Papadopoulos: The importance of plasma effects on electron-cyclotron maser emission from flaring loops, Astron. Astrophys. 112, 377, 1982.

[55] Shawhan, S.D. and D.A. Gurnett: Polarization measurements of auroral kilometric radiation by DE-1, Geophys. Res. Lett. 9, 913, 1982.

[56] Strangeway, R.J.: Wave dispersion and ray propagation in a weakly relativ-istic electron plasma: Implications for the generation of auroral kilometric radiation, J. Geophys. Res. 90, 9675, 1985.

[57] Trubnikov, B.A.: In Plasma physics and the problem of controlled thermo-nuclear reactions, ed. M.A. Leontovich, Pergamon Press Inc., New york, 1959.

[58] Ungstrup, E., A. Bahnsen, H.K. Wong, M. André, and L. Matson: Energy source and generation mechanism for AKR, J. Geophys. Res. 95, 5973, 1990.

[59] Wu, C.S. and L.C. Lee: A theory of the terrestrial kilometric radiation, Astrophys. J. 230, 621, 1979.

[60] Zarka, P.: The auroral radio emissions from planetary magnetospheres: what do we know, what don't we know, what do we learn from them? Adv. Space Res. 12, 99. 1992.

4

Generation of Emissions by Fast Particles in Stochastic Media

G.D. Fleishman

National Radio Astronomy Observatory, Charlottesville, VA 22903
gfleishm@nrao.edu

Abstract. We demonstrate the potential importance of small-scale turbulence in the generation of radio emission from natural plasmas. This emission, being reliably detected and interpreted, probes small-scale turbulence in remote sources in the most direct way. The radiation emitted is called "diffusive synchrotron radiation" because it is related to shaking the electron distribution when passing through randomly distributed small-scale plasma inhomogeneities caused in turbulence. The emissivity is calculated and shown to be in the observable range. A further effect of inhomogeneities is transition radiation arising from fast particles which interact with small-scale density inhomogeneities. This emission process generates continuum emission below synchrotron and is applicable to some solar radio bursts. It serves for probing number densities. The effect of inhomogeneities on coherent emissions is either broadening or splitting of the spectral peaks generated by the electron cyclotron maser mechanism.

Key words: Solar radio bursts, random inhomogeneities, turbulent radiation, diffuse synchrotron radiation

4.1 Introduction

Interaction of charged particles with each other and/or with external fields results in emission of electromagnetic radiation. This article considers the emission arising as fast (nonthermal) particles move through media with random inhomogeneities. The nature of these inhomogeneities might be rather arbitrary. One of the simplest examples of inhomogeneities is a distribution of the microscopic particles (atoms or molecules) in an amorphous substance, so the medium is inhomogeneous at microscopic scales (of the order of the mean distance between particles), perhaps remaining uniform (on average) at macroscopic scales.

More frequently, however, real objects are inhomogeneous on macroscopic scales as well. The irregularities might be related to the interfaces between

G.D. Fleishman: *Generation of Emissions by Fast Particles in Stochastic Media*, Lect. Notes Phys. **687**, 87–104 (2006)
www.springerlink.com

inhomogeneities, variations of the elemental composition, temperature, density, electric, and magnetic field.

Random inhomogeneities of any of these parameters may strongly affect in various ways the generation of electromagnetic emission. For example, the presence of the density inhomogeneities implies that the dielectric permeability tensor is a random function, as well as the refractive indices of the electromagnetic eigen-modes. As a result, the eigen-modes of the uniform medium are not the same as the eigen-modes of the real inhomogeneous medium. The irregularities of the electric and magnetic fields affect primarily the motion of the fast particle (although the effect of the field fluctuations on the dielectric tensor exists as well).

Below we consider one of many emission processes appearing due to or affected by small-scale random inhomogeneities, namely, *diffusive synchrotron radiation* arising as fast particles are scattered by the small-scale random fields. This emission process is of exceptional importance since current models of many astrophysical objects (see, e.g., [6, 7] and references therein) imply generation of rather strong small-scale magnetic fields. The effect of the inhomogeneities on other emission processes is discussed briefly as well.

4.2 Statistical Methods in the Theory of Electromagnetic Emission

The trajectories of charged particles and the fields created by them are random functions as the particles move through a random medium. Thus, the use of appropriate statistical methods is required to describe the particle motion and the related fields.

4.2.1 Spectral Treatment of Random Fields

For a detailed theory of the random fields we refer to a monograph [Toptygin, 20] and mention here a few important points only. To be more specific, let us discuss some properties of the random magnetic fields. Assume that total magnetic field is composed of regular and random components $\boldsymbol{B}(\boldsymbol{r},t) = \boldsymbol{B}_0(\boldsymbol{r},t) + \boldsymbol{B}_{st}(\boldsymbol{r},t)$, such as $\boldsymbol{B}_0(\boldsymbol{r},t) = \langle \boldsymbol{B}(\boldsymbol{r},t) \rangle$ and $\langle \boldsymbol{B}_{st}(\boldsymbol{r},t) \rangle = 0$, where the brackets denote the statistical averaging. Note, that the method of averaging depends on the problem considered.

The statistical properties of the random field might be described with a (infinite) sequence of the multi-point correlation functions, the most important of which is the (two-point) second-order correlation function

$$K_{\alpha\beta}^{(2)}(\boldsymbol{R},T,\boldsymbol{r},\tau) = \langle B_{st,\alpha}(\boldsymbol{r}_1,t_1)B_{st,\beta}(\boldsymbol{r}_2,t_2) \rangle , \tag{4.1}$$

where $\boldsymbol{R} = (\boldsymbol{r}_1 + \boldsymbol{r}_2)/2$, $\boldsymbol{r} = \boldsymbol{r}_2 - \boldsymbol{r}_1$, $T = (t_1 + t_2)/2$, and $\tau = t_2 - t_1$.

Since the regular and random fields are statistically independent ($\langle B_0 B_{st} \rangle = B_0 \langle B_{st} \rangle = 0$), each of them satisfies the Maxwell equations separately. In particular $\nabla \cdot \boldsymbol{B}_{st} = 0$, so only two of three vector components of the random field are independent.

For a statistically uniform random field the Fourier transform of the correlator $K_{\alpha\beta}^{(2)}(\boldsymbol{r}, \tau)$ over spatial and temporal variables $\boldsymbol{r}$ and τ gives rise to the spectral treatment of the random field

$$K_{\alpha\beta}(\boldsymbol{k}, \omega) = \int \frac{d\boldsymbol{r}d\tau}{(2\pi)^4} e^{i(\omega\tau - \boldsymbol{kr})} K_{\alpha\beta}^{(2)}(\boldsymbol{r}, \tau) . \tag{4.2}$$

In the case of isotropic turbulence we easily find

$$K_{\alpha\beta}(\boldsymbol{k}, \omega) = \frac{1}{2} K(\boldsymbol{k}) \delta(\omega - \omega(\boldsymbol{k})) \left(\delta_{\alpha\beta} - \frac{k_\alpha k_\beta}{k^2} \right) , \tag{4.3}$$

which, in particular, satisfies the the Maxwell equation $\nabla \cdot \boldsymbol{B}_{st} = 0$, since the tensor structure of the correlator is orthogonal to the $\boldsymbol{k}$ vector: $k_\alpha K_{\alpha\beta} = 0$.

Although the spectral shape of the correlators is not unique and may substantially vary depending on the situation, we will adopt for the purpose of the model quasi-power-law spectrum of the random measures:

$$K(\boldsymbol{k}) = \frac{A_\nu}{(k_{min}^2 + k^2)^{\nu/2+1}}, \quad A_\nu = \frac{\Gamma(\nu/2+1)k_{min}^{\nu-1}\langle A^2 \rangle}{3\pi^{3/2}\Gamma(\nu/2 - 1/2)} , \tag{4.4}$$

where ν is the spectral index of the turbulence, and the spectrum $K(\boldsymbol{k})$ is normalized to d^3k:

$$\int_0^{k_{max}} K(\boldsymbol{k}) d^3k = \langle A^2 \rangle , \text{ for } k_{min} \ll k_{max}, \ \nu > 1 , \tag{4.5}$$

where $\langle A^2 \rangle$ is the mean square of the corresponding measure of the random field, e.g., $\langle B_{st}^2 \rangle$, $\langle E_{st}^2 \rangle$, $\langle \Delta N^2 \rangle$ etc.

4.2.2 Emission from a Particle Moving along a Stochastic Trajectory

The intensity of the emission of the eigen-mode σ

$$\mathcal{E}_{\boldsymbol{n},\omega}^\sigma = (2\pi)^6 \frac{\omega^2 n_\sigma(\omega)}{c^3} |(\boldsymbol{e}_\sigma \cdot \boldsymbol{j}_{\omega,\boldsymbol{k}})|^2 \tag{4.6}$$

depends on the trajectory of the radiating charged particle since the Fourier transform $\boldsymbol{j}_{\omega,\boldsymbol{k}}$ of the corresponding electric current has the form:

$$\boldsymbol{j}_{\omega,\boldsymbol{k}} = Q \int_{-\infty}^{\infty} \boldsymbol{v}(t) \exp(i\omega t - i\boldsymbol{kr}(t)) \frac{dt}{(2\pi)^4} , \tag{4.7}$$

where Q is the charge of the particle. For the stochastic motion of the particle, we have to substitute (4.7) into (4.6) and perform the averaging of the corresponding expression:

$$
\mathcal{E}_{n,\omega}^{\sigma} = \frac{Q^2\omega^2 n_\sigma(\omega)}{4\pi^2 c^3}
$$

$$
\times \mathrm{Re} \int_{-T}^{T} dt \int_{0}^{\infty} d\tau e^{i\omega\tau} \langle e^{-i\boldsymbol{k}[\boldsymbol{r}(t+\tau)-\boldsymbol{r}(t)]} (\boldsymbol{e}_\sigma^* \cdot \boldsymbol{v}(t+\tau))(\boldsymbol{e}_\sigma \cdot \boldsymbol{v}(t)) \rangle ,
$$
(4.8)

where $2T$ is the total time at which the emission occurs, and $\boldsymbol{e}_\sigma$ is the polarization vector of the eigen-mode σ.

It is convenient to perform the averaging denoted by the brackets with the use of the distribution function of the particle(s) $F(\boldsymbol{r},\boldsymbol{p},t)$ at the time t and the conditional probability $W(\boldsymbol{r},\boldsymbol{p},t;\boldsymbol{r}',\boldsymbol{p}',\tau)$ for the particle to transit from the state $(\boldsymbol{r},\boldsymbol{p})$ to the state $(\boldsymbol{r}',\boldsymbol{p}')$ during the time τ. For statistically uniform random field we obtain:

$$
\mathcal{E}_{n,\omega}^{\sigma} = \frac{Q^2\omega^2 n_\sigma(\omega)}{4\pi^2 c^3}
$$

$$
\times \mathrm{Re} \int_{-T}^{T} dt \int_{0}^{\infty} d\tau e^{i\omega\tau} \int d\boldsymbol{r} d\boldsymbol{p} d\boldsymbol{p}' (\boldsymbol{e}_\sigma^* \cdot \boldsymbol{v}')(\boldsymbol{e}_\sigma \cdot \boldsymbol{v}) F(\boldsymbol{r},\boldsymbol{p},t) W_{\boldsymbol{k}}(\boldsymbol{p},t;\boldsymbol{p}',\tau) .
$$
(4.9)

Then, the integration over τ gives rise to the temporal Fourier transform of W, so the spectrum of emitted electromagnetic waves is expressed via the spatial and temporal Fourier transform of the distribution function of the particle in the presence of the random field.

4.2.3 Kinetic Equation in the Presence of Random Fields

The conditional probability W, which substitutes the particle trajectory in the presence of random fields, can be obtained from the kinetic Boltzmann-equation:

$$
\frac{\partial f}{\partial t} + \boldsymbol{v} \cdot \frac{\partial f}{\partial \boldsymbol{r}} + \boldsymbol{F}_L \cdot \frac{\partial f}{\partial \boldsymbol{p}} = 0 ,
$$
(4.10)

where $\boldsymbol{F}_L = Q\boldsymbol{E} + \frac{Q}{c}(\boldsymbol{v} \times \boldsymbol{B})$ is the Lorentz force, while the electric $(\boldsymbol{E})$ and magnetic $(\boldsymbol{B})$ fields contain in the general case both regular and random components. Let us express the Lorentz force as a sum of these two components explicitly:

$$
\boldsymbol{F}_L = \boldsymbol{F}_R + \boldsymbol{F}_{st} .
$$
(4.11)

Accordingly, we'll seek a distribution function in the form of the sum of the averaged (W) and fluctuating (δW) components:

$$
f(\boldsymbol{r},\boldsymbol{p},t) = W(\boldsymbol{r},\boldsymbol{p},t) + \delta W(\boldsymbol{r},\boldsymbol{p},t) .
$$
(4.12)

The equation for the averaged component W can be derived from (4.10) applying the Green's function method [Toptygin, 20]:

$$\frac{\partial W}{\partial t} + v \cdot \frac{\partial W}{\partial r} - (\mathbf{\Omega} \cdot \vec{\mathcal{O}})W = \left(\frac{Qc}{\mathcal{E}}\right)^2 \tag{4.13}$$

$$\times \int_0^\infty d\tau \mathcal{O}_\alpha T_{\alpha\beta}[\mathbf{\Delta r}(\tau), \tau]\mathcal{O}_\beta W[r - \mathbf{\Delta r}(\tau), p - \mathbf{\Delta p}(\tau), t - \tau],$$

where

$$\mathbf{\Omega} = \frac{Q\mathbf{B}c}{\mathcal{E}} \tag{4.14}$$

is a vector pointing in the direction of the magnetic field and whose magnitude equals the rotation frequency of the charged particle with energy $\mathcal{E}$,

$$T_{\alpha\beta}(r,\tau) = \langle B_{st,\alpha}(r_1,t_1)B_{st,\beta}(r_2,t_2)\rangle = \frac{\langle B_{st}^2\rangle}{3}\left\{\psi(r)\delta_{\alpha\beta} + \psi_1(r)\frac{r_\alpha r_\beta}{r^2}\right\}. \tag{4.15}$$

To derive (4.13) we transformed the terms with the magnetic field using the following property of the scalar triple product:

$$\frac{Q}{c}(v \times B) \cdot \frac{\partial}{\partial p} = -\frac{QB}{c} \cdot \left[v \times \frac{\partial}{\partial p}\right] = -\frac{QBc}{\mathcal{E}} \cdot \mathcal{O}, \qquad \mathcal{O} = \left[v \times \frac{\partial}{\partial v}\right], \tag{4.16}$$

where $\mathcal{O}$ is the operator of the angular variation of the velocity.

The equation (4.13) is rather general and can be applied to the study of both emission by fast particles and particle propagation in the plasma [Toptygin, 20]. Further simplifications of equation (4.13) can be done by taking into account some specific properties of the problems considered. The theory of wave emission involves a fundamental measure called the *coherence length* (or the formation zone) that refers to that part of the particle path where the elementary radiation pattern is formed.

The coherence length is much larger than the wavelength for the case of relativistic particles, e.g., the coherence length for synchrotron radiation in the presence of the uniform magnetic field is $l_s = R_L/\gamma = Mc^2/(QB)$, where R_L is the Larmor radius, $\gamma = \mathcal{E}/Mc^2$ is the Lorentz-factor of the particle. Length l_s is by the factor of γ^2 larger than the corresponding wave length. The effect of magnetic field inhomogeneity on the elementary radiation pattern is specified by the ratio of the spatial scale of the field inhomogeneity and the coherence length. If the scale of inhomogeneity is much larger than the coherence length, the effect of the inhomogeneity is small and can typically be discarded. However, if the magnetic field changes noticeably at the coherence length, the inhomogeneity affects the emission strongly, so the spectral and angular distributions of the intensity and polarization of the emission can be remarkably different from the case of the uniform field.

This means, in particular, that in the presence of magnetic turbulence with a broad distribution over the spatial scales, the large-scale spatial irregularities should be considered like the regular field, while the small-scale fluctuations should be properly taken into account as the random field. Since the variation of the particle speed (momentum) over the correlation length of the small-scale random field is small, then we can adopt $\Delta p(\tau) = 0$, $\Delta r(\tau) = v\tau$ in the right-hand-side of (4.13). Then, the kinetic equation takes the form

$$\frac{\partial W}{\partial t} + v \cdot \frac{\partial W}{\partial r} - (\Omega \cdot \vec{\mathcal{O}})W = \frac{\langle B_{st}^2 \rangle}{3} \left(\frac{Qc}{\mathcal{E}}\right)^2 \mathcal{O}^2 \int_0^\infty d\tau\, \psi(v\tau) W(r - v\tau, p, t - \tau) ,$$

(4.17)

4.2.4 Solution of Kinetic Equation

Let us outline the solution of the kinetic equation (4.17) for the averaged distribution function W. First of all, the stochastic field has to be split into large-scale $(\widetilde{B})$ and small-scale (B_{st}) components. To see this, consider a purely sinusoidal spatial wave of the magnetic field with the strength B_0 and the wavelength $\lambda_0 = 2\pi/k_0$. If the wavelength λ_0 is less than the coherence length l_{s0} calculated for the emission in a uniform field B_0, $l_{s0} \sim Mc^2/(QB_0)$, this wave represents the small-scale field, whose spatial inhomogeneity is highly important for the emission; in the opposite case, $\lambda_0 \gtrsim l_{s0}$, it is a large-scale field.

The splitting is less straightforward when the random field is a superposition of the random waves with a quasi-continuous distribution over the spatial scales. Let us consider the effect provided by a random magnetic field corresponding to a small range Δk in the spectrum (4.4) on the charged particle trajectory. The energy of this magnetic field is

$$\delta\mathcal{E}_{st} \sim K(k)k^2 \Delta k .$$

(4.18)

The corresponding non-relativistic "gyrofrequency" is $\delta\omega_{st} \sim Q\sqrt{\delta\mathcal{E}_{st}}/(Mc)$. For a truly random field, when the harmonics with k and $k+dk$ are essentially uncorrelated, we can arbitrarily select the value Δk to be small enough to satisfy $\delta\omega_{st} \ll kc$ for any k, so all the independent field components represent the small-scale field. However, in a more realistic case the Fourier components of the random field with similar yet distinct k are typically correlated, so they disturb the particle motion coherently and Δk in estimate (4.18) cannot be arbitrarily small any longer. Accordingly, all components of the random field with $k \lesssim \delta\omega_{st}/c$ (where $\delta\omega_{st}$ is calculated for the smallest allowable Δk) must be treated as a large-scale field.

The large-scale field $\widetilde{B}$ together with the regular field B_0 specifies the vector Ω in the left-hand-side of equation (4.17):

$$\Omega = \frac{Q(B_0 + \widetilde{B})c}{\mathcal{E}} .$$

(4.19)

For the analysis of the emission process (and, respectively, for the solution of the kinetic equation (4.17)), we treat the large-scale field (which is the sum of the regular and large-scale stochastic fields) as uniform, $(\Omega = const)$;

the actual inhomogeneity might be taken into account by averaging the final expressions of the emission if necessary.

Equation (4.17) has been solved in the presence of a uniform magnetic field and small-scale random magnetic fields [cf. 13]:

$$W_{\boldsymbol{k}} = \frac{1}{p^2}\delta(p - p_0)\exp\left[-i\frac{\omega v}{c}\left(1 - \frac{\omega_{pe}^2}{2\omega^2}\right)\tau\right]w(\boldsymbol{\theta}_0, \boldsymbol{\theta}, \tau), \qquad (4.20)$$

where

$$w(\boldsymbol{\theta}_0, \boldsymbol{\theta}, \tau) = \frac{x}{\pi\sinh z\tau}\exp\Bigg[-x(\theta^2 + \theta_0^2)\coth z\tau$$
$$+ 2x\boldsymbol{\theta}\boldsymbol{\theta}_0\sinh^{-1}z\tau - \frac{(\boldsymbol{\theta} - \boldsymbol{\theta}_0)\cdot(\boldsymbol{n}\times\boldsymbol{\Omega})}{2q} - \frac{\Omega_\perp^2\tau}{4q}\Bigg], \quad (4.21)$$

$$\boldsymbol{\theta}_0 = \frac{\boldsymbol{v}_0}{v} - \boldsymbol{n}\left(1 - \frac{\theta_0^2}{2}\right), \qquad \boldsymbol{\theta} = \frac{\boldsymbol{v}}{v} - \boldsymbol{n}\left(1 - \frac{\theta^2}{2}\right), \qquad (4.22)$$

$$x = (1 - i)\left(\frac{\omega}{16q}\right)^{\frac{1}{2}}, \qquad z = (1 - i)(\omega q)^{\frac{1}{2}}, \qquad (4.23)$$

$$q(\omega, \theta) = \pi\left(\frac{Qc}{\mathcal{E}}\right)^2\int d\boldsymbol{k}' K(\boldsymbol{k}')\delta[\omega - (\boldsymbol{k} - \boldsymbol{k}')\boldsymbol{v}]. \qquad (4.24)$$

If there is only a random field and no regular field, the function w reads

$$w(\boldsymbol{\theta}_0, \boldsymbol{\theta}, \tau) = \frac{x}{\pi\sinh z\tau}\exp\left\{-x(\theta^2 + \theta_0^2)\coth z\tau + 2x\boldsymbol{\theta}\boldsymbol{\theta}_0\sinh^{-1}z\tau\right\}, \quad (4.25)$$

while in the opposite case, when there is no random field, we have

$$w(\boldsymbol{\theta}_0, \boldsymbol{\theta}, \tau) = \delta(\boldsymbol{\theta} - \boldsymbol{\theta}_0 + [\boldsymbol{n}\times\boldsymbol{\Omega}]\tau)\exp\left\{\frac{i\omega\tau}{2}\left[\theta_0^2 - \boldsymbol{\theta}_0\cdot[\boldsymbol{n}\times\boldsymbol{\Omega}]\tau + \frac{\Omega_\perp^2\tau^2}{3}\right]\right\}. \tag{4.26}$$

Calculation of the emission with the use of this distribution function leads evidently to the standard expressions of synchrotron radiation in the uniform magnetic field.

Finally, the distribution function of the free particle (moving without any acceleration), which does not produce any emission in the vacuum or uniform plasma, is

$$w^0(\boldsymbol{\theta}_0, \boldsymbol{\theta}, \tau) = \delta(\boldsymbol{\theta} - \boldsymbol{\theta}_0)\exp\left\{\frac{i\omega\tau}{2}\theta^2\right\}. \qquad (4.27)$$

4.3 Emission from Relativistic Particles in the Presence of Random Magnetic Fields

4.3.1 General Case

Let us consider the energy emitted by a single particle (regardless of polarization) based on the general expression (4.9), i.e., we take the sum of (4.9) over the two orthogonal eigen-modes:

$$
\mathcal{E}_{n,\omega} = \frac{Q^2 \omega^2 \sqrt{\varepsilon(\omega)}}{2\pi^2 c^3}
$$
(4.28)
$$
\times \operatorname{Re} \int_{-T}^{T} dt \int_{0}^{\infty} d\tau e^{i\omega\tau} \int d\boldsymbol{r}\, d\boldsymbol{p}\, d\boldsymbol{p}'[\boldsymbol{n} \times \boldsymbol{v}'] \cdot [\boldsymbol{n} \times \boldsymbol{v}] F(\boldsymbol{r},\boldsymbol{p},t) W_{\boldsymbol{k}}(\boldsymbol{p},t;\boldsymbol{p}',\tau) ,
$$

where we neglected the difference between the two refractive indices in the magnetized plasma and adopted $n_\sigma(\omega) \approx \sqrt{\varepsilon(\omega)}$.

In the presence of a statistically uniform and stationary magnetic field the emitted energy (4.28) is proportional to the time (on average), although the intensity of emission at a given direction $\boldsymbol{n}$ depends on time since the angle between the instantaneous particle velocity $\boldsymbol{v}(t)$ and $\boldsymbol{n}$ changes with time as described by the dependence of the function $F(\boldsymbol{r},\boldsymbol{p},t)$ on time t. This kind of the temporal dependence is not of particular interest, e.g., it represents periodic pulses provided by the rotation of the particle in the uniform magnetic field, so it is more convenient to proceed with time-independent intensity of radiation emitted into the full solid angle

$$
I_\omega = \frac{Q^2 \omega^2}{2\pi^2 c} \sqrt{\varepsilon(\omega)} \operatorname{Re} \int_0^\infty d\tau \exp\left[\frac{i\omega\tau}{2\gamma^2}\left(1 + \frac{\omega_{pe}^2 \gamma^2}{\omega^2}\right)\right]
$$
(4.29)
$$
\times \int d^2\theta\, d^2\theta'(\boldsymbol{\theta}\boldsymbol{\theta}')(w(\boldsymbol{\theta},\boldsymbol{\theta}',\tau) - w^0(\boldsymbol{\theta},\boldsymbol{\theta}',\tau)) ,
$$

where $w^0(\boldsymbol{\theta},\boldsymbol{\theta}',\tau)$ is the distribution function of the free particle (4.27), which does not contribute to the electromagnetic emission (since the Vavilov-Cherenkov condition cannot be fulfilled in the plasma or vacuum). Then, calculation of (4.29), described in detail in [21], results in

$$
I_\omega = \frac{8Q^2 q'(\omega)}{3\pi c} \frac{\gamma^2 \Phi_1(s_1,s_2,r)}{[1 + \gamma^2(\omega_{pe}/\omega)^2]} + \frac{Q^2 \omega}{4\pi c\gamma^2}\left(1 + \frac{\gamma^2 \omega_{pe}^2}{\omega^2}\right)\Phi_2(s_1,s_2,r) ,
$$
(4.30)

where $\Phi_1(s_1,s_2,r)$ and $\Phi_2(s_1,s_2,r)$ stand for the integrals:

$$
\Phi_1(s_1,s_2,r) = \frac{6|s|^4}{s_1 s_2} \operatorname{Im} \int_0^\infty dt \exp(-2st) \times
$$
(4.31)
$$
\times \left\{ \coth t \exp\left[-2rs^3\left(\coth t - \sinh^{-1} t - \frac{t}{2}\right)\right] - \frac{1}{t}\right\} ,
$$

$$\Phi_2(s_1, s_2, r) = 2r|s|^2 \text{Re} \int_0^\infty dt \, \frac{\cosh t - 1}{\sinh t} \times \tag{4.32}$$

$$\times \exp\left[-2st - 2rs^3\left(\coth t - \sinh^{-1} t - \frac{t}{2}\right) - i\phi\right],$$

which depend on the dimensionless parameters s_1, s_2, r:

$$s = s_1 - is_2 = \frac{e^{-i(\frac{\pi}{4}+\frac{\phi}{2})}}{4\sqrt{2}\gamma^2}\left(\frac{\omega}{|q(\omega)|}\right)^{\frac{1}{2}}\left(1 + \frac{\omega_{pe}^2\gamma^2}{\omega^2}\right), \tag{4.33}$$

$$r = 32\gamma^6\left(\frac{\Omega_\perp}{\omega}\right)^2\left(1 + \frac{\omega_{pe}^2\gamma^2}{\omega^2}\right)^{-3}. \tag{4.34}$$

The parameter s depends on the rate of scattering of the particle by magnetic inhomogeneities $q(\omega)$, which has the form

$$q(\omega) = \frac{\sqrt{\pi}\Gamma(\nu/2)\omega_{st}^2\omega_0^{\nu-1}}{3\Gamma(\nu/2-1/2)\gamma^2(\alpha^2+\omega_0^2)^{\nu/2}} + i\frac{(\nu-1)\omega_{st}^2\alpha}{3\gamma^2\omega_0^2(1+(\nu-1)\alpha^2/\omega_0^2)}, \tag{4.35}$$

for power-law distribution of magnetic irregularities over scales: $P(k) = 4\pi K(\boldsymbol{k})k^2 \propto k^{-\nu}$ at $k \gg k_{min} = \omega_0/c$, where $\omega_{st}^2 = Q^2\langle B_{st}^2\rangle/(Mc)^2$ is the square of the cyclotron frequency in the random magnetic field, $\alpha = (a\omega/2)\left(\gamma^{-2} + \omega_{pe}^2/\omega^2\right)$, and a is a factor of the order of unity.

In the general case the integrals (4.31, 4.32) cannot be expressed in terms of elementary functions. However, there are convenient asymptotic expressions of these integrals. In particular, if $r|s|^3 \gg 1$ and $r|s|^3 \gg |s|$ we have

$$\Phi_1 \approx -\frac{|s|^4}{s_1 s_2}(2\pi - 3\phi), \qquad \Phi_2 \approx 2^{\frac{1}{3}}3^{\frac{1}{6}}\Gamma\left(\frac{2}{3}\right)r^{\frac{1}{3}}, \tag{4.36}$$

while for $|s| \gg 1$, $r \ll 1$, the functions Φ_1 and Φ_2 contain exponentially small terms:

$$\Phi_1 \approx 1 - \frac{3\pi^{\frac{1}{2}}r^{\frac{1}{4}}|s|^4}{2^{\frac{5}{4}}s_1 s_2}\exp\left(-\frac{8}{3}\sqrt{\frac{2}{r}}\right), \tag{4.37}$$

$$\Phi_2 \approx \frac{r}{32|s|^2} + 2^{\frac{1}{4}}\pi^{\frac{1}{2}}r^{\frac{1}{4}}\exp\left(-\frac{8}{3}\sqrt{\frac{2}{r}}\right). \tag{4.38}$$

Complementary, for $s \ll 1$ and $rs^3 \ll 1$ we obtain

$$\Phi_1 \approx 6s\frac{1-rs^2}{1+r^2s^4}, \qquad \Phi_2 \approx \frac{rs}{2}\frac{1+rs^2}{1+r^2s^4}. \tag{4.39}$$

4.3.2 Special Cases

The radiation intensity (4.30) depends on many parameters, allowing many different parameter regimes. It is clear that in the absence of the random fields we arrive at standard expressions for synchrotron radiation in a uniform magnetic field. Let us consider here a few interesting cases when the presence of small-scale random field results in a considerable change of the emission.

Weak Random Magnetic Inhomogeneities Superimposed on Regular Magnetic Field

Consider the case of weak magnetic irregularities with a broad (power-law) distribution over spatial scales, with $B_\perp^2 \gg \langle \tilde{B}^2 \rangle$, so that

$$\omega_c \gg \tilde{\omega}_{st} \gg \omega_0 . \tag{4.40}$$

Here ω_c is the gyro-frequency, and $\tilde{\omega}_{st}$ is the gyro-frequency related to the total random field $\langle \tilde{B}^2 \rangle^{\frac{1}{2}}$.

Radiation from highly relativistic particles [Toptygin and Fleishman, 21] is mainly specified by the regular field, since either $|s| \gg 1$ or $r|s|^3 \gg 1$. However, at high frequencies, where synchrotron radiation decreases exponentially, the spectrum is controlled by the small-scale field: the spectral index of radiation is equal to the spectral index of the random field, see Fig. 4.1.

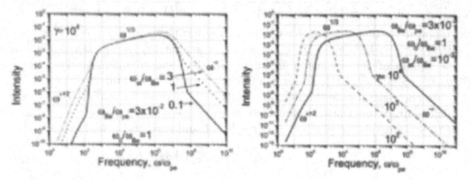

Fig. 4.1. Spectra of radiation by a relativistic particle with $\gamma = 10^4$ for differing value of the random magnetic field (*left*) and with different γ in the presence of weak random magnetic field $\langle B_{st}^2 \rangle / B_0^2 = 10^{-4}$ (*right*)

However, a more interesting regime, which has not been considered so far, takes place for moderately relativistic (and non-relativistic) particles moving in a dense plasma (the case typical for solar and geospace plasmas), when synchrotron radiation is known to be exponentially suppressed according to (4.37), (4.38) by the effect of plasma density (Razin-effect [5, 19]) at all frequencies. The contribution of the small-scale random field, which we refer to as *diffusive synchrotron radiation*, in this conditions takes the form:

$$I_\omega = \frac{2^{\nu+1}\Gamma(\nu/2)(\nu^2 + 7\nu + 8)}{3\sqrt{\pi}\Gamma(\nu/2 - 1/2)(\nu + 2)^2(\nu + 3)} \frac{Q^2}{c} \frac{\omega_{st}^2 \omega_0^{\nu-1} \gamma^{2\nu}}{\omega^\nu \left(1 + \omega_{pe}^2 \gamma^2/\omega^2\right)^{\nu+1}} . \tag{4.41}$$

It is important that this radiation decreases with the increase of the plasma density (plasma frequency) much more slowly (as a power-law, $\sim \omega_{pe}^{-\nu}$) than synchrotron radiation. As a result, the diffusive synchrotron radiation can

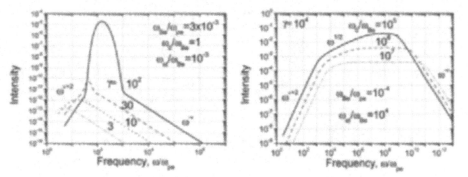

Fig. 4.2. Spectra of radiation by a relativistic particle with different γ in a dense plasma in the presence of weak random magnetic field $\langle B_{st}^2 \rangle / B_0^2 = 10^{-6}$ (*left*), and with $\gamma = 10^4$ in small-scale random magnetic field (*right*). If ω_0 is big enough (e.g., $\omega_0/\omega_{ce} = 10^7$ in the figure) the spectral region provided by multiple scattering, $\omega^{1/2}$, disappears

dominate the entire spectrum even if the random field is much weaker than the regular field, as is evident from the left part of Fig. 4.2: the emission by particles with $\gamma \lesssim 10$ is defined exclusively by the small-scale field.

Small-Scale Magnetic Field

Consider an extreme case, which might be relevant in the physics of cosmological gamma-ray bursts, when there is only a small-scale random magnetic field but no (very weak) regular field, so that $\omega_0 \gg \omega_{st}$ [4]. Now, the parameter q depends substantially on ω_0, the particle motion is similar to the random walk, so the radiation spectrum is similar to some extent to bremsstrahlung provided by multiple scattering of the fast particle by randomly located Coulomb centers. In particular, the spectrum of diffusive synchrotron radiation can contain a flat region (as standard bremsstrahlung) and a $\propto \omega^{\frac{1}{2}}$ region (like bremsstrahlung suppressed by multiple scattering), Fig. 4.2 right. However, at sufficiently high frequencies ($\omega > \omega_0 \gamma^2$), the flat spectrum gives way to a power-law region $\propto \omega^{-\nu}$ typical for the diffusive synchrotron radiation. We should note, that the spectrum depends significantly on the energy of radiating particle (compare the left part in Fig. 4.3). For low-energy particles some parts of the spectrum (e.g., flat region) might be missing.

4.3.3 Emission from an Ensemble of Particles

The results presented in the previous section can be directly applied to mono-energetic electron distributions, which can be generated in the laboratory, but are rare exceptions in nature (e.g., in astro- and geo-plasmas). Natural particle distributions can frequently be approximated by power-laws, say, as function of the dimensionless parameter γ:

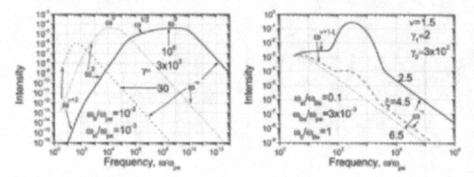

Fig. 4.3. Spectra of radiation by a relativistic particle with different $\gamma = 30$, $3 \cdot 10^3$, 10^6 in the presence of small-scale random magnetic field (*left*). Emissivity by fast electron ensemble with different energetic spectra ($\xi = 2.5$, 4.5, 6.5) for the case of dense plasma, $\omega_{ce}/\omega_{pe} = 3 \cdot 10^{-3}$ (*right*)

$$dN_e(\gamma) = (\xi - 1)N_e\gamma_1^{\xi-1}\gamma^{-\xi}, \quad \gamma_1 \le \gamma \le \gamma_2 , \tag{4.42}$$

where N_e is the number density of relativistic electrons with energies $\mathcal{E} \ge mc^2\gamma_1$, and ξ is the power-law index of the distribution. Evidently, the intensity of incoherent radiation produced by the ensemble (4.42) of electrons from the unit source volume is

$$P_\omega = \int I_\omega dN_e(\gamma) . \tag{4.43}$$

Hard Electron Spectrum

As we will see, the radiation spectrum produced by an ensemble of particles differs for hard ($\xi < 2\nu + 1$) and soft ($\xi > 2\nu + 1$) distributions of fast electrons over energy. Let us consider first the case of hard spectrum [Toptygin and Fleishman, 21], which is typical, e.g., for supernova remnants and radio galaxies. Assuming the small-scale field to be small compared with the regular field, we may expect the contribution of diffusive synchrotron radiation to be noticeable only in those frequency ranges where synchrotron emission is small.

In particular, at low frequencies $\omega \ll \max(\omega_{pe}^2/\omega_{c\perp}, \omega_{pe}\sqrt{\omega_{pe}\gamma_1/\omega_{c\perp}})$, synchrotron radiation is suppressed by the effect of density. Diffusive synchrotron radiation is produced by relatively low energy electrons at these frequencies, and each electron produces the emission according to (4.41), which peaks at $\omega \sim \omega_{pe}\gamma$. Evaluation of the integral (4.43) gives rise to

$$P_\omega \simeq \frac{(\xi-1)\Gamma\left(\frac{\nu}{2}\right)(\nu^2+7\nu+8)}{3\sqrt{\pi}\Gamma\left(\frac{\nu}{2}-\frac{1}{2}\right)(\nu+2)^2(\nu+3)} \frac{e^2N_e\gamma_1^{\xi-1}}{c} \frac{\omega_{st}^2\omega_0^{\nu-1}}{\omega_{pe}^\nu}\left(\frac{\omega}{\omega_{pe}}\right)^{\nu+1-\xi}$$

$$\tag{4.44}$$

in agreement with Nikolaev and Tsytovich [14]. The spectrum can either increase or decrease with frequency depending on the spectral indices ν and ξ.

This expression holds for $\omega \gg \omega_{pe}\gamma_1$. If there are no particles with $\gamma < \gamma_1$, the spectrum at even lower frequencies drops as

$$P_\omega = \frac{(\xi-1)2^{\nu+1}\Gamma(\nu/2)(\nu^2+7\nu+8)}{3(\xi+1)\sqrt{\pi}\Gamma(\nu/2-1/2)(\nu+2)^2(\nu+3)} \frac{e^2 N_e}{c} \frac{\omega_{st}^2\omega_0^{\nu-1}\omega^{\nu+2}}{\gamma_1^2\omega_{pe}^{2\nu+2}}. \quad (4.45)$$

At high frequencies $\max(\omega_{pe}^2/\omega_{c\perp}, \omega_{pe}\sqrt{\omega_{pe}\gamma_1/\omega_{c\perp}}) \ll \omega \ll \omega_{c\perp}\gamma_2^2$, where the effect of density is not important, the spectrum is specified by standard synchrotron radiation. However, at higher frequencies, $\omega \gg \omega_{c\perp}\gamma_2^2$, the intensity of synchrotron radiation decreases exponentially, and the contribution of diffusive synchrotron radiation dominates again. Adding up contributions from all particles described by (4.41) at these frequencies, we obtain

$$P_\omega = \frac{2^{\nu+1}(\xi-1)\Gamma(\frac{\nu}{2})(\nu^2+7\nu+8)}{3\sqrt{\pi}(2\nu-\xi+1)\Gamma(\frac{\nu}{2}-\frac{1}{2})(\nu+2)^2(\nu+3)} \frac{e^2 N_e\gamma_1^{\xi-1}}{c} \frac{\omega_{st}^2\omega_0^{\nu-1}\gamma_2^{2\nu-\xi+1}}{\omega^\nu}. \quad (4.46)$$

Thus, power-law spectrum of relativistic electrons with a cut-off at the energy $\mathcal{E} = mc^2\gamma_2$ produces diffusive synchrotron radiation at high frequencies, whose spectrum shape is defined by the small-scale field spectrum. Remarkably, the corresponding flattening in the synchrotron cut-off region has recently been detected in the optical-UV range for the jet in the quasar 3C273 [8], which would imply the presence of a relatively strong small-scale field there in agreement with the model of Honda and Honda [6].

Although formally the spectrum (4.46) is valid at arbitrarily high frequencies, there is actually a cut-off related to the minimal scale of the random field l_{min}. Accordingly, the largest frequency of the diffusive synchrotron radiation is about $\omega_{max} \sim (c/l_{min})\gamma_2^2$.

Soft Electron Spectrum

Let us now turn to the case of sufficiently soft electron spectra, $\xi > 2\nu + 1$, which are typical, e.g., in many solar flares. The contribution of synchrotron radiation is described by the standard expression, $P_\omega \propto \omega^{-\alpha}$, $\alpha = (\xi-1)/2$, which is steeper for soft spectra than the spectrum of diffusive synchrotron radiation, $P_\omega \propto \omega^{-\nu}$. Hence, for soft electron spectra, diffusive synchrotron radiation can dominate even at $\omega < \omega_{c\perp}\gamma_2^2$. The spectrum of diffusive synchrotron radiation has the same shape as before but its level is defined by lower-energy electron contribution:

$$P_\omega = \frac{2^{\nu+1}(\xi-1)\Gamma\left(\frac{\nu}{2}\right)(\nu^2+7\nu+8)}{3\sqrt{\pi}(\xi-2\nu-1)\Gamma\left(\frac{\nu}{2}-\frac{1}{2}\right)(\nu+2)^2(\nu+3)} \frac{e^2 N_e\gamma_1^{2\nu}}{c} \frac{\omega_{st}^2\omega_0^{\nu-1}}{\omega^\nu}. \quad (4.47)$$

At low frequencies, $\omega \ll \omega_{pe}\gamma_1$, the radiation is still specified by expression (4.44), see the corresponding curves on the right in Fig. 4.3. One may note that in the case of soft electron spectra, the emission produced by the electron ensemble is similar to the emission from a mono-energetic electron distribution with $\gamma = \gamma_1$, which is the main difference between the cases of hard and soft electron spectra.

Let us estimate the ratio of the diffusive synchrotron radiation intensity to the synchrotron radiation intensity. For simplicity, we neglect factors of the order of unity, assume $\omega_0 = \omega_{ce}$ and $\gamma_1 \sim 1$, and introduce frequency $\omega_* = \omega_{pe}^2/\omega_{ce}$ $(\omega \equiv (\omega/\omega_*)\omega_*)$, where synchrotron radiation has a peak, then

$$\frac{P_{sh}}{P_{syn}} \sim \frac{\omega_{st}^2}{\omega_{ce}^2}\left(\frac{\omega_{pe}}{\omega_{ce}}\sqrt{\frac{\omega}{\omega_*}}\right)^{\xi - 2\nu - 1}. \qquad (4.48)$$

Evidently, this ratio increases with frequency, so that diffusive synchrotron radiation can become dominant well before the frequency reaches $\omega_{c\perp}\gamma_2^2$. Moreover, in the case of dense plasma, $\omega_{pe} \gg \omega_{ce}$, diffusive synchrotron radiation can dominate at all frequencies under the condition

$$\frac{\omega_{st}^2}{\omega_{ce}^2}\left(\frac{\omega_{pe}}{\omega_{ce}}\right)^{\xi - 2\nu - 1} > 1, \qquad (4.49)$$

even if the random field is small compared with the regular field $\omega_{st} \ll \omega_{ce}$. On top of this, the radiation spectrum produced from the dense plasma depends critically on the highest energy of the accelerated electrons. Indeed, if $\gamma_2 \ll \omega_{pe}/\omega_{ce}$ (e.g., $\gamma_2 = 4$ in Fig. 4.4), then the radiation spectrum is entirely set up by the small-scale field, in spite of its smallness ($\langle B_{st}^2\rangle/B_0^2 = 10^{-4}$ in Fig. 4.4). Evidently, the standard synchrotron emission increases and becomes observable as far as γ_2 increases.

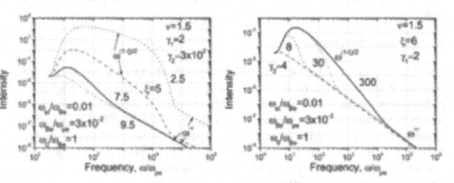

Fig. 4.4. *Left*: Same as in Fig. 4.3, right, for less dense plasma, $\omega_{ce}/\omega_{pe} = 3 \cdot 10^{-2}$. The contribution from the uniform field (synchrotron radiation) decreases for softer electron spectra (i.e., as ξ increases). *Right*: Emissivity by fast electron ensemble with ($\xi = 6$) from dense plasma ($\omega_{ce}/\omega_{pe} = 3 \cdot 10^{-2}$) in the presence of weak magnetic inhomogeneities $\langle B_{st}^2\rangle/B_0^2 = 10^{-4}$ for different high-energy cut-off values γ_2. When γ_2 is small enough, the uniform magnetic field does not affect the radiation spectrum

Diffusive Synchrotron Radiation from Solar Radio Bursts?

According to microwave and hard X-ray observations of solar flares, the energetic spectra of accelerated electrons are frequently rather soft [9, 15]. Consequently, the diffusive synchrotron radiation can dominate the microwave emission for dense enough radio sources. Nevertheless, as a rule the microwave emission from solar flares meets reasonable quantitative interpretation as synchrotron (gyrosynchrotron) radiation by moderately relativistic electrons (moving in non-uniform magnetic field of the coronal loop). An example of microwave burst produced by gyrosynchrotron emission is given in Fig. 4.5, left.

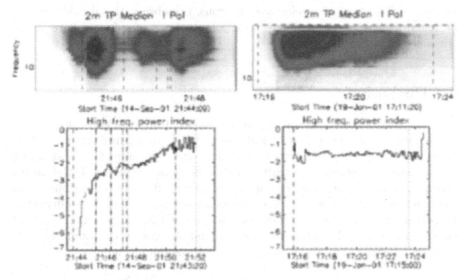

Fig. 4.5. Two microwave bursts recorded by Owens Valley Solar Array in the range 1–18 GHz with 40 spectral channels and 4 sec temporal resolution. The first one (*left bottom*) displays evident spectral hardening with time, while the second one shows remarkable constancy of the high-frequency spectral slope (courtesy of D.E. Gary)

Note, that the high-frequency spectral index δ (the radio flux is fitted by a power-law, $F \propto f^\delta$, at high frequencies, Fig. 4.5, bottom left) decreases in value with time. Such spectral evolution typical for solar microwave bursts is well-understood in the context of the energy-dependent life time of electrons against the Coulomb collisions. Indeed, higher energy electrons have longer life times, which results in spectral hardening of the trapped electron population [Melrose and Brown, 12], and, respectively, hardening of the produced gyrosynchrotron radiation as observed by Melnikov and Magun [11].

However, if microwave emission is produced by diffusive synchrotron radiation as fast electrons interact with small-scale magnetic (and/or electric) fields, then the radio spectrum is specified by the spectrum of random fields rather then of fast electrons. Thus, no spectral evolution (related to electron distribution modification) is expected. Indeed, there is a minority of solar microwave bursts, which do not show any spectral evolution (e.g., no spectral hardening). An example of such a burst, demonstrating constancy in time of the high-frequency spectral index, is shown in Fig. 4.5 at the right. Curiously, the spectral index is $\delta = -1.5$ to -1.7 in agreement with standard models [Vainshtein et al., 22] and measurements of the turbulence spectra, e.g., in interplanetary [see, e.g., Toptygin, 20] and interstellar [Cordes et al., 1] space.

Although it has not been firmly proven so far, such microwave bursts are possibly produced by diffusive synchrotron radiation mechanism. Since (to be dominant) this mechanism requires relatively dense plasma at the source site and soft spectra of accelerated electrons, the observational evidence can be found from analysis of simultaneous observations of soft and hard X-ray emissions from the same flares.

4.4 Discussion

The analysis presented demonstrates the potential importance of small-scale turbulence in the generation of radio emission from natural plasmas. This emission, being reliably detected and interpreted, provides the most direct measurements of small-scale turbulence in the remote sources. The diffusive synchrotron radiation is only one of the observable effects of the turbulence on the radio emission. Indeed, the presence of density inhomogeneities affects the properties of bremsstrahlung, because the Fourier transform of the square of the electric potential produced by background charges in a medium depends on spatial distribution of the charges through the double sum:

$$| \varphi_{q_0,q} |^2 \propto \left\langle \sum_{A,B} e^{-iq(R_A - R_B)} \right\rangle = \left(N + (2\pi)^3 \frac{|\Delta N|_q^2}{V} \right) , \qquad (4.50)$$

where R_A and R_B are the radius-vectors of the particles A and B, respectively, $|\Delta N|_q^2$ is the spectrum of the inhomogeneity, and V is the volume of the system. In statistically uniform media the positions of various particles are uncorrelated and this double sum equals the total number of particles N. However, the macroscopic inhomogeneities make the positions correlated, so the double sum deviates from N. The second term in (4.50) gives rise to *coherent bremsstrahlung*, which in a certain spectral range dominates the incoherent bremsstrahlung [Platonov et al., 17].

Another important radiation process in the turbulent plasma is *transition radiation* arising when fast particles interact with small-scale density inhomogeneities of the background plasma (see Platonov and Fleishman [18], and references therein), whose potential importance for ionospheric conditions has been pointed out long ago by Yermakova and Trakhtengerts [23] (see also the discussion in LaBelle and Treumann [10]). This emission process, giving rise to enhanced low-frequency (at frequencies lower than the accompanying synchrotron emission) continuum radio emission, has recently been reliably confirmed in a subclass of two-component solar radio bursts [3, 16]. This finding is of particular importance for diagnostics of the number density, the level of small-scale turbulence, and the dynamics of low-energy fast particles in solar flares.

In addition, the turbulence can also affect the coherent emissions from unstable electron populations [Fleishman et al., 2], e.g., providing strong broadening (or splitting) of the spectral peaks generated by electron cyclotron maser (ECM) emission. The typical bandwidth of the broadened ECM peaks and its distributions are found to be quantitatively consistent with those observed for narrowband solar radio spikes [Fleishman et al., 2].

Acknowledgements

The National Radio Astronomy Observatory is a facility of the National Science Foundation operated under cooperative agreement by Associated Universities, Inc. This work was supported in part by the Russian Foundation for Basic Research, grants No. 03-02-17218, 04-02-39029. I am very grateful to T. S. Bastian for his numerous comments on this paper.

References

[1] Cordes, J.M., M. Ryan, J.M. Weisberg, D.A. Frail, and S.R. Spangler: Nature 354, 121 (1991).
[2] Fleishman, G.D.: Astrophys. J. 601, 559 (2004).
[3] Fleishman, G.D., G.M. Nita, and D.E. Gary: Astrophys. J. 620, 506 (2005).
[4] Fleishman, G.D.: Astro-ph/0502245 (2005).
[5] Ginzburg, V.L. and S.I. Syrovatsky: Ann. Rev. Astron. Astrophys. 3, 297 (1965).
[6] Honda, M. and Y.S. Honda: Astrophys. J. 617, L37 (2004).
[7] Jaroschek, C. H., H. Lesch, and R. A. Treumann: Astrophys. J. 618, 822 (2005).
[8] Jester, S., H.-J. Röser, K. Meisenheimer, and R. Perley: Astron. Astrophys. 431, 477 (2005).
[9] Kundu, M.R., S.M. White, N. Gopalswamy, and J. Lim: Astrophys. J. Suppl. 90, 599 (1994).
[10] LaBelle, J. and R. A. Treumann: Space Sci. Rev. 101, 295 (2002).
[11] Melnikov, V.F. and A. Magun: Solar Phys. 178, 153 (1998).
[12] Melrose, D.B. and J.C. Brown: Monthly Notic. Royal Astron. Soc. London 176, 15 (1976).
[13] Migdal, A.B.: Dokl. Akad. Nauk SSSR (in Russian) 96, 49 (1954); Phys. Rev. 103, 1811 (1956).
[14] Nikolaev, Yu.A. and V.N. Tsytovich: Phys. Scripta 20, 665 (1979).

[15] Nita, G.M., D.E. Gary, and J. Lee: Astrophys. J. 605, 528 (2004).

[16] Nita, G.M., D.E. Gary, and G.D. Fleishman: Astrophys. J. 629, L65 (2005).

[17] Platonov, K.Yu., I.N. Toptygin, and G.D. Fleishman: Uspekhi Fiz. Nauk (in Russian) 160, 59 (Engl. transl.: Sov. Phys. Uspekhi, 33, 289) (1990).

[18] Platonov, K.Yu. and G.D. Fleishman: Physics-Uspekhi 45, 235 (2002).

[19] Razin, V. A.: Radiofizika 3, 584 (1960).

[20] Toptygin, I.N.: *Cosmic rays in interplanetary magnetic fields* (Dordrecht-Holland, D. Reidel, 1985) 387 pp.

[21] Toptygin, I.N. and G.D. Fleishman: Astroph. Space Sci. 13, 213 (1987).

[22] Vainshtein, S.I., A.M. Bykov, and I.N. Toptygin: *Turbulence, Current Sheets, and Shocks in Cosmic Plasma* (The Fluid Mechanics of Astrophysics and Geophysics, Vol. 6, Langhorne: Gordon and Breach Science Publ., 1993) 398 pp.

[23] Yermakova, E.N. and V.Yu. Trakhtengerts: Geomagn. Aeron. (Engl. Transl.) 21, 56 (1981).

5

Auroral Acceleration and Radiation

R. Pottelette[1] and R.A. Treumann[2,3]

[1] CETP/CNRS, 4 av. de Neptune, 94107 St. Maur des Fossés Cedex, France
raymond.pottelette@ipsl.cetp.fr
[2] Ludwig-Maximilians Universität München, Sektion Geophysik, Theresienstr.
37-41, 80333 München, Germany,
art@mpe.mpg.de
[3] Department of Physics and Astronomy, Dartmouth College, Hanover, New
Hampshire 03755, USA

Abstract. A brief review is given of the recent achievements in understanding the connection between processes in the generation of auroral acceleration and processes taking place at the tailward reconnection site. It is shown that most of the acceleration in the aurora is due to local field-aligned electric potentials which are located in vertically narrow double layers along the magnetic field of the order of ∼10 km and which are the site of preferential excitation of phase space holes of kilometer size extension along the magnetic field which by themselves sometimes represent local potential drops and accelerate electrons and ions antiparallel to each other such that the energy modulation of the electron and ion energy fluxes are in antiphase. Auroral kilometric radiation observations suggest that these structures may be the elementary radiation sources which build up the entire spectrum of the auroral kilometric radiation. Leaving open the very generation mechanism of the double layer whether produced by locally applied shear flows as recently suggested in the literature (and reviewed here as well) we argue that the field aligned current generator responsible for the production of the initial auroral current system is non-local but is related to reconnection in the tail. The field aligned currents are interpreted as the closure currents required to close the recently observed electron Hall current system in the ion diffusion region at the tail reconnection site. Such a model is very attractive as it does not need any other secondary current disruption mechanism. Coupling to the ionosphere may be provided by kinetic Alfvén waves emanating from the Hall reconnection region as surface waves and generating local shear flow when focussing close to the ionosphere and transforming into shear-kinetic Alfvén waves. A main problem still remains in how the decoupling of the two hemispheres observed in the aurora is produced at reconnection site. Multiple reconnection would be one possible solution.

Key words: Auroral processes, auroral particle acceleration, double layers, auroral kilometric radiation, elementary radiation sources, electron holes, reconnection, generator region

R. Pottelette and R.A. Treumann: *Auroral Acceleration and Radiation*, Lect. Notes Phys. **687**, 105–138 (2006)
www.springerlink.com

5.1 Introduction

For many years it has been a mystery why electrons and ions are accelerated in the auroral zone. The idea of Störmer [40] that cosmic ray particles spiral in along Earth's magnetic field to excite the atoms and ions of the upper atmosphere and ionosphere and generate the aurora has readily been proven wrong. Cosmic rays are too energetic to cause anything like an aurora. For them the Earth's magnetic field is practically invisible. They impinge on the ionosphere isotropically causing showers of secondary elementary particles and become absorbed in the atmosphere at altitudes way below those of the aurora. Solar energetic particles on the other hand are not energetic enough to reach anywhere deep enough into the ionosphere. According to their magnetic hardness they are excluded from the inner magnetosphere, while when flowing in from their forbidden zone boundary they are still too energetic to cause substantial aurora. Moreover, they appear at too low latitudes to be associated with auroral phenomena. In addition, aurora is located most frequently on the antisolar side of the magnetosphere even though there is also aurora on the dayside under the cusp and driven by particles accelerated in magnetopause reconnection. These dayside aurorae complete the auroral oval.

The energy range required for electrons to become effective in generating aurorae lies between several 100 eV and several keV. The electron distribution in the magnetospheric tail, i.e. the plasma sheet, is at the lower end of this range. It thus can serve as the source for the auroral electron distribution. However, it must become additionally accelerated by some secondary acceleration process which up till now has only been badly identified. There is a high probability that this process does not consist of one unique step but that the acceleration of electrons takes place in a primary step followed by a sequence of a few or several secondary acceleration steps until the electrons have enough energy to play their role in the aurora. Presumably these two steps are located in spatially separate regions of the magnetosphere. Since aurorae occur during magnetic substorms and storms which are believed to be the result of violent reconnection going on in the magnetospheric tail current sheet, it is reasonable to assume that the first step of acceleration is directly related to the tailward reconnection site. The second step, on the other hand, has been found to occur much closer to the ionosphere at altitudes of ∼2000–8000 km.

Both steps act in concert but from the point of view of the aurora the second step is the more interesting one. It is also related to the generation of the famous auroral kilometric radiation which, under special plasma conditions, is emitted from the acceleration region. In the present paper we briefly review the processes in the second step acceleration region adding some founded speculations on the processes in the first step acceleration region.

5.2 Morphology of the Auroral Acceleration Region

During the past decade various spacecraft have passed the auroral zone at different altitudes. The lowest flying spacecraft was the Swedish satellite Freja (at a nominal altitude of ~2000 km), followed by the NASA satellite FAST (at nominal altitude ~4000 km) and the Swedish satellite Viking (at ~8000 km altitude). Also, Geotail and Polar have passed magnetic field lines connected to the auroral zone at much larger distances from Earth thus providing occasional information about auroral region-magnetosphere connections. According to these observations we may divide the auroral-magnetosphere connection into two different regions: the lower magnetospheric auroral region and the tail auroral region. We will first discuss the former as the processes taking place there have recently been illuminated best.

Traditionally this region is divided into a sequence of field-aligned current elements: the upward current and the downward current regions, respectively. Since the auroral field-aligned currents are carried mainly by electrons with the ions contributing only little, the upward current region is the region where electrons from the magnetosphere flow down into the ionosphere. Traditionally this region is called the "inverted-V" a name adapted from the Λ-like shape of the electron energy fluxes versus time – or space – during satellite crossings of the auroral upward current region. An example is shown in Fig. 5.1 for a relatively long-lasting "inverted-V" event which in the low time resolution looks rather like a step function with sharp onset in the electron energy flux around a few keV at the low latitude boundary of the event followed by a long-lasting plateau and a correspondingly sharp dropout of the flux at the high latitude boundary. In high time resolution this increase is rather gradual indicating that the potential drops on a somewhat longer spatial distance. The whole event lasts for roughly 30 s, a very short time which for a spacecraft velocity of ~6 km s^{-1} corresponds to the short horizontal distance of only ~180 km in the corresponding northern hemispherical auroral passage of FAST. We will return to this particular fact in Sect. 5.6 below. In fact, the length of the "inverted-V" event is marked even better by the ion energy fluxes than by the electron fluxes. This is very clearly seen in Fig. 5.1 where the sharp increase in the ion energy accompanies the changes in energetic electron energy flux.

Before entering into a discussion of the entire sequence of data during this short passage we note that this short "inverted-V" event is a section of a longer auroral disturbance shown in Fig. 5.2 in very low time resolution. This entire event lasts for ~5 min, corresponding to a horizontal distance of ~1800 km on the northern hemispherical auroral region in the altitude range between 4000 km and 3500 km. It covers a latitudinal range of roughly 5 degrees in invariant latitude and a longitudinal range of roughly 1 hour in magnetic local time (or 15 degrees) in the pre-midnight sector. Thus the entire event is located fully on the northern hemisphere. This is important to remember for our later discussion. However, the self-sustained short time section shown in Fig. 5.1 which is an "inverted V" on its own shows that the full event consists

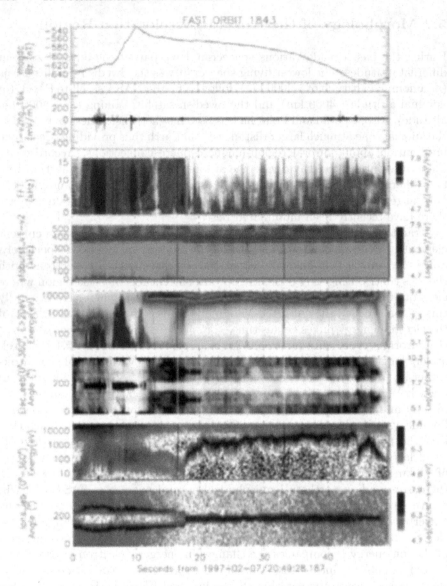

Fig. 5.1. An overview of data from FAST orbit 1843 during a passage of the region above an aurora on the northern hemisphere

of a sequence of several sub-events more or less well covered by the spacecraft orbit, each of them a separate "inverted V". Inspection of the overall event thus suggests that it consists of a mixture of a sequence of more or less well developed upward and downward current regions, and in order to obtain a true impression of its nature it would be necessary to investigate its higher

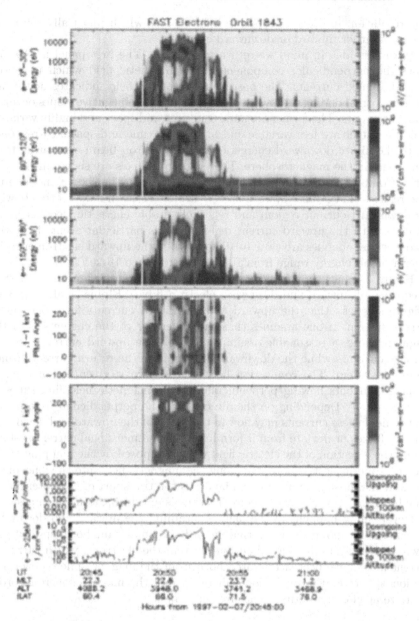

Fig. 5.2. The whole auroral disturbance on FAST orbit 1843 as seen in the electron flux. Shown are the downward (*upper panel*), the perpendicular (*second panel*), the upward electron flux (*third panel*), and pitch angle distributions selected for two electron energy ranges (*fourth and fifth panels*) in very low time resolution overview (courtesy C. W. Carlson, UCB)

time resolution in order to separate it into parts which physically represent regions of either upward or downward currents.

With this idea in mind we return to Fig. 5.1. The first panel in Fig. 5.1 shows the perpendicular component of the magnetic field which is caused by field aligned currents. The positive slope of the field indicates downward currents corresponding to upward electron flux, while negative slopes occur in upward currents. The former are narrow in time and space and highly variable, while the latter are less variable and broader in time and space, corresponding to the dilute downward energetic auroral electron beam emanating from somewhere in the magnetosphere. These electron fluxes are shown in the fifth panel. Combined with the pitch angle distribution in the sixth panel one indeed realizes that intense upward highly variable electron fluxes correlate with the downward current region, and relatively stable energetic electron fluxes correlate with the upward current region. In this particular event the downward electron energies are close to 10 keV, while the upward electron energies cover a broad energy range from 10 eV upward up to 10 keV.

Figure 5.3 shows a schematic of the auroral current system inferred from a sequence of upward and downward electron fluxes [Elphic et al., 11]. Such a figure suggests that the upward and downward currents form one closed current system. If one assumes that the generator of the currents is in the magnetosphere, a reasonable assumption, then the upward currents are the primary currents while the downward currents are the return currents which close the system. The connection between the two is given by ionospheric Pedersen currents flowing perpendicular to the magnetic field flux tubes in the ionosphere. Depending on the direction of the perpendicular ionospheric electric field these currents may flow in the plane of the upward and downward electron fluxes or deviate from it forming a three-dimensional current system. In fact the direction of the electric field will be imposed by the mapping of the magnetospheric electric potential along the magnetic field into the ionosphere which in the most general case will be 3d as the transport of the potentials is done by shear Alfvén waves. Figure 5.3 thus is an idealization which assumes that everything happens in the plane containing the upward and downward currents. It is, however, clear that in order to maintain both upward and downward currents in a closed current system the electrons, which carry the current must have been accelerated both from the magnetosphere down into the ionosphere and from the ionosphere up into the magnetosphere in order to arrive at closed and divergence free currents.

5.2.1 Upward Current Region

The upward current region or downward electron acceleration region is in our view the primary acceleration region. Ignoring for the moment the primary magnetospheric acceleration processes, the data suggest that it contains a warm energetic primary electron beam of a few keV energy and several 100 eV velocity spread or temperature. This beam is not necessarily Maxwellian.

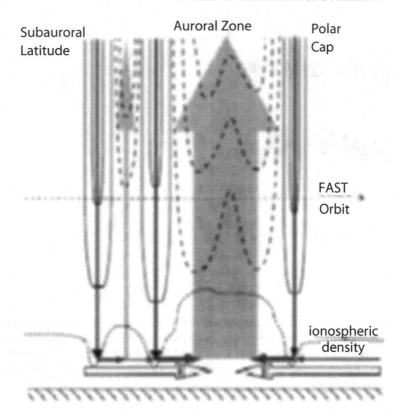

Fig. 5.3. Schematic of the auroral field-aligned current system inferred from a path of FAST across a sequence of downward and upward electron fluxes [Elphic et al., 11] (Reprinted with permission of the American Geophysical Union)

Figure 5.4 shows on its right a cut through the electron distribution function (for FAST orbit 1773) along the magnetic field. The auroral beam sits on a broad plateau in the distribution function. The increase at low velocities (energies) is due to electrons <10 eV. The beam has keV energy and has a shape which varies in time. In addition there are a few reflected electrons flowing upward. The top right panel shows that high energy beams are cooler than low energy beams, a fact which is barely understood.

The important observation in the upward current region is that the ions in panel seven and eight of Fig. 5.1 with entry in the upward current region are violently accelerated upward to energies comparable to the downward electron beam. The antiparallel acceleration of electrons and ions in this region is an absolutely convincing indication not only of the field-aligned currents but also of the presence of a strong field-aligned electric potential as indicated in Fig. 5.3. The narrowness of the ion energy flux in addition shows that the ions are of cold ionospheric origin. The entire upper ionosphere in the upward

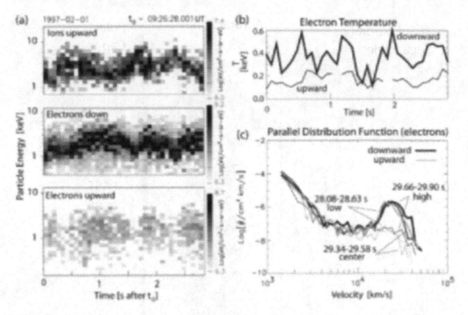

Fig. 5.4. Time variation of particles during a short time interval in the upward current region. (a) particle energy flux for upward ions and upward and downward electrons, (b) electron temperatures, (c) the primary downward parallel electron distribution function [Pottelette et al., 35] (Reprinted with permission of the European Geophysical Society)

current region is transformed into a field-aligned cold upward ion beam. In fact, the high time resolution data of Fig. 5.4 (upper left panel) show in addition that deep in the upward current region the variations of the electron and ion beam energies is strictly anti-correlated suggesting that the electron and ion beams have passed simultaneously an upward directed electric field layer [Pottelette et al., 35]. The variation in energy in both cases amounts to a few keV; hence, the parallel potential difference across the layer is of the order of a few kV. This is sketched schematically in Fig. 5.5.

5.2.2 Downward Current Region

Marklund et al. [21, 22] have called the downward current region the "black aurora" in order to credit the absence of any auroral phenomena. Clearly, if no electrons precipitate no aurora will be excited. The ionospheric electrons sucked up along the magnetic field into the magnetosphere are collisionless. The principle of this is contained in Fig. 5.3 for the return current.

The main property of the downward current region is the broad distribution of upflowing ionospheric electrons. Since no magnetospheric ions are available there are no corresponding downflows of ions. From Fig. 5.1 it is obvious that the electron distribution here is grossly different from that in

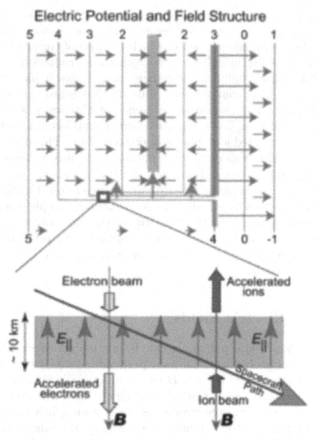

Fig. 5.5. Idealized representation of the field-aligned potential structure in the upward current region under conditions when the beam passes the bottom of the potential. In the lower part of the figure a small portion of the structure is zoomed in showing the spacecraft path across the parallel field layer. When electrons pass from above across the layer they become accelerated while ions are accelerated by the same amount when passing from below

the upward current region. While in the upward current region the parallel distribution is centered around keV beam energies and for larger pitch angles the distribution is a ring shell or horseshoe distribution [Chiu and Schulz, 8], in the downward current region the electron distribution is parallel or at best conical with maximum phase space density at low ionospheric energies. It is clear that the two different kinds of electron distributions will lead to entirely different behaviour with respect to the excitation of instability and radiation. The expectation is that the upward current region exhibits stronger instability than the downward current region, and that radiation is emitted mainly from the upward current region. This is basically true, and therefore we concentrate on the conditions in the upward current region. Since both regions, however, are not independent we will nevertheless occasionally have to refer to the downward current region which because of the special conditions there is highly unstable as well while not generating radiation.

There is a simple qualitative argument that makes it plausible why the upward current region should radiate while the downward current region will not. This argument is based on the presence or absence of a dense low energy background plasma. The downward current region contains the relatively dense upward accelerated ionospheric electron plasma. This plasma when becoming unstable with respect to some plasma instability is dense enough for transporting plasma waves and dissipating their free energy in some way. In the highly dilute plasma of the upward current region on the other hand any free excess energy cannot be easily dissipated because no plasma background is available that could accept the heat. Hence the only way to get rid of any excess energy is to radiate it away into free space. In principle this is the reason why the upward current region serves as a source of extraordinarily intense radio emission.

5.3 Parallel Electric Fields

One may ask how static potentials can be maintained along the magnetic field. There is no problem to maintain perpendicular electric fields. By the Lorentz transformation $E = -v \times B$ they simply imply that the magnetized plasma is in motion perpendicular to the magnetic field with convection speed $v = E \times B/B^2$ seen by any observer at rest or motion relative to the magnetic field frame. Parallel electric potential drops will, on the other hand, be extinguished readily by the thermal motion of electrons along the magnetic field on the timescale of the inverse plasma frequency. Hence, parallel potentials in collisionless plasma will necessarily exist only for limited time or otherwise under forced conditions. This can be seen from the generalized Ohm's law holding in a two or multi-component plasma:

$$E + v \times B - \eta j = \frac{1}{\omega_{pe}^2 \epsilon_0} \left[\frac{\partial j}{\partial t} + \nabla \cdot \left(jv + vj - \frac{1}{en} jj \right) \right]$$
$$+ \frac{1}{en} \left(j \times B - \nabla \cdot \mathsf{P}_e + F_{\mathrm{pmf}} \right) \qquad (5.1)$$

Here v is the bulk plasma velocity, $\mathbf{j}$ current density, η the resistivity which in collisionless plasma is zero unless it is caused by anomalous processes, P_e the electron pressure tensor, and F_{pmf} a possible ponderomotive force exerted by turbulence on the plasma.

It is clear that even in this simple plasma model which does not take into account any kinetic processes parallel electric fields could be maintained by pressure gradients, pressure anisotropies, ponderomotive forces, time variations and spatial variations of current densities and bulk flows. For an anisotropic pressure tensor $P_e = p_\perp I + (p_\parallel - p_\perp)BB/B^2$, for instance, a parallel electric field $enE_\parallel = -\nabla_\parallel p_\parallel + (p_\parallel - p_\perp)\nabla_\parallel \log B$ is obtained showing that parallel pressure gradients and magnetic field inhomogeneity may generate a parallel field. Because of the weak (logarithmic) dependence on the magnetic field, the main contribution comes from the parallel pressure gradient. However, these fields are altogether very weak and unable to produce the wanted parallel potential drops unless strong pressure variations become possible or very large distances are available. Both conditions are not given in the auroral region.

Similarly, the other terms in Ohm's law will produce only weak parallel fields and are thus incapable of producing the required potential drops. The final possibility is an anomalous resistance η_{an} caused by a strongly turbulent wave field. It can be shown quite generally that such an anomalous resistance is given by $\eta_{an} = \nu_{an}/\epsilon_0\omega_{pe}^2$ with the anomalous electron collision frequency $\nu_{an} = \alpha\omega_{pe}W_{turb}/nk_BT_e$. It is proportional to the turbulent wave energy density W_{turb} normalized to the thermal energy density nk_BT_e, and α is some numerical coefficient which for any turbulent interaction of interest is of the order of 0.01 [see, e.g., Sagdeev, 38]. For smaller α anomalous resistance plays no role anyway. Usually the turbulent wave energy densities are much below thermal energy density, the ratio being of the order of 10^{-8}. However, even if the ratio reaches 10^{-5} anomalous resistance is unimportant in producing parallel potential drop via Ohm's law. Stronger average incoherent wave intensities are highly improbable.

One may thus safely conclude that "classical" anomalous resistance is not responsible for the generation of field-aligned potential drops of the size needed in aurorae. Here by classical we mean that ordinary incoherent plasma turbulence is responsible for anomalous collisions. The sole possibility remaining is that some coherent process is responsible for the observed potentials. Of course, formally one can also define an anomalous collision frequency in these cases, but this makes sense only from a global point of view and to some extent hides the microscopic physics.

5.3.1 Electrostatic Double Layers

Since large scale double layers in collisionless plasma cannot be sustained, any average large scale potential difference will consist of a sequence of microscopic potential differences adding up to the observed one. The maximum sustainable electric potential drop across an electrostatic shock in collisionless plasma can be easily estimated from $|j_\parallel E_\parallel| \approx d(nk_BT_e)/dt$. Assuming that the shock is quasi-stationary and causes a density depression Δn by evacuating plasma in a region of length Δx while moving at speed v_s, the potential drop will be given approximately by

$$|\Delta\Phi| = \Delta x|E_\parallel| \approx \frac{\Delta n}{n}\frac{n}{n_b}\frac{v_s}{v_b}\left(\frac{k_BT_e}{e}\right) = \frac{\Delta n}{n}\frac{nv_s}{j_\parallel}k_BT_e \qquad (5.2)$$

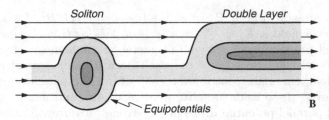

Fig. 5.6. Two types of electrostatic potential structures. The first is typical for a (two-dimensional) soliton in parallel direction, the second S-type or U-type potential structure is that of a double layer or "electrostatic shock". Only the latter one has a finite potential difference across it while the net potential difference across the soliton is zero [after Treumann and Baumjohann, 41]

Typical shock velocities are of the order of the ion acoustic speed c_{ia}. Hence for keV beam electrons the velocity ratio is of the order of $< 10^{-2}$. Assuming a nearly complete evacuation of the plasma $\Delta n \sim n$ as is frequently observed in the upward current region, the total potential difference is $\Delta \Phi < 10^{-2}(n/n_b)$ kV. Since the beam to plasma density ratio is $\sim 10^{-2} - 10^{-3}$ the potential difference can reach values of ~ 1 kV, implying that a few of these strong double layers suffice to account for the necessary potential drops in the aurora. Alternatively, if the density depression is much less, many very weak double layers are required. These will be slightly asymmetric, soliton-like structures or phase space holes.

The question of how strong double layers are generated has not yet been ultimately solved. It has early on been argued that the field-aligned current in large-amplitude shear Alfvén waves would generate them in regions of anomalous resistance produced self-consistently by current driven ion-cyclotron waves as shown by Lysak and Dum [20]. Such shear Alfvén waves can be launched from a tailward magnetospheric source as a kinetic Alfvén wave and transform into a shear wave in the low density upper auroral ionosphere. However, as discussed above, sufficient anomalous resistances seem highly improbable even for large ion-cyclotron wave amplitudes. Still the picture that the shear wave carries the current responsible for the double layer may be true.

Another possibility is that the double layers evolve from large amplitude ion-acoustic waves excited in the presence of two electron populations of different densities and temperatures, as Berthomier et al. [3] assume to be present in the aurora. Figure 5.7 shows the pseudo-potential for ion-acoustic waves and the range when double layers may occur. In the potential such double layers have an S-shaped or U-shaped equipotential structure as seen on the right in Fig. 5.7. The two legs of the U are along the magnetic field while the bottom of the U connects two magnetic field lines thereby giving rise to the parallel electric field. This is clearly localized. The pseudo-potential solutions for the two electron population driven ion-acoustic waves show that such

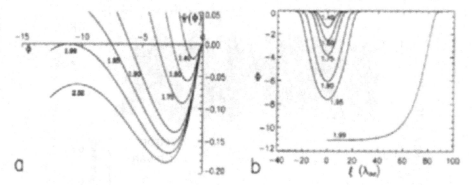

Fig. 5.7. a: The pseudo-potential for the transition from ion-acoustic solitons to ion-acoustic double layers. **b**: The two different forms of solitons and double layers. These structures are solutions of the stationary Korteweg-deVries equation. The pseudo-potential indicates the regions where solutions exist. In the negative closed domain the solutions are soliton-like, in the negative open domain the solutions are nonsymmetric potential structures which correspond to a net electric potential drop across the structure and thus a double layer as shown on the right [after Berthomier et al., 3] (Reprinted with permission of the American Geophysical Union). Note that the lengths are normalized to the fast electron Debye length

microscopic double layers are solutions of the stationary nonlinear equation describing large-amplitude ion acoustic waves.

A real measurement of a double layer potential in the upward current region by the FAST spacecraft is shown in Fig. 5.8. In the case in question the double layer (negative) field-aligned electric field is about 1.5 V/m and points upward implying that electrons will be accelerated earthward. The average electron energy changes by an amount of 5–6 keV (from about 4 keV to about 10 keV) during the crossing implying that the total potential drop in the double layer was about 5-6 kV which, with the above value of the electric field yields a parallel extension of the double layer of roughly 30-40 km in this case. From Fig. 5.8 in this case we would conclude that the entire structure survives for at least 10–20 s or longer than the crossing time of the spacecraft. Thus the interesting question is not simply how the double layer is generated but why it can survive for long times.

So far neither of these questions can be answered. The stationary theory mentioned above does not give any hint of how these microscopic double layers evolve and survive. Numerical simulations in 1d and 2d have been performed during the past decade as well as more recently, showing that such double layers are highly time variable structures. In these simulations by Goldman et al. [15] and Newman et al. [27] the double layer is either generated by simply imposing a large density depression on the plasma or by artificially putting a U shaped potential into the plasma as recently done by Singh et al. [39]. The latter simulations are of considerable interest. They describe the potential structure as being generated by a shear flow on the top of the box. The model

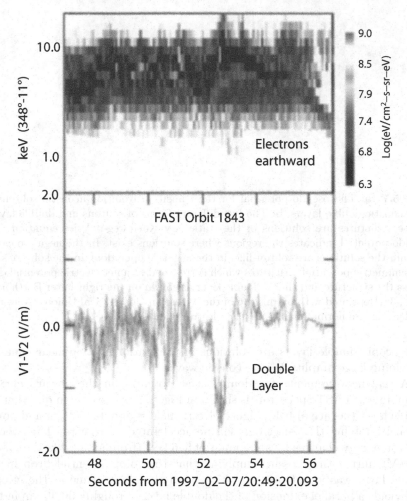

Fig. 5.8. *Top*: Downward auroral electron energy flux in reaction to the presence of a double layer potential. *Bottom*: The measured microscopic double layer potential during an inverted-V crossing by FAST. The large unipolar negative excursion of the parallel electric field indicates the presence of a field-aligned potential difference of U shape or S shape. The field points upward in this case accelerating electrons down as seen in the upper part of the figure where the electron energy jumps up by the amount of a few keV. Note the change in the character of the field fluctuations before and after crossing the double layer (Reprinted with permission of the American Geophysical Union partly adapted from Pottelette and Treumann [33])

is sketched in Fig. 5.9. The simulation is shown in Fig. 5.10. The simulations of Singh et al. [39] come already close to reality in that a shear flow imposed at the magnetospheric edge of the box transfers its transverse potential drop along the magnetic field into a parallel potential drop resembling a double

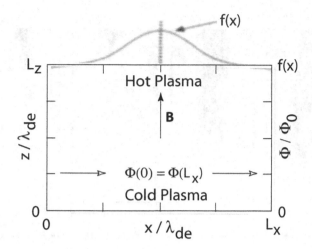

Fig. 5.9. Initial conditions for the simulation of the evolution of a double layer along the magnetic field **B** in the simulation box. The bell-shaped potential prescribed at the top of the box corresponds to a local convective shear flow (as in a shear Alfvén wave). A hot dilute electron plasma (current) is injected at the top, while at the bottom a cold dense plasma is sitting. In this configuration a double layer evolves [after Singh et al., 39] (Reprinted with permission of the European Geophysical Society)

layer. The density shows a region of strongly diluted plasma evolving at the bottom of the box where the cold plasma is heated and pulled out of the region of parallel potential drop with it being accelerated at the boundaries of the double layer. Moreover, the hot plasma injected from the top is seen to become more dense with time and possibly even more energetic because of acceleration in the field-aligned potential drop. The field-aligned potential layer becomes unstable at later times with respect to density fluctuations and resolves into a number of smaller scale double layers.

Even though such half-self-similar simulations are a step in the direction of real progress, the self-consistent evolution of a double layer in the magnetospheric geometry has not yet been shown in simulations. For this to happen one needs to include the source of the shear flow which in the current simulations is arbitrary. Probably one also needs to include the correct converging magnetic field geometry in addition to the prescription of the magnetospheric source which can be either a shear flow, i.e. a converging or diverging convection electric field at some altitude far above the ionosphere as in the case of Singh et al. [39], or Alfvénic wave pulses of the kind of kinetic Alfvén waves generated somewhere in the magnetosphere by some mechanism like reconnection or other kinds of instabilities. In addition the plasma must carry field aligned currents consisting of downward energetic electrons injected from the magnetospheric source as well as the local auroral plasma environment. In an Alfvénic pulse such currents are naturally included. However, how they are produced remains to be an unresolved question. So far simulations of this kind are out of reach.

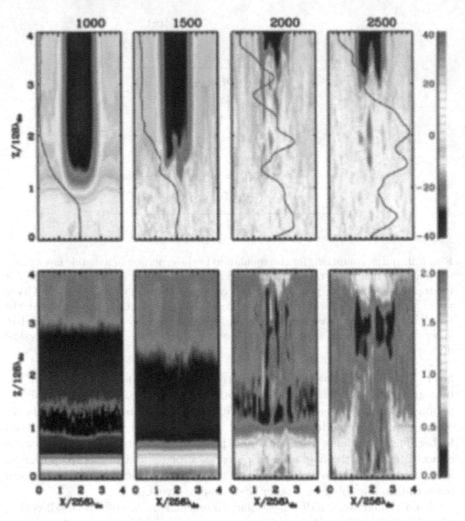

Fig. 5.10. Evolution of a one-sided shear flow driven double layer in the upward current region. The different columns show the potential structure (*top row*) and plasma density (*bottom row*) for four different simulation times. Initially after very short time a U shaped potential evolves with the shear flow potential drop mapped to a parallel electric field. At later times the laminar double layer is gradually destroyed and assumes a filamentary structure [from Singh et al., 39] (Reprinted with permission of the European Geophysical Society)

5.3.2 Microscopic Structures – Phase Space Holes

The above discussed double layer simulations can be taken as the macroscopic background scenario for microscopic processes in the auroral plasma. They show already that the macroscopic picture is not entirely stable to fluctuations and that it decays in time into smaller scale structures unless it is driven permanently from the outside. Such a permanent drive is, however, unrealistic as the source processes deep in the magnetosphere are by themselves highly variable in time. This time variation will be imposed on the input and will cause the auroral processes to be variable as well on two scales, the scale of the driver and the intrinsic scale of the processes taking place in the auroral region.

Field-aligned potential differences can also be generated as the superposition of very many small-scale small potential drops. This can be realized by long chains of quasi-solitary structures along the magnetic field. Figure 5.6 left-hand structure shows recurring potential structures with zero net potential drop. However, small asymmetries in these potentials may exist giving rise to very small field-aligned electric fields which can be maintained in the plasma by small charge separations on scales of the Debye length. Such structures are known as Bernstein-Green-Kruskal (BGK) modes after Bernstein et al. [2] who first investigated them using kinetic plasma theory.

BGK modes are the result of the nonlinear interaction between plasma waves and particles on the Debye scale. This scale is not covered by fluid theory (Korteweg-deVries modes, Nonlinear Schrödinger equation, and other nonlinear fluid equations) and hence the resulting phase space structures cannot be described by these fluid theories.

The BGK modes represent holes in phase space (see Fig. 5.13). This figure shows on its left the electron distribution function in phase space. The ordinate is the velocity, the abscissa is one-dimensional space. The electron density in phase space is shown in gray scale. Electron holes are not only localized structures in real configuration space, but in addition they are connected with local deformations of the particle distribution function in phase space, so called phase space holes. The particle distribution splits into a trapped and a passing distribution. In addition, the size of the hole increases with increasing hole amplitude [Muschietti et al., 25]. This is in contrast to soliton solutions of the fluid equations where the size decreases with amplitude in order to keep the invariant number of "trapped plasmons" constant. Different types of holes are possible depending on the form of the potential. Figure 5.14 shows a model of a top-hat potential hole. Because plasmas consist of two kinds of particles, ions and electrons, there are two types of holes which can evolve in the plasma: ion holes and electron holes, respectively. Ion holes are depressions in the ion distribution, they are negatively charged entities which move across the plasma. Electron holes on the other hand are positively charged structures. Hence, ion holes possess converging parallel electric fields, while near electron holes the electric fields diverge.

The theory of phase space holes has been given by Dupree [10]. A first numerical simulation of ion holes by Gray et al. [16] in a current carrying plasma is shown in Fig. 5.11. The simulation is one-dimensional with space on the ordinate and time on the abscissa. The local particle density at each point and time is ray scale coded (see the gray scale bar to the left of the figure).

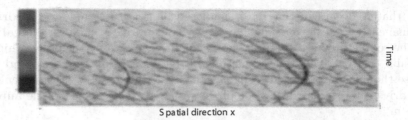

Time

S patial direction x

Fig. 5.11. One-dimensional numerical particle simulation of the evolution of ion holes in a plasma carrying a field aligned current [adapted from Gray et al., 16]. The abscissa is along the magnetic field, while the ordinate is time. The colour code gives the density normalized to background. Here the ion holes, which are the dark structures, result from an ion-acoustic instability driven by the strong current. The ion density depression in the holes is nicely visible. Interestingly the two big holes change the direction of their motion during evolution (with permission of the American Geophysical Union)

Depression of ion density in the holes is clearly seen as dark lines occurring on the light background. These ion holes were generated by ion acoustic instability evolving in a plasma carrying a parallel current as required for the auroral region. In the simulation they occur as density depressions which during their evolution change their velocity from being initially parallel to the field to ultimately moving antiparallel. This behaviour can be understood from the dynamics of the ion acoustic waves from which they evolve. These move initially at ion-acoustic speed less than the electron drift velocity in the direction of the electrons which carry the current. However, since ion holes are effectively negative charges they are reflected from electrons and attracted to ions, and as a result their velocity rotates until they move with the ions.

It seems that electron holes can evolve more easily than ion holes because the higher mobility of the electrons allows them to react faster in the presence of fluctuating electric fields. Recent one-dimensional simulations of electron holes by Newman et al. [27, 29] and Goldman et al. [14, 15] have indeed shown that electron holes are quite a natural nonlinear phenomenon in dilute plasmas under many different conditions. This has been confirmed by observations in the auroral region by Carlson et al. [5] and Ergun et al. [5], in the tail of the magnetosphere by Cattell et al. [7], in the magnetosheath by Pickett et al. [31] and in the bow shock by Bale et al. [1].

These observations have identified electron holes as bipolar electric field signatures in the downward current region. Simulation studies have demonstrated that these bipolar structures are indeed caused by the two stream instability [14]. This is shown in Fig. 5.12 where in the initial phase of the two stream instability electron holes evolve. However, when the nonlinear state is reached, these holes decay very soon into intense whistlers. This is not entirely a property of the simulation. Very frequently the observations in the downstream region contain very intense saucer emissions in the VLF which

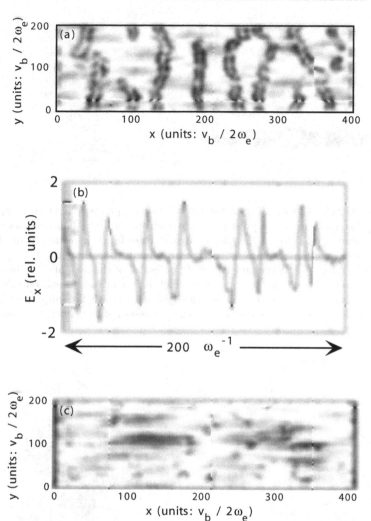

Fig. 5.12. Two-dimensional particle in cell simulation of the evolution of electron holes in a plasma out of the two stream instability [adapted from Goldman et al., 14]. The first panel shows the evolution of bipolar electric field signatures typical for an electron hole. The second panel shows the bipolarity of the electric field signatures when crossing the holes along the x direction. At later times in the two-dimensional simulation the holes decay into whistlers. This is shown in the third panel (Reprinted with permission of the American Geophysical Union)

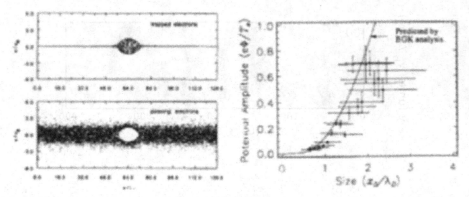

Fig. 5.13. *Left*: The splitting of the distribution in phase space into trapped and passing distributions in the presence of an electron hole. The hole appears as a real hole of missing particles in the passing distribution function, while in the trapped distribution function it contains particles of energy less than the hole potential. *Right*: The relation between the maximum potential of the hole and the hole width. The hole width increases with potential which is in contrast to soliton solutions of the fluid equations where the soliton width decreases with the amplitude [adapted from Muschietti et al., 25] (Reprinted with permission of the American Geophysical Union)

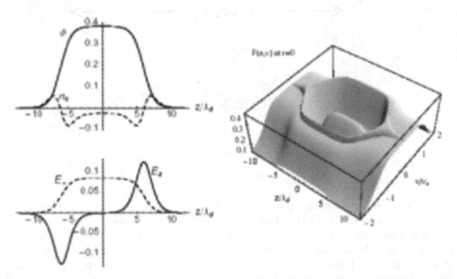

Fig. 5.14. Model of a so-called stretched electron hole, showing the potential in cross section, the corresponding electric field, and the phase space structure of the hole. Interestingly such holes possess a maximum in the density of the trapped electrons in the center of the hole, and a sharp edged boundary consisting of lowest energy passing particles. These concentrations of particles cause peaks in the electric field [adapted from Muschietti et al., 26] (Reprinted with permission of the European Geophysical Society)

originate from a very narrow source region [see the discussion in LaBelle and Treumann, 18] located in the downstream current region and obviously related to the most intense electron holes. One may speculate that the chain of two stream instability-electron holes-saucer emission in this region serves as the most efficient way of dissipation of the free energy contained in the downward current. Since the current cannot feed the whistler directly it first generates electron holes which are too slow to transport energy away. Hence, they decay into oblique whistlers which are fast enough to distribute the free energy over the plasma in a wide region away from the source region.

The interest in electron holes is manifold. They are not only interesting micro-structures in the auroral plasma. They divide the electron distribution into two separate parts as suggested by Bernstein et al. [2] and first shown in the model of Muschietti et al. [25], a passing energetic and accelerated distribution and a trapped distribution of electrons which have local parallel speeds in the electron hole frame smaller than the potential of the hole such that they can be trapped. Hence the holes cool the small trapped part of the electron distribution and accelerate the passing part. Holes thus contribute substantially to particle acceleration and VLF whistler radiation. As usual the really interesting physics takes place on microscopic scales and very short times.

5.3.3 Relation between Holes and Double Layers

Goldman et al. [15] investigated the time evolution of phase space holes which are driven in the presence of a strong microscopic double layer. The idea behind such an investigation is that double layers once immersed into or evolving in the plasma will necessarily accelerate particles along the magnetic field. They will generate a current, and they will reflect particles from the location of the double layer. In such a situation one can expect that a beam-beam instability between incoming and reflected electrons will evolve and generate electron holes. This happens indeed, as shown in Fig. 5.15 which plots the time history of the evolution of holes in the vicinity of an artificially implanted double layer by digging a deep depletion into the plasma density distribution. Note that this indeed causes a double layer as described above when mentioning the simulations of Singh et al. [39], since there the application of a shear potential caused a deep density depletion to evolve.

The space-time plot of the time history shows that the double layer serves as the generator of a large number of phase space holes. Initially many electron holes are generated by an electron-electron instability moving at very high velocity to the right into the high potential side. These holes are small amplitude. At later time a chain of holes begins to escape from the double layer to the left into the low potential side at much lower speed. This speed is comparable to the velocity of the double layer accelerated ion beam. The holes in this chain have much larger amplitude than the original right moving electron holes. These large-amplitude holes quasi-periodically interrupt the

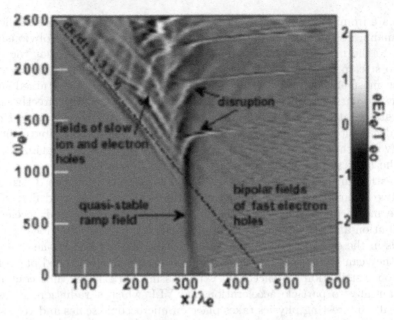

Fig. 5.15. Time history of the evolution of phase space holes in a plasma containing a strong double layer. The double layer is the dark nearly vertical region in this space-time plot of the electric potential across the simulation box [adapted from Goldman et al., 15]. One observes that early in time fast and small amplitude electron holes escape from the double layer to the right. At later times a chain of slow large amplitude phase space holes escapes from the double layer to the left. This chain consists of tripolar electric field structures which are probably a mixture of ion and electron holes, the former being generated by a kinetic two-stream instability, the latter resulting from an electron-electron instability driven by initial and reflected electrons (Reprinted with permission of the European Geophysical Society)

double layer until it has again reformed. The chain consists of ion holes and electron holes and apparently has tripolar electric field signature.

This can be realized from inspection of the evolution of the chain. In fact, the chain starts as a slow left-moving ion hole which hooks up an electron hole that tries to move to the right but is braked by the formation of another ion hole such that it practically comes to rest and stands in front of the double layer for comparably long time while growing. It is thus nearly stationary in the frame of the double layer. When its amplitude becomes comparable to that of the double layer it becomes strong enough to break through the potential barrier of the double layer, speed up, and escape to the right with high velocity to join the group of small fast moving electron holes (features labelled as "disruptions" in Fig. 5.15). This happens several times with other electron holes as well such that the region to the right of the double layer which is the low potential side becomes highly turbulent containing a mixture of large amplitude ion and electron holes. This whole process corrugates the double layer, broadening it, structuring it into smaller pieces, and due to overall momentum conservation letting it move to the left in the direction into which the ion holes move, i.e. into the direction the ion beam is moving. In real space this implies that such a corrugated double layer will be moving slowly upward in the upward current region while being a site of accumulation of electron holes which try to pass through the potential barrier. The whole turbulent dynamics is basically confined to the low potential side of the double layer.

The ion holes are the result of a two-stream current instability, while the electron holes are generated by an electron-electron instability which involves ion hole accelerated or reflected electrons. Thus, chains of this kind act as strong secondary accelerators of particles. They probably exist in the upward current region as has been demonstrated by Pottelette and Treumann [33] and is in contrast to the bipolar holes which have been found in the downward current region. This is an important observation as the upward current region is the source region of the auroral kilometric radiation which is fed by accelerated electrons. In contrast, in the downward current region no radiation is emitted. There, in the sufficiently dense plasma, the free energy is transferred through the bipolar electron holes into the fast escaping whistlers forming the saucer emissions. In the dilute upstream current region the transfer takes place through electron holes in the chains directly into the auroral kilometric radiation.

Figure 5.16 shows the relation between the evolution of the double layer and chain of holes at different phases of the simulation and the electron and ion phase space distributions, shown by cuts taken at various places along the

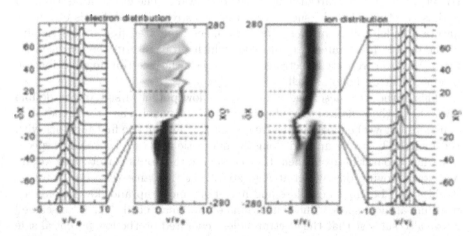

Fig. 5.16. The relation between the evolution of the double layer and the holes and the deformation of the electron and ion distributions in the simulation box [adapted from Goldman et al., 15] (Reprinted with permission of the European Geophysical Society)

potential structure. On the left part of the figure the broadening of the electron distribution with time is nicely displayed. Behind the ramp of the double layer it consists of a thin modulated electron beam and a broad reflected electron component. The inward cuts in this distribution are the electron holes passing through the effectively heated electron plasma. The evolution of the distribution is very well monitored by the 16 cuts through the distribution at different places. In front of the ramp the electron distribution is a broad bump with several small maxima. Behind the ramp it has a narrow beam bump on the right and a flat, broad, hot component on the left. The ion distribution in front of the ramp consists of two well separated beams of equal heights. Behind the ramp these merge to become only one broad beam.

5.3.4 Holes in the Upward Current Region

Our main interest concerns the upward current region since this is the source region of AKR. In light of the above discussion of the plasma dynamics the upward current region is rather distinct from the downward current region. The near absence of a dense background plasma here [McFadden et al., 23] changes the character of the double layer and phase space holes. The main plasma components are the cold fast ion beam moving upward along the magnetic field, and the ~keV downward auroral electron beam. As described earlier, this beam has a broad pitch angle distribution forming a ring or horseshoe distribution with empty loss cone. This distribution is the result of the combined action [8] of the reflecting magnetic mirror force in the converging geomagnetic field and the presence of the upward parallel double layer electric field which accelerates the electrons downward. The electron distribution is thus effectively hot. In such a case the two-stream instability is of kinetic nature [Kindel and Kennel, 17], and the holes generated first in the vicinity of the double layer ramp are ion holes which propagate in the ion acoustic mode. However, reflected electrons together with the electron ring distribution produce electron holes as well.

This is the case described above on the low potential side of the double layer. Both these holes are visible in Fig. 5.1 in panel c as broadband VLF emissions below the local plasma frequency f_{pe}. They move first together with the ion holes basically upward along the magnetic field at relatively low speed, but when separating from them they corrugate the double layer ramp locally and start moving downward at high speed. At the same time the distorted double layer begins to oscillate and may both move up and down along the field line at slow speed. The simulations of Goldman et al. [14, 15] discussed above also suggest that the electron holes generated on the low potential side of the double layer stay for a relatively long time in the proximity of the double layer apparently moving together with it.

This is the plasma situation to which we will refer in the next section which deals with the generation of the auroral kilometric radiation.

5.4 Auroral Kilometric Radiation

Auroral kilometric radiation is the extremely intense electromagnetic radiation emitted from the "inverted-V" region during auroral and substorm activity Gurnett [21]. It radiates away to free space up to several percent of the total substorm energy, an enormous amount which cannot be explained by any incoherent radiation mechanism. Its gross spectrum is shown in Fig. 5.17. It has a sharp low frequency cut-off near the cyclotron frequency at the upper end of its source region.

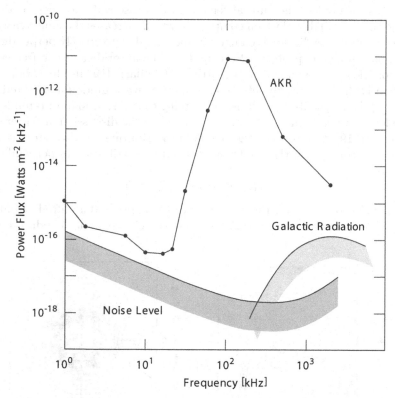

Fig. 5.17. The average spectrum of auroral kilometric radiation as registered in the first complete observational work on this radiation [after Gurnett, 21]) (Reprinted with permission of the American Geophysical Union)

Many theories have been proposed to explain the generation of AKR. However, the only theory which survived is that of a weakly relativistic coherent maser radiation [Wu and Lee, 42]. It is reviewed in depth in the article of Louarn [19] in this book. The essence of that theory is that even very weakly relativistic auroral electrons with energies of a few keV with distribution function deviating from the beam (or Maxwellian) distribution in a very dilute plasma invert the absorption coefficient of the plasma. The plasma then radiates in concert like a maser emitting coherent and therefore very intense radiation. This is a purely linear process and therefore rules out any nonlinear much less efficient mechanisms. However, the original idea that the maser worked at the boundary of the empty loss cone turned out to be illusory. There are not enough particles available to feed sufficient energy into the radiation.

A more realistic mechanism was found when it was realized that strong parallel double layer electric fields in cooperation with the reflecting mirror force (see Figs. 5.18 and 5.19) lift the auroral beam electron distribution into an excited level by generating a ring or horseshoe distribution of the kind actually measured in the auroral region. In this case radiation can be driven by the entire available electron component which increases the emissivity by a large factor. This radiation is emitted in the x-mode into strictly perpendicular direction posing the problem of escape from the auroral cavity to free space. This problem is described in the article of Louarn [19] in this book. The solution is that the cavity acts as an imperfect wave guide for EM and TM modes such that at its boundaries a coupling to external modes: o-mode and z-modes as well as x-modes become possible [see the discussion and literature by Louarn, 19]. In any case, the particles contributing to wave growth must satisfy two conditions: the condition of inverted distribution in perpendicular direction, viz.

$$\partial f(v_{\parallel}, v_{\perp})/\partial v_{\perp}|_{v_{\parallel}=v_r} > 0 \tag{5.3}$$

which is the condition for the presence of free energy and a "lifted" to higher energy level distribution function, and the particle resonance condition

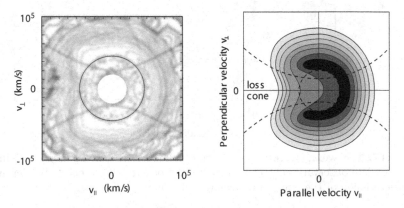

Fig. 5.18. *Left*: Measured [after Delory et al., 9, with permission of the American Geophysical Union] and *Right*: model horseshoe electron distribution in the auroral region. The measured distribution consists of the spread electron beam moving down along the field line and a plateau which is produced by the quasilinear interaction of the distribution with the self-generated VLF waves. The loss cone on the left in both distributions is practically empty

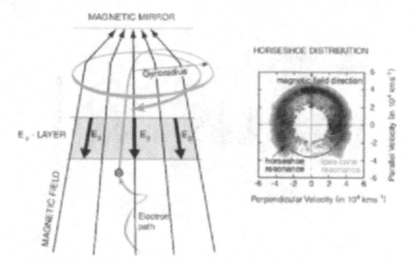

Fig. 5.19. *Left*: Schematics of the mechanism of generation of horseshoe distributions in a converging magnetic field. The converging field reflects the particles while the parallel electric potential accelerates the particles. These gain in this way perpendicular energy. The result is the distribution on the *Right*: which is a horseshoe distribution. The formal analytical treatment of this process has been given by Chiu and Schulz [8]

$$k_\parallel v_r/2\pi = f - f_{ce}/\gamma_r \qquad (5.4)$$

where $k_\parallel$ is the parallel wave number, f the wave frequency, f_{ce} the local electron cyclotron frequency, and $\gamma_r = [1 - v^2/c^2]^{-\frac{1}{2}}$ the relativistic energy factor. Clearly, for strictly perpendicular radiation $k_\parallel = 0$, the resonance line becomes a circle in the $(v_\parallel, v_\perp)$-plane which can be positioned along the nearly circular horseshoe distribution where the derivative in the above equation is positive and maximum. This is the simple idea of emission in the x-mode. A theory like this generates very broadband emission unless the distribution is deformed locally very strongly.

Inspection of the high-resolution emission spectrum however demonstrates that strongest emission is generated in very narrow bands which by themselves drift on variable speed across the frequency-time spectrogram. Such situations are shown in Figs. 5.20 and 5.21 for two different cases: when the narrow band emission is standing and when it is moving. Figure 5.21 in addition shows the observation of two such interacting narrow band AKR sources of high intensity. In this case the two move initially at different velocities. When interacting, they obviously repel each other and move together at the same speed to higher frequencies downward along the magnetic field. This can be interpreted as both emission regions being of same polarity and thus not mixing.

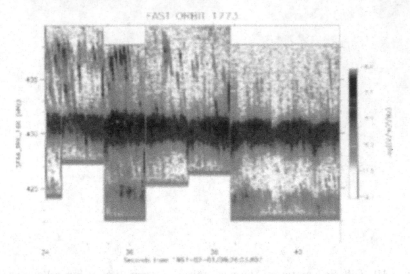

Fig. 5.20. An example of a narrow emission band in AKR which in this case is stationary in space and frequency

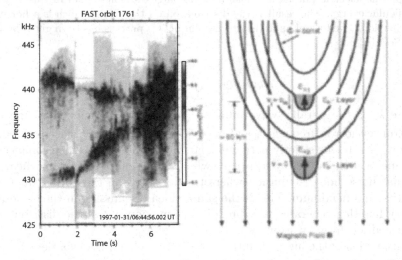

Fig. 5.21. *Left*: Two narrow band AKR emission lines moving across the frequency-time spectrogram approach each other, collide and move together down along the magnetic field at same speed. *Right*: The explanation of this case as two electric field structures of similar amplitude and equal polarity approaching each other [adapted from Pottelette et al., 32] (Reprinted with permission of the American Geophysical Union)

Another very interesting observation can be made from these two figures. They show that the emission of AKR is not continuum but consists of the superposition of a large number of "elementary radiation events" (ERE's) the emitted radiation of which superimposes to generate the apparently continuous AKR spectrum. The nature of these elementary radiators has not yet been determined. However, in light of the above discussion of the dynamics of the auroral plasma it seems quite natural to assume that the elementary radiators are the electron holes which are generated in the plasma and move on the background of the horseshoe distribution. If this is true then the same theory applies as before with two modifications:

- the electron holes must generate steep gradients on the distribution in perpendicular velocity which implies that they must be bent in phase space;
- the emission from each electron hole is then not necessarily perpendicular as this is no longer required by the resonance condition. It will, however, be inclined, and the strongest emission is detected when the spacecraft will be in the emission cone of the elementary radiators.

The existence of strong emission lines which move together may suggest that many such radiators have become attracted to a certain region and move approximately together over relatively long times. This is possible when realizing that the electron holes are effectively positive charges on the electron background which implies that they will interact in a certain way with the main electron component becoming attracted by the bulk of the electrons and thus trapped in the electron distribution while at the same time growing in amplitude as their absolute depth will be conserved. This can lead to a collection of holes in a more narrow space region which then acts as the radiation source. The questions related to these problems have not yet been solved and are subjects of ongoing research.

5.5 The Tail Acceleration Region

We now come to the discussion of the ultimate source of the auroral electron beam. This discussion will be rather brief as there is no consensus about this hot problem yet. It seems, however, reasonable to assume that processes in the tail of the magnetosphere provide the primary energy source since aurora are most prominent during substorms which are directly related to reconnection in the tail current sheet. It is thus natural to consider the structure of the reconnection site and to speculate about the reconnection site being the ultimate energy source of aurorae.

It has recently been shown by Øieroset et al. [30] that the reconnection diffusion site in the tail current sheet is the site of separate electron and ion dynamics. On the small scale of the ion diffusion length $c/2\pi f_{pi}$ the motion of electrons and ions becomes different. The ions in the plasma carrying the tail current are effectively non-magnetic on this scale while the electrons are still tied to the magnetic field.

This difference gives rise to the generation of Hall currents since the magnetized electrons follow the convective motion of the magnetic field from the

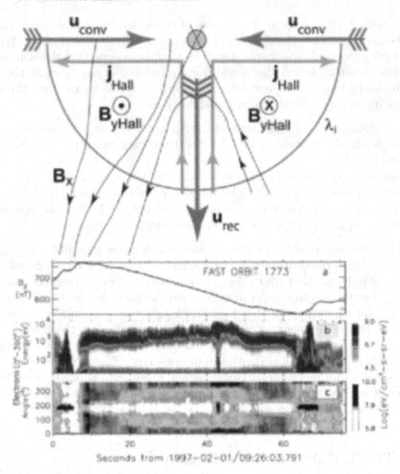

Fig. 5.22. Schematic of the earthward half of the ion diffusion region in the magnetospheric tail current sheet during reconnection and its relation to the auroral electron flux regions

tail lobe into the tail current sheet. These Hall currents must close somewhere. However, since such a closure cannot be done locally, the only possibility for them to close is through field-aligned currents. Inflow of electrons on the boundary far from the current sheet and outflow from the boundary in the current sheet are required. This is demonstrated in Fig. 5.22 which shows a schematic of the connection of a part of the ion diffusion region in the tail and the auroral electron fluxes which have been measured during one particular path of FAST.

In this figure the electron inflows and outflows in the ion diffusion region in reconnection are seen as large arrows, while the corresponding Hall current is indicated in oppositely directed lines. The electron speed is essentially the inward-convection and outward reconnection-jet speed, respectively, to both

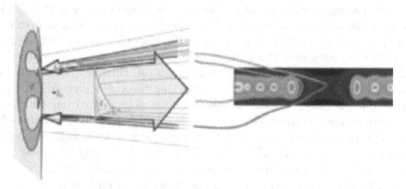

Fig. 5.23. The schematic connection between tail reconnection and the auroral region. Tail reconnection is shown as a simulation result for the magnetic field x-point region, when one dominant x-point has evolved and several secondary x-points are visible in the tail current sheet. In this case the current sheet is very thin, of the order of the ion skin depth, such that the entire current sheet is affected by reconnection. The Hall coupling of the reconnection region to the auroral zone causes the coupling of the reconnection to the processes in the aurora. It is not yet clear in which way this really happens

sides of the diffusion region. The electron outflow is related to the downward auroral electron beam fluxes, while the electron inflow is fed by or at least related to the ionospheric outflow of electrons in the downward current region. In this picture the downward currents are the Hall currents emanating from the near-Earth reconnection site in the tail.

Figure 5.23 shows the speculative geometry how activity in the tail connects to the auroral ionosphere. This picture does not take into account fast particles generated during reconnection in the tail since these particles are only a fraction of the entire electron distribution. It also suppresses the real process of coupling between the reconnection site and the ionosphere. This is a process which depends on the real conditions, on the propagation speed of the currents and particles and how fast particles can be extracted from the ionosphere to feed the Hall current in the tail reconnection region.

The coupling might be produced by kinetic Alfvén waves which are generated in the Hall region with transverse scales just of the ion inertial scale. When these waves, which are surface waves, move down to the ionosphere they constitute a current pulse as well as a transverse electric field pulse which may cause the required shear flow potential at the topside ionosphere which, as has been discussed here, will drive a U-shaped potential in the ionosphere, causing the double layer and, depending on its polarity, extracting electrons out of the ionosphere.

An important problem in this connection is that in all available models the magnetospheric tail reconnection site is so large that it should affect both hemispheres in the same way. Hence aurorae should be symmetrical with respect to the equator if a coupling like the one proposed here will exist. This cannot be the case for the auroral processes. The observations of FAST and other spacecraft show that in most cases the closure is nearly local. At first glance this suggests that the auroral field-aligned current system should be local and should have little in common with a tail current system generated in reconnection. One might therefore argue that the above picture is incorrect as it requires a global closure of the magnetic field-aligned currents in the ionosphere similar to those closure processes which have been predicted for decades in the literature [for a review of the various auroral field-aligned current system generator mechanisms see Borovsky, 4]. Though this may sometimes be indeed the case, there are possibilities for small scale reconnection in the tail being restricted solely to one hemisphere and even to part of it when reconnection is multiple [as has recently been argued by Pottelette and Treumann, 34] or when, as has been observed recently with Cluster [as can be concluded from observations by Runov et al., 36, 37], the current sheet in the tail bifurcates into narrow current layers.

The mechanism of such a bifurcation is not understood yet. It is probably related to the preference of the current layer to generate conditions which are in favor of reconnection to develop on a very fast time scale. For this to happen it is required that the current sheet is very thin. This, however, causes again problems with the topology of the magnetic field which becomes very complicated. In principle bifurcation can become possible only when the initially two-dimensional magnetic field and current structure develops into a three-dimensional configuration. If this is the case, as most recent numerical simulations of reconnection have shown, then a model of the connection between tail reconnection in the near Earth tail and auroral processes can be developed which is restricted solely to small scale auroral phenomena taking place on one hemisphere only in the auroral region.

5.6 Conclusions

We have given a brief overview of the current state of the art of our knowledge in the fundamental acceleration, radiation, and source processes in the nighttime auroral field-aligned current system. After a quarter of a century in situ observations from S3-3 to Viking, Freja, Polar, FAST, Geotail and finally Cluster a stage has been reached on which we can firmly conclude that our understanding of the auroral processes has stepped up from a purely descriptive one to a semi-quantitative understanding of the dominant acceleration processes of charged particles, both electrons and ions, in aurorae, of the structure of the auroral current system that consists of upward and downward field-aligned currents which close in the ionosphere through Pedersen currents on surprisingly small scales, on the radiation mechanism in auroral kilometric radiation, and on the generation of the parallel potential drops. We know by now that indeed such parallel potential drops are generated which cause tremendous density decreases in the ionosphere while being highly dynamic. Usually in an auroral acceleration region not only one such drop exists but several are present along the magnetic field. Their longitudinal scale is of the order of not more than few 10 km. They thus comprise quasi-stationary but small-scale double layers containing potential drops of order 100 eV to ~ 1 keV, in rare cases few keV. These double layers move along the magnetic field and can interact with each other. They are the sources of electron holes, small scale structures in the electric field and electron phase space distribution of enormous dynamics. These electron holes may themselves contain smaller potential drops which erase the large double layer potential and at the same time cause out of phase variability in the acceleration of electrons and ions. Moreover, these electron holes seem to be the very sources of the auroral kilometric radiation, serving as elementary radiation sources. Thus they are of enormous dynamical importance in the auroral processes and in analogous applications under astrophysical conditions.

Nevertheless, a number of open questions still remain which in near future will have to be attacked by multi-spacecraft missions and numerical simulations of the auroral and magnetospheric processes. One of the most interesting of these problems is the relation of the auroral processes and the current generator processes in the geomagnetic tail. These are most probably related to reconnection in the tail current sheet under thinning conditions during substorms. A direct relation has been argued for in this review which is based on the realization that the reconnection Hall current system in the tail current sheet must be closed by field aligned currents which connect down to the ionosphere. Many aspects of such a model are in agreement with observation. However, one basic property is still barely understood: this is the fact that aurorae are very obviously local phenomena on one hemisphere only. How this can happen when reconnection is the driving force has still to be clarified. Since reconnection is the most probably generator one may thus ask how reconnection can be imagined being restricted to one hemisphere only. We have suggested ideas on a possible resolution of this puzzle only without going into more detail since models of this kind are still under evolution.

References

[1] Bale, S. J., et al.: Astrophys. J. Lett. 575, L25 (2002).
[2] Bernstein, I.B., J.M. Greene, and M.D. Kruskal, Phys. Rev. 108, 546 (1957).
[3] Berthomier, M., R. Pottelette, and M. Malingre: J. Geophys. Res. 103, 4261 (1998).
[4] Borovsky, J.E.: J. Geophys. Res. 98, 6101 (1993).
[5] Carlson, C.W., et al.: Geophys. Res. Lett. 25, 2017 (1998).
[6] Cattell, C., et al.: Geophys. Res. Lett. 26, 425 (1999).
[7] Cattell, C., et al.: J. Geophys. Res. 110, A01211 (2005).
[8] Chiu, L. and M. Schulz: J. Geophys. Res. 83, 629 (1978).
[9] Delory, G.T., et al.: Geophys. Res. Lett. 25, 2069 (1998).

[10] Dupree, T.: Phys. Fluids 26, 2460 (1983).

[11] Elphic, R., et al.: Geophys. Res. Lett. 25, 2033 (1998)

[12] Ergun, R.E., et al.: Geophys. Res. Lett. 25, 2041 (1998a).

[13] Ergun, R.E., et al.: Geophys. Res. Lett. 25, 2025 (1998b).

[14] Goldman, M.V., M.M. Oppenheim, D.L. Newman: Geophys. Res. Lett. 13, 1821 (1999).

[15] Goldman, M.V., D.L. Newman, and R.E. Ergun: Nonlin. Proc. Geophys. 10, 37 (2003).

[16] Gray P., et al.: Geophys. Res. Lett. 17, 1609 (1990).

[17] Kindel, J.F. and C.F. Kennel: J. Geophys. Res. 76, 3055 (1971).

[18] LaBelle, J. and R.A. Treumann: Space Sci. Rev. 101, 295 (2002).

[19] Louarn, P.: this volume (2005).

[20] Lysak, R.L. and C.T. Dum: J. Geophys. Res. 88, 365 (1983).

[21] Marklund, G., T. Karlsson, and J. Clemmons: J. Geophys. Res. 102, 17509 (1997).

[22] Marklund, G., et al.: Nature 414, 724 (2001).

[23] McFadden, J.P., et al.: Geophys. Res. Lett. 25, 2045 (1998).

[24] McFadden, J.P., C.W. Carlson, and R.E. Ergun: J. Geophys. Res. 104, 14453 (1999).

[25] Muschietti, L., et al.: Geophys. Res. Lett. 26, 1093 (1999).

[26] Muschietti, L., et al.: Nonlin. Proc. Geophys. 9, 101 (2002).

[27] Newman, D.L., et al.: Phys. Rev. Lett. 87, 255001 (2001).

[28] Newman, D.L., M.V. Goldman, and R.E. Ergun: Phys. Plasmas 9, 2337 (2002).

[29] Newman, D.L., et al.: Comp. Phys. Comm. 164, 122 (2004).

[30] Øieroset, M., et al.: Nature 412, 414 (2001).

[31] Pickett, J.S., et al.: Ann. Geophys. 22, 2515 (2004).

[32] Pottelette, R., R. A. Treumann, and M. Berthomier: J. Geophys. Res. 106, 8465 (2001).

[33] Pottelette, R. and R. A. Treumann: Geophys. Res. Lett. 32, L12104, doi:10.1029/2005GL022547 (2005a).

[34] Pottelette, R. and R. A. Treumann: Geophys. Res. Lett. 32, submitted (2005b).

[35] Pottelette, R., R. A. Treumann, and E. Georgescu: Nonlin. Proc. Geophys. 11, 197 (2004).

[36] Runov, A., et al.: Geophys. Res. Lett. 30, 1036, doi: 10.1029/2002GL016136 (2003).

[37] Runov, A., et al.: Ann. Geophys. 22, 2535 (2004).

[38] Sagdeev, R.Z.: Rev. Mod. Phys. 51, 1 (1979).

[39] Singh, N., et al.: Nonlin. Proc. Geophys. 12, in press (2005)

[40] Störmer C.: Ergebn. kosm. Physik 1, 1 (1931).

[41] Treumann, R. A. and W. Baumjohann: *Advanced Space Plasma Physics* (Imperical College Press, London, 1997).

[42] Wu, C.S. and L.C. Lee: Astrophys. J. 230,621 (1979).

Part II

High-Frequency Waves

6

The Influence of Plasma Density Irregularities on Whistler-Mode Wave Propagation

V.S. Sonwalkar

Department of Electrical and Computer Engineering, University of Alaska
Fairbanks, Fairbanks, AK, 99775, USA
ffvss@uaf.edu

Abstract. Whistler mode (W-mode) waves are profoundly affected by Field-Aligned Density Irregularities (FAI) present in the magnetosphere. These irregularities, present in all parts of the magnetosphere, occur at scale lengths ranging from a few meters to several hundred kilometers and larger. Given the spatial sizes of FAI and typical wavelength of W-mode waves found in the magnetosphere, it is convenient to classify FAI into three broad categories: large scale FAI, large scale FAI of duct-type, and small scale FAI. We discuss experimental results and their interpretations which provide physical insight into the effects of FAI on whistler (W) mode wave propagation. It appears that FAI, large or small scale, influence the propagation of every kind of W-mode waves originating on the ground or in space. There are two ways FAI can influence W-mode propagation. First, they provide W-mode waves accessibility to regions otherwise not reachable. This has made it possible for W-mode waves to probe remote regions of the magnetosphere, rendering them as a powerful remote sensing tool. Second, they modify the wave structure which may have important consequences for radiation belt dynamics via wave-particle interactions. We conclude with a discussion of outstanding questions that must be answered in order to determine the importance of FAI in the propagation of W-mode waves and on the overall dynamics of wave-particle interactions in the magnetosphere.

Key words: Field aligned irregularities, whistler propagation, ducting, plasma inhomogeneity, ionosphere

6.1 Introduction

The whistler mode is a cold plasma wave mode with an upper cutoff frequency at the plasma frequency (f_{pe}) or cyclotron frequency (f_{ce}), whichever is lower. Waves propagating in whistler mode (W-mode) are found in all regions of the Earth's magnetosphere. They are also found in the magnetospheres of other planets. These waves may originate in sources residing outside the magnetosphere, such as lighting or VLF transmitters, or they may originate within the magnetosphere as a result of resonant wave-particle interactions. W-mode

V.S. Sonwalkar: *The Influence of Plasma Density Irregularities on Whistler-Mode Wave Propagation*, Lect. Notes Phys. **687**, 141–191 (2006)
www.springerlink.com

waves have been detected on every spacecraft carrying a plasma wave receiver and at numerous ground stations [see, e.g., 48, 51, 54, 74, 107, 115].

W-mode waves are important because they influence the behavior of the magnetosphere and partly because they are used as experimental tools to investigate the upper atmosphere. W-mode waves and their interactions with energetic particles have been a subject of interest since the discovery of the radiation belts. These interactions establish high levels of ELF/VLF waves in the magnetosphere and play an important role in the acceleration, heating, transport, and loss of energetic particles in the magnetosphere via cyclotron and Landau resonances [see, e.g., 56, 67, 78, 104]. Propagation, including reflection, refraction, guiding, and scattering, determines the extent to which whistler mode energy can remain trapped in the magnetosphere and influence the nature of wave-particle interactions. Plasma density irregularities in turn determine to a large extent the nature of whistler mode propagation. The role of plasma density irregularities on whistler mode propagation and their effects on the overall dynamics of the magnetosphere is the subject of this paper.

The magnetosphere is a highly structured magnetoplasma containing field aligned irregularities (FAI) ranging in size from meters to hundreds of kilometers in the cross-B_0 direction [30, 39, and references therein] where B_0 is the geomagnetic field. These irregularities are present in all parts of the magnetosphere. Plasma density and density structures are believed to play an important role in many physical processes at low and high latitudes: wave particle interactions, mode conversion, particle acceleration and precipitation, including auroral precipitation, visible and radar aurora, and substorm activity [e.g., 66, 105, 113]. On the practical side, these irregularities are important because they contribute to the fading of high frequency trans-ionospheric signals and to the degradation of ground-satellite communication. Forecasting and specification of these irregularities is a major component of space weather programs.

Though the subject of this paper is to understand the role of FAI on whistler mode propagation, the knowledge of these irregularities themselves is incomplete and is a subject of ongoing research [e.g., 30, 41]. Moreover, a wealth of information about FAI has actually been obtained by studying their effects on plasma wave propagation. Thus the research on W-mode waves, and in general on plasma waves, and plasma density irregularities are intimately tied to one another.

Evidence that FAI play an important role in the propagation of W-mode waves has been accumulating since the 1950s and 1960s [e.g., 26, 58, 60, 65, 112, and references therein]. In these earlier studies, W-mode propagation in the presence of FAI was considered mainly to explain the occurrence rates of lightning-generated whistlers. In recent years it has been realized that FAI may play a role more important in the overall physics of the magnetosphere than initially envisioned. For example, lighting-generated whistlers refracting through large scale ionospheric horizontal density gradients in the topside ionosphere may lead to generation of plasmaspheric hiss, believed to be primarily responsible for the slot region in the radiation belts [44, 79, 118]. On the other hand, experimental observations of burst electron precipitation by lightning generated whistlers, which propagate through FAI called ducts, indicate that discrete whistlers may play a significant role in the loss rates for the radiation belts in the mid-to-low latitudes within the plasmasphere [20, 134].

The general subject of whistler mode wave generation, propagation, and their interaction with energetic particles is a broad subject and is discussed in various books [40, 54] and review papers [48, 51, 74, 115]. Plasma density irregularities also affect waves propagating in free space R-X and L-O modes and in Z-mode, which are discussed elsewhere [e.g., 14, 15, 30, 31, 123].

We focus in this review paper on particular aspects of whistler mode propagation motivated by the following questions:

- How is whistler mode propagation affected by FAI of various scale sizes?
- What is the significance of FAI influence on W-mode propagation from the magnetosphere to the ground and vice versa?
- What are the implications of W-mode wave refraction or scattering by FAI for the generation of new kinds of waves?
- What are the implications for mode conversion?
- What are the implications for wave-particle interaction?

We shall find that:

- FAI are responsible for providing W-mode waves accessibility to various regions in the magnetosphere otherwise not reachable in a smooth magnetosphere;
- FAI guide, reflect, and scatter W-mode waves and, in general, modify wave structure, which may have profound effects on wave-particle interactions and the general dynamic equilibrium of the radiation belt particles.

It is impossible in this brief review to provide a comprehensive review of the vast amount of literature available on these topics. Our approach is to provide physical insight into various ways FAI may affect W-mode propagation with the help of qualitative explanations and a few illustrative examples.

This chapter is organized as follows: Section 6.2 briefly describes magnetospheric plasma and field aligned irregularities; Section 6.3 provides a theoretical background on the propagation of whistler mode waves; Section 6.4 presents ground and spacecraft observations of whistler mode waves illustrating the role played by field aligned irregularities in the propagation of these waves, and Sect. 6.5 follows with a discussion and concluding remarks.

6.2 Magnetospheric Plasma Distribution: Field Aligned Irregularities

The Earth's magnetosphere is described in many texts and monographs [e.g., 66, 69]. Here we briefly review aspects of magnetospheric cold plasma

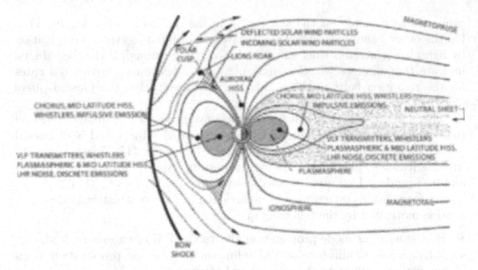

Fig. 6.1. Schematic showing in the noon-midnight meridian various regions and features of the magnetosphere. Also shown are the locations where whistler-mode waves of various types are observed

distribution important for understanding how various magnetospheric boundaries and density irregularities affect the whistler mode propagation. Figure 6.1 shows schematically various regions of the magnetosphere in the noon-midnight meridian and the locations where whistler mode waves of various types are observed.

Immediately surrounding the Earth is the non-conducting atmosphere roughly 60–80 km thick, transparent to the propagation of radio waves. The next layer above the neutral atmosphere is called the ionosphere which extends up to ~1000 km, completely encircles the Earth, and then merges into the magnetosphere. The boundary between the ionosphere and the neutral atmosphere below it is called the Earth-ionosphere boundary and the region between this boundary and the Earth is called the Earth-ionosphere waveguide. The ionosphere is a conductive medium, in which a small fraction (<0.1%) of the atmospheric constituents are ionized due to solar UV and X-ray radiation. The ionospheric plasma is horizontally stratified, with distinct D, E, F_1, and F_2 regions. The electron density increases from about 10^3 cm^{-3} in the D region to a peak of about 10^6 cm^{-3} in the F_2 region at about 300 km. Primary ions in the D and E regions (60–150 km) are NO$^+$, O$_2^+$ and N$_2^+$, in the F region (300 km) are O$^+$, and at altitudes greater than 500 km are H$^+$ and He$^+$. The topside ionosphere starts at the F2 layer peak and extends upward with decreasing density to a transition height, around 500–1000 km, where O$^+$ ions become less numerous than H$^+$ and He$^+$.

The magnetospheric plasma above the ionosphere contains the plasmasphere, a doughnut shaped relatively cool (<1 eV) high density ($\sim 10^2$–10^3 cm^{-3})

plasma consisting mainly of electrons and H^+ ions, and a smaller population of He^+ and O^+ ions. The plasma in this region corotates with the Earth, and can also flow along the field lines from one hemisphere to the other. It is believed that cold plasma of ionospheric origins flows along the field lines to fill the plasmasphere. The cold plasma density measured at the geomagnetic equator gradually decreases with increasing distance from the Earth up to the boundary called plasmapause at which the electron density drops by as much as one or two orders of magnitude. The plasmapause is roughly aligned with the geomagnetic field and is located anywhere from $\sim$2 to 8 R_E in the equatorial plane, depending on the magnetic activity. The location of plasmapause is often near $L = 4-5$ during quiet geomagnetic conditions. The distribution of plasma within the plasmasphere can be roughly described by a diffusive equilibrium model [Angerami and Thomas, 3]. The plasmapause can be considered as the boundary between the plasma of ionospheric origin that corotates with the Earth and the plasma that is influenced by the magnetospheric electric fields. Outside the plasmasphere the plasma density continues to decrease, generally following a R^{-N} drop off in density, where $N \sim 3 - 5$.

The magnetosphere is dominated by the Earth's magnetic field. The geomagnetic field within the inner magnetosphere, the low geomagnetic latitude ($< 60°$) region within about 5-6 R_E from the Earth, can be approximated by a centered dipole. Further out, however, the dipole geometry is distorted and the geomagnetic field lines are compressed into a form that departs significantly from a dipole field. In addition to cold plasma, the inner magnetosphere is also populated by energetic particles, mainly protons and electrons with energies extending from approximately 100 eV up to hundreds of MeV. These particles, which form the Earth's radiation belts, are magnetically trapped in the geomagnetic field and execute a helical gyro motion around the field lines. They bounce back and forth between the two conjugate hemispheres and slowly drift in longitude across the field lines. These particles are believed to be responsible for various wave particle interactions that occur in the magnetosphere.

Superposed on the magnetospheric plasma described above are cold plasma field aligned density irregularities. In these irregularities, the plasma density varies in the direction perpendicular to the geomagnetic field ($\mathbf{B_0}$). They are called field aligned irregularities because relative to their cross-$\mathbf{B_0}$ field extent they extend long distances in the direction of the $\mathbf{B_0}$. The spatial scale size of these irregularities in the cross-$\mathbf{B_0}$ direction ranges from 0.1 m to 100 km and the density change ranges from a few percent at shorter spatial scales to 500% at longer spatial scales [see Fejer and Kelley, 39].

Past whistler mode wave observations, discussed in Sect. 4, suggest that it is convenient to classify FAI into two broad categories based on their effects on wave propagation: (1) large scale and (2) small scale.

- Large scale FAI have spatial scale sizes (L_{FAI}) in the cross-B_0 direction large relative to the whistler mode wavelength. These in general slowly refract waves without significant reflections.
- Small scale FAI have spatial scale sizes in the cross-B_0 direction small relative to the whistler mode wavelength. These in general drastically affect wave propagation over short distances through reflection, scattering, and mode conversion.

Budden [19] has given a general condition under which the slowly varying approximation (W.K.B.) holds for wave propagation in an inhomogeneous and anisotropic medium (when W.K.B. holds, wave propagation can be described by ray theory). Bell et al. [9] have given a condition under which the W.K.B. condition fails for whistler mode propagation and when the wave undergoes strong reflection, scattering and mode conversion. In general the W.K.B. condition holds when the medium can be considered to be slowly varying. A medium can be considered slowly varying if the rate of change of wavelength over one wavelength is much smaller than unity. A medium varies rapidly and W.K.B. fails if this rate is greater than unity. Assuming the whistler mode refractive index to be roughly proportional to the square root of the local electron density, and ignoring its dependence on other factors such as wave normal direction, the following approximate relations can be used to determine if a given FAI is large or small scale.

Large Scale FAI, W.K.B. is satisfied

Ray theory holds :
$$\frac{\Delta N_e \lambda}{N_e L_{FAI}} \ll 1 \quad (6.1)$$

Small Scale FAI, W.K.B. is not satisfied

Ray theory fails :
$$\frac{\Delta N_e \lambda}{N_e L_{FAI}} \geq 1 \quad (6.2)$$

where λ is the W-mode wavelength, N_e is the background electron density, and ΔN_e is the magnitude of the density perturbation over the FAI spatial scale size L_{FAI}.

The large scale FAI contain all those irregularities with cross-B scale sizes much larger than the typical whistler mode wave length. For typical ionospheric and magnetospheric parameters and for waves in the very low frequency range (VLF), FAI with cross-B spatial sizes of the order of 1–10 km or larger in the ionosphere or of the order of a few hundred km in the equatorial region, and with typical density changes from a few percent to a few tens of percent, fall into this category. These irregularities refract or bend W-mode ray paths without significantly changing the character of W-mode waves. Ducts are large scale FAI that extend over a significant fraction of the length of a field line. Ducts can guide W-mode energy along a field line over long distances. When ducts extend from one hemisphere to other and are capable of carrying lighting generated whistlers they are often called "whistler ducts." There are other duct-type FAI. The plasmapause can be considered as a one-sided duct. Other examples of duct-type FAI include equatorial anomaly and field-aligned density drop-offs in the plasmasphere.

The small scale FAI contain all those irregularities with cross-B scale sizes comparable or smaller than typical W-mode wavelength. These FAI typically have spatial scale sizes of the order of 10 m–100 m or larger and typical density changes of a few percent. The small scale FAI can scatter W-mode waves in all directions and may profoundly change their character.

FAI are found in all parts of the magnetosphere including ionosphere, plasmasphere including plasmapause region, high latitude auroral and polar regions [see, e.g., Carpenter et al., 30, and references therein]. FAI are caused by various processes including plasma instabilities, particle precipitation, and plasma drifts [cf., 30, 39, 66, 105, and references therein]. Information on these irregularities has come to us by various means including Langmuir probes [e.g., Kletzing et al., 70], passive wave measurements [e.g., Persoon et al., 89, San-tolík et al., 98], and active wave experiments such as ground based ionosondes and radars [e.g., Kelley, 66], topside sounders [e.g., 13, 15, 84], Very Large Array radio interferometer [Jacobson and Erickson, 63], radio beacon from geostationary satellites [Jacobson et al., 64], and the recent RPI instrument on the IMAGE satellite [e.g., 14, 30, 94, 123].

6.3 Propagation of Whistler-Mode Waves

The salient features of plasma wave propagation in a magnetoplasma can be explained by the classical cold plasma magneto-ionic theory found in detail in several textbooks like Budden [19] and Stix [124]. This theory takes into account the motion of electrons and ions but ignores their thermal motion. For frequencies much higher than ion characteristic frequencies, ion motion can also be ignored. The Appleton-Hartree equation describes wave propagation in these limits. This equation predicts four plasma wave modes for a cold plasma at frequencies near the electron cyclotron (f_{ce}) and plasma frequencies (f_{pe}). These modes are the (1) free space $L-O$ mode, (2) free space $R-X$ mode, (3) Z mode and (4) whistler mode. Plasma waves belonging to each of these modes have commonly been observed in the magnetosphere as reviewed by Gurnett and Inan [48] or Sonwalkar [115]. Figure 6.2 shows the frequency ranges of these four wave modes in the magnetosphere near the equatorial and polar regions, respectively. The details of the whistler mode wave propagation are discussed next.

6.3.1 Propagation of Plane Whistler-Mode Waves in Uniform Magnetoplasma

The whistler mode is named after lightning generated "whistlers" which prop-agate in this mode [Helliwell, 54]. The whistler mode propagates at frequencies

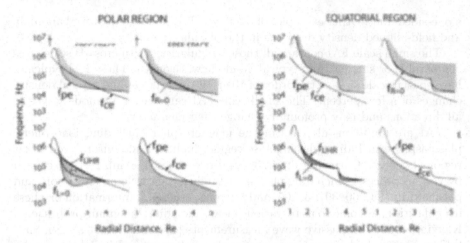

Fig. 6.2. Frequency range of the four most important plasma wave modes at frequencies close to characteristics electron frequencies for a representative plasma density profile near the equatorial plane. The plasma density, proportional to f_{pe}^2, is usually larger near the equator than over the polar region [adapted from Gurnett and Inan, 48] (Reprinted with permission of American Geophysical Union)

below either f_{pe} or f_{ce}, whichever is lower. As a result of this, the maximum frequency of W-mode waves observed in the magnetosphere generally decreases with altitude (Fig. 6.2). The typical frequency range of W-mode waves found in the magnetosphere is from a few kHz to a few tens of kHz, and therefore they are also called VLF (Very Low Frequency) waves. W-mode is a slow wave mode with refractive index (n) always greater than unity. It has a right hand circular polarization when propagating parallel to the magnetic field. Waves propagating in this mode are found throughout the magnetosphere, including the equatorial and polar regions, polar cusp, near the magnetopause, and in the magnetosheath and the bow shock.

W-mode wave propagation in a magnetoplasma is both anisotropic and dispersive. A study of the W-mode refractive index as a function of frequency and wave normal direction provides insight into the nature of W-mode wave propagation. A simplified expression for the Appleton-Hartree refractive index equation for a wave at frequency f and wave normal angle θ with respect to $\mathbf{B}_0$ under quasi-longitudinal approximation $(\sin^2 \theta / \cos \theta) < (2f_{pe}^2 / 3f f_{ce})$ is given by

$$n^2 = 1 + \frac{f_{pe}^2}{f f_{ce} \cos \theta - f^2} \simeq \frac{f_{pe}^2}{f f_{ce} \cos \theta - f^2} \tag{6.3}$$

The expression is modified for both large wave normal angles and for frequencies close to and below the lower hybrid frequency, when ion effects become important. The expression shows that the refractive index is always larger than unity and thus, unlike other cold plasma wave modes, the W-mode does not have a $n = 0$ cutoff. The expression also shows that n depends strongly on plasma and cyclotron frequencies and the wave normal direction. It also shows that there is an angle $\theta = \theta_{RES} = \cos^{-1}(f/f_{ce})$, called the "resonance cone angle", at which the refractive index becomes infinity. In practice, when a W-mode wave normal angle approaches θ_{RES}, the refractive index becomes very large, and the wave becomes almost electrostatic and may suffer Landau damping in a thermal plasma. Using the full expression for refractive index as given by the Appleton-Hartree equation [see, e.g., Budden, 19], it can be shown that when a fixed frequency W-mode wave approaches a region where $f = f_{pe}$, $\theta_{RES} \rightarrow 0$ and reflection or damping occurs.

At frequencies well below f_{pe} and f_{ce} and near $f_{LHR} \simeq (f_{ce}f_{ci})^{1/2}$, ion effects on propagation become evident. For frequencies below f_{LHR}, there is no resonance and whistler mode signals propagating with wave normal angles near $\theta = 90°$ reflect when $f = f_{LHR}$ [Kimura, 68]. This kind of reflection is called magnetospheric reflection (MR). A quasi-electrostatic mode (generally whistler mode signals propagating with large wave normal angles) with a lower cutoff at f_{LHR} is frequently observed in the ionosphere and inner magnetosphere for $R \leq 6R_E$, where R is the geocentric distance and R_E the Earth's radius. In the literature it is often called as lower hybrid (LH) emission.

It is clear from equation (6.3) that a magnetoplasma is an *anisotropic* (n depends on θ) and dispersive (n depends on f) medium. Figure 6.3 shows a polar plot of n versus θ for three different cases: (1) $f < f_{LHR}$, (2) $f_{LHR} < f < f_{ce}/2$, and (3) $f > f_{ce}/2$. The complete expression for refractive index given by the Appleton-Hartree equation is used to plot this figure. Since there is an azimuthal symmetry around $\mathbf{B_0}$, the *refractive index surfaces* are generated by the revolution of $n - \theta$ curves around $\mathbf{B_0}$. As shown in Fig. 6.3, the refractive index surface undergoes a topological change, from a closed to an open one, as

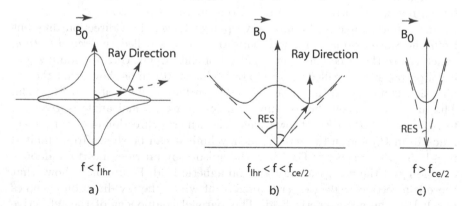

Fig. 6.3. The refractive index as a function of wave normal direction for (a) $f < f_{lhr}$, (b) $f_{lhr} < f < f_{ce}/2$, and (c) $f > f_{ce}/2$. The ray direction is perpendicular to the refractive index surface and in general is in the direction different from the wave normal direction as illustrated in (a) and (b)

frequency increases in value from below f_{LHR} to a value higher than f_{LHR}. In a closed refractive index surface ($f < f_{LHR}$), shown in Fig. 6.3a, the refractive index values remain finite for all values of the wave normal angle θ. In an open refractive index surface, the refractive index tends to infinity at a certain angle called the *resonance angle*, θ_{RES}; the corresponding surface of revolution is called *resonance cone*. For frequencies $f_{LHR} < f < f_{ce}/2$, as shown in Fig. 6.3b, the refractive index surface is concave downward for smaller values of θ and then concave upward for large values of θ. Refractive index surface undergoes another topological change as the frequency increases beyond $f_{ce}/2$ and becomes concave upward for all wave normal angles. These changes in refractive index surface topology have important consequences for wave propagation in the magnetosphere, as discussed in the next subsection.

As Fig. 6.3b and Fig. 6.3c show, for $f > f_{LHR}$, the refractive index $n \to \infty$ as $\theta \to \theta_{RES}$. At wave normal angle close to resonance cone angle, the refractive index becomes large and W-mode waves become short wavelength quasi-electrostatic waves and may suffer Landau damping. For cold magnetospheric plasma with temperatures $<1\,\text{eV}$, Landau damping may become significant when the refractive index values are in a few hundred to a few thousand range. For VLF frequencies, these values correspond to wavelengths of the order of meters to tens of meters [Sonwalkar et al., 123].

A key property of the whistler mode waves resulting from its rather large refractive index values is that its phase velocity can become comparable to velocity of energetic electrons in the magnetosphere. The whistler mode waves are therefore capable of strong resonant interactions with the hot electron plasmas of the magnetosphere. For whistler mode waves at frequencies $f_{LHR} < f < \min(f_{pe}, f_{ce})$ the refractive index value can vary over two to three orders of magnitude, leading to wavelength variation ranging from tens of meters to tens of kilometers. Whistler mode waves can therefore be refracted or scattered by plasma density irregularities of various spatial scale sizes.

In an anisotropic medium, a wave packet travels in a direction different from the wave normal direction. This direction is called the *ray direction*, which is also the direction of propagation of energy. The corresponding velocity of propagation is called the group ray velocity. It can be shown that the ray direction is normal to the refractive index surface as illustrated in Fig. 6.3a. The angle α between the ray direction and wave normal angle is given by $\tan \alpha = (-1/n)(\partial n/\partial \theta)$. At low frequencies, the ray direction does not depart much from $\mathbf{B_0}$ and in the zero frequency limit it can be shown to be limited to $\sim 19.3°$ with respect to $\mathbf{B_0}$. Thus the anisotropy provides a certain amount of guiding of the energy along the geomagnetic field. Figure 6.3b shows that there is a certain angle greater than zero at which the ray direction becomes parallel to the geomagnetic field. The parallel component of the refractive index is minimum at this angle. This angle is called Gendrin angle [Gendrin, 43]. At the Gendrin angle the group velocity is parallel to $\mathbf{B_0}$ and becomes approximately independent of frequency. This has important consequences for the nonducted propagation and wave particle interactions of W-mode waves [57, 61].

6.3.2 Propagation of Whistler-Mode Waves in the Magnetosphere

As discussed in Sect. 6.2, the magnetosphere is an inhomogeneous medium with its plasma density, composition, and the geomagnetic field varying from point to point. In most instances, however, the medium properties change slowly over many wave lengths of a plasma wave mode, and thus locally one can still consider propagation of plasma waves in terms of the propagation modes described in Sect. 6.3.1 for an infinite homogeneous plasma. This approximation, called the ray approximation, enables one to derive ray equations which are used to describe long distance propagation in a slowly varying magnetoplasma [Budden, 19]. The ray approximation is valid even when FAI with scale sizes larger than W-mode wave lengths are present. The ray approximation fails for W-mode propagation across the plasma boundaries, such as the Earth-ionosphere boundary, or the small size FAI where the plasma densities vary rapidly over a wavelength of the W-mode wave. In carrying out ray tracing simulations on modern computers, a realistic model of the magnetosphere incorporating various plasma density gradients such as the plasmapause, ducts, and the vertical and horizontal density gradients in the ionosphere are employed.

One way to classify sources of W-mode waves is to consider their locations:

- sources below the Earth-ionosphere boundary, such as VLF transmitters and lighting, and
- sources above the Earth-ionosphere boundary, such as the naturally occurring VLF emissions which originate in the magnetosphere.

It is easier to understand W-mode propagation from the first type of sources because we generally know their locations, and can approximate their radiation as fixed frequency radiation (e.g. VLF transmitters) or impulsive radiation (e.g., lightnings). In contrast, propagation of W-mode waves from sources within the magnetosphere is difficult to analyze because we do not know their locations or natures, i.e. if they are point or extended sources, and their initial wave normal direction of propagation.

Propagation of W-mode Waves in the Magnetosphere: Waves Originating in a Source Below the Ionosphere

W-mode propagation in the magnetosphere from a source below the ionosphere is illustrated schematically in Fig. 6.4a-c. The free space electromagnetic radiation from the source spreads in the Earth-ionosphere cavity, some of it reaching the receiver labelled R_x. A fraction of this radiation penetrates the ionosphere as W-mode waves, the only mode possible in the ionosphere at VLF, and propagates in the magnetosphere by various ray paths labelled A, B, B', etc. We can understand this propagation by first considering propagation in the horizontally stratified ionosphere and then in the bulk of the magnetosphere.

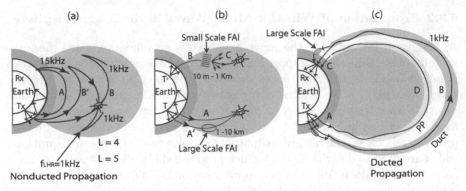

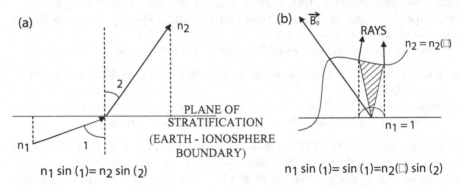

Fig. 6.4. An illustration of various ray paths for whistler mode wave propagation in the magnetosphere for a source on the ground (e.g. VLF transmitter) or in the atmosphere (e.g. lightning discharge) below the ionosphere: (**a**) nonducted propagation in a smooth magnetosphere, (**b**) nonducted propagation in the presence of large scale and small scale FAI, and (**c**) ducted propagation

Fig. 6.5. (**a**) An illustration of Snell's law at a boundary between two media with different refractive indices. (**b**) An illustration of the transmission cone at the lower boundary of the ionosphere

Propagation through the Horizontally Stratified Ionosphere

The typical free space wavelength at W-mode frequencies of 1–30 kHz ranges from 300 to 10 km. Since in the lower ionosphere refractive index varies rapidly with vertical distance in less than a wavelength, the usual assumption that the medium is slowly varying is not valid. Under these circumstances waves undergo partial reflections and mode coupling the analysis of which requires full wave solutions. However, Helliwell [54] has shown that to a good approximation the Earth-ionosphere boundary can be assumed sharp and Snell's law can be applied.

Figure 6.5a–b schematically shows the application of Snell's law at the Earth-ionosphere boundary. Snell's law states that the component of refractive index parallel to the boundary between the two media is conserved. This law is also applicable to anisotropic media. Consider the air-ionosphere boundary at the lower edge of the ionosphere, as shown in Fig. 6.5b. The diagram shown in this figure is called *Poeverlein-diagram* in which the refractive index surfaces are drawn to two sides of the boundary and the component of the refractive index parallel to the boundary is matched to determine the wave normal angle of the transmitted wave [Poeverlein, 91]. For a wave incident on the ionosphere from below, $n_1 = 1$. Since the whistler mode refractive index n_2 in the ionosphere is large compared to $n_1 = 1$, all waves that enter the ionosphere from below with different wave normal angles, have their wave normal angles bent sharply toward the vertical so that they lie within the shaded region, called the *transmission cone*. The angular width of the transmission cone depends on the refractive index in the ionosphere and thus on the plasma frequency, gyrofrequency, and the direction of the geomagnetic field (magnetic dip angle). In general the transmission cone half width is of the order of a few degrees and therefore any W-mode wave originating in the atmosphere (e.g. lightning, ground transmitter signals) should enter the magnetosphere with an essentially vertical wave normal angle.

For a whistler wave incident on the Earth-ionosphere boundary from above, the reverse problem exists. If the wave normal angle of the wave incident from above is inside the transmission cone, it can propagate through the ionosphere into the earth-ionosphere wave guide. If it is outside the transmission cone, the wave will be reflected back into the magnetosphere. Such reflection is called total internal reflection. Because the transmission cone is very narrow, most of the whistler mode energy incident on the Earth-ionosphere boundary from above remains trapped in the magnetosphere.

Various aspects of the ionospheric plasma, such as small and large scale FAI present in the topside ionosphere, profoundly affect the subsequent propagation of W-mode waves. Highly collisional D-region plasma can lead to a 10 to 20 dB reduction in wave energy that passes through this region (upward or downward). A consequence of D-region absorption is that many magnetospheric wave phenomena are better observed at nighttime when the D-region absorption is minimum, and about 10 dB less than that at daytime [Helliwell, 54].

Nonducted Propagation through the Magnetosphere

In the absence of any FAI, the W-mode propagation takes place in nonducted mode, so called to distinguish it from ducted mode described next. This non-ducted propagation is illustrated by rays labelled A, B, and B′ in Fig. 6.4a. The initial wave normal angle of a whistler mode wave entering the ionosphere is nearly vertical at all latitudes. Spatial gradients in the refractive index tend to rotate the wave normal towards the magnetic field direction but not as rapidly as the field direction itself rotates. Consequently, the wave normal angle increases steadily towards the resonance cone angle. Eventually these rays undergo reflections at low altitudes. Figure 6.4a shows that a ray labelled A at frequency $f > f_{LHR,max}$, where $f_{LHR,max}$ is the maximum lower hybrid frequency along the ray path, will propagate to the other hemisphere and will undergo total internal reflection at the Earth-ionosphere boundary. A ray such as B or B′ at a low frequency will undergo magnetospheric reflection. In general, W-mode waves injected from the ground will remain trapped in the magnetosphere. Also, rays can reach any given location by more than one path after undergoing magnetospheric reflections, such as the rays B and B′ arriving at the satellite as shown in the Fig. 6.4a.

Figure 6.4b shows effects of large scale and small scale FAI on the W-mode propagation. A large scale FAI, satisfying condition (1) and thus W.K.B., can bend the upgoing ray such that at a given location rays may arrive by more than one path, each starting at a slightly different latitude (rays A and A′ in Fig. 6.4b). A W-mode wave (ray B in Fig. 6.4b) incident on small scale irregularity can scatter into W-mode waves of both small and large wave normal angles (rays collectively labelled C in Fig. 6.4b).

FAI with scale sizes comparable or smaller than W-mode wavelengths can scatter electromagnetic W-mode waves into quasi-electrostatic W-mode propagating with wave normal angles close to resonance cone angle θ_{RES}. In general, when condition (2) is satisfied, W.K.B. fails and strong scattering of W-mode waves occurs. The excited quasi-electrostatic waves are cut off at f_{LHR} and are a type of lower hybrid (LH) wave. A number of mechanisms, both linear and nonlinear, have been proposed for producing LH waves from W-mode waves [6, 7, 8, 45, 86, 133]. Figure 6.6 schematically shows the linear mode conversion mechanism first proposed by *Ngo and Bell* [1985]. This mechanism is attractive because it is simple and because it explains most experimental observations better than others do, as shown by Bell and Ngo [7].

Ducted Propagation through the Magnetosphere

Figure 6.4c shows W-mode propagation guided by a duct (ray B) and by the plasmapause (ray D). Some of the energy injected from the ground (rays A in Fig. 6.4c) can couple into whistler mode duct or other duct type FAI such as plasmapause and be guided to the other hemisphere with relatively small wave normal angles. At the other hemisphere, part of the energy can propagate down to the Earth and part may be reflected back into the magnetosphere, some of it getting coupled into the same duct.

Because the refractive index surface goes through a topological change (see Fig. 6.4) at $f/f_{ce} = \frac{1}{2}$, conditions for ducting also change at that frequency [22, 54, 112]. Snell's construction shows that the density gradients on both sides of a crest tend to rotate the wave normal toward the geomagnetic field direction. For frequencies below $f_{ce}/2$, the refractive index surface is concave downward. This geometry requires a density enhancement (crest) for ducting. For frequencies above $f_{ce}/2$, the refractive index surface is concave upward. This geometry requires a density depletion (trough) for ducting. However,

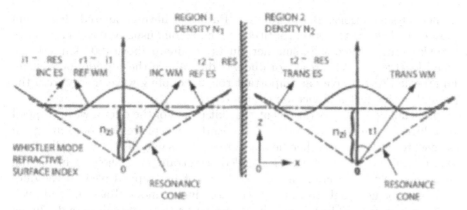

Fig. 6.6. A geometric description of linear mode conversion of whistler mode waves. An incident small wave-normal ($\theta_{i1} < \theta_R$) whistler mode wave (labelled INC WM in region 1) is partly reflected as a small wave normal angle wave (labelled REF WM in region 1) and as a quasi-electrostatic wave with a large wave-normal angle close to resonance cone (labelled REF ES in region 1) and partly transmitted as a small wave-normal whistler mode wave (labelled TRANS WM in region 2), and as a quasi-electrostatic wave with a large wave normal angle close to resonance cone (labelled TRANS ES in region 2). The linear mode conversion is possible because W-mode refractive index has multiple solutions for the parallel component (n_{zi}) of the refractive index in the $f_{LHR} \leq f < f_{ce}$ range. Thus W-mode waves scattered by this mechanism exhibit a low frequency cutoff at f_{LHR}. The mechanism of mode conversion also works when the incident wave is at large wave normal angle as shown by dotted arrow ($\theta_{i}1 \sim \theta_{RES}$) [adapted from Bell and Ngo, 7] (Reprinted with permission of American Geophysical Union)

since this mode of ducting requires $f > f_{ce}/2$, the wave frequency must be ~ 0.5 MHz or higher in order to be ducted all the way down to the ionosphere. Thus this type of ducting does not apply to whistlers. We conclude that whistlers received on the ground require enhancement ducts and that ducting should be effective up to one half of the minimum electron gyrofrequency along the path. Alternatively, a ducted whistler propagating along a certain geomagnetic field line will show an upper cutoff at half the gyrofrequency at a point where the field line crosses the equator. Both ground and spacecraft observations of whistlers confirm this upper frequency cutoff [Carpenter et al., 28]. Whistler dispersion, time delay as a function of frequency, depends on both the path length and the electron density along the path.

Several factors including duct L shell, width, and enhancement/depletion determine whether or not the duct will trap waves incident from below the ionosphere. Helliwell [54] has discussed in detail the trapping of rays in ducts as a function of density enhancement/depletion, the scale size (gradients), and the initial wave normal angle.

The coupling of wave energy into and out of a duct may also depend on how low the duct extends downward in altitude [e.g., James, 65]. Using numerical simulations, Bernhardt and Park [16] have estimated that ducts may extend down to 300 km altitude at night but usually terminate above 1800 km during the day. In summer, ducts terminate above 1000-km altitude at all local times. As a result of ducts ending above the topside ionosphere, large scale FAI may play an important role in coupling wave energy from the ground to the duct and vice versa [James, 65].

Depending on the conditions at the exit points of the duct a wave trapped in a duct can undergo reflections at the Earth-ionosphere boundary and then propagate back to the other hemisphere in the same duct. Such reflections lead to multiple traces of whistler, called echo train, each showing increasing dispersion. After reflection, a whistler may not be trapped back in the duct and can propagate subsequently in the non-ducted mode [Rastani et al., 93]. In the case of several ducts, lightning energy from the same lightning discharge may propagate along several ducts and can be detected on the ground station as a multi-component whistler.

Ducts and duct-type large scale field aligned density drop offs in the magnetosphere can also affect non-ducted propagation [Edgar, 37]. In such cases the ducts are not able to trap the waves, but can significantly modify their ray paths such that signals such as MR whistlers show distinctive signatures.

Equatorial anomaly and plasmapause are other examples of duct-type irregularities. Equatorial anomaly, in which electron concentration during the daytime is depleted near the equatorial ionosphere and is enhanced in two regions on either side of the equator, also acts like a duct [Sonwalkar, 115]. It can be shown that ducts need not have density gradients on both sides to guide the waves from hemisphere to hemisphere. Thus plasmapause, where the density decreases sharply with increasing radial distance, provides an excellent one-sided whistler duct [61]. Plasmapause can also trap nonducted W-mode waves in certain cyclic trajectories [Thorne et al., 131]. The ducts need not go from one hemisphere to another. They may start and terminate in the same hemisphere. Such ducts have been found at high latitude and have been used to explain the dispersion of impulsive auroral hiss [Siren, 109].

Propagation of W-Mode Waves in the Magnetosphere: Waves Originating in a Source in the Magnetosphere

Whistler mode emissions generated near the magnetic equator including mid-latitude hiss, discrete, quasi-periodic, periodic emissions, and chorus are also believed to be generated by gyroresonance instabilities, sometimes called whistler mode instabilities [Sonwalkar, 115, and references therein]. The energy sources in most of these cases are the energetic electrons ($\sim$1–100 keV) trapped in the magnetosphere. In the case of chorus, the energy sources are 5–150 keV electrons injected in the dayside region beyond the plasmapause during the period of heightened magnetic activity. The W-mode emissions in the auroral region include auroral hiss which may be produced via Cherenkov mechanism by precipitating beams of electrons with energies greater than

10 keV [LaBelle and Treumann, 74]. Recently, for the first time, we have a manmade source of W-mode waves on-board the IMAGE satellite [94, 123]. The Radio Plasma Imager (RPI) on the IMAGE satellite is capable of radiating in W-mode during the low altitude portion of its orbit.

The general principles of W-mode propagation for a source in the magnetosphere are similar to those discussed for the case of a ground source. The principal difference between these two types of sources is that the waves injected from the ground enter the ionosphere vertically, whereas the waves injected into the magnetosphere from space may start with a wide range of initial wave normal angles.

Figure 6.7a schematically illustrates the propagation of whistler mode signals from a source location in the equatorial plane within the magnetosphere. Most of the rays, such as A and B, from a magnetospheric source cannot reach a ground receiver because they either undergo total internal reflection at the Earth-ionosphere boundary or they undergo LHR reflections in

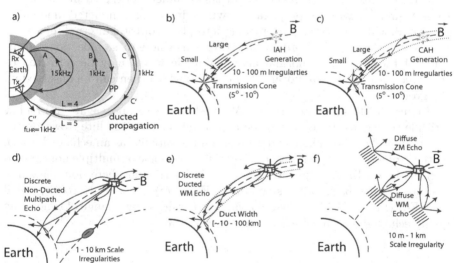

Fig. 6.7. An illustration of various ray paths for whistler mode wave propagation in the magnetosphere from a source in the magnetosphere: (**a**) source at the equator, ducted and nonducted propagation; (**b**) source in the auroral region, nonducted propagation followed by scattering by small scale FAI and propagation to the ground; (**c**) source in the auroral region, ducted propagation followed by scattering by small scale FAI and propagation to the ground; (**d**) source on a low altitude satellite, reflection from the ionosphere leading to an echo; echoes arriving from multiple paths resulting from propagation through a large scale FAI; (**e**) source on a low altitude satellite, ducted echoes, (**f**) source on a low altitude satellite, W-mode and Z-mode diffuse echoes resulting from scattering by small scale FAI [adapted from Sonwalkar et al., 123] (Rerpinted with permission of American Geophysical Union)

the magnetosphere. However, in the presence of ducts, some of the rays, such as C, can be guided to the low altitude ionosphere with their wave normal at small angles with respect to the local vertical; these rays can be observed at a ground receiver after propagating within the Earth-ionosphere wave guide [Helliwell, 54]. The ducted signals that have backscattered from the ionosphere-waveguide boundary can propagate back into the duct or can propagate as nonducted signals [Rastani et al. 93]. In the topside ionosphere, FAI may refract or scatter the downcoming W-waves, which may play a crucial role in determining subsequent paths of these waves. The waves can also reach the ground directly from the source if the source is at low altitudes and radiates such that some of the wave energy reaches the Earth-ionosphere boundary with sufficiently small wave normal angles with respect to the local vertical [Thompson and Dowden, 132]. Other modes of guiding energy from a magnetospheric source to the ground include propagation along the plasmapause [61] and a sub-protonospheric mode [92, 110].

Figures 6.7b and 6.7c show propagation from a source located on auroral field lines and generating waves at large wave normal angles. It is believed that auroral hiss (AH) is generated in this manner. As shown in the figure, these W-mode waves can propagate down in ducted or nonducted mode and, in general, will be reflected back by magnetospheric or total internal reflection. However, if there are FAI present in the regions where these reflections occur, some of the large wave normal angle W-mode waves may be refracted or scattered into small wave normal angle waves which then can be observed on the ground.

Figures 6.7d-f show various ways W-mode waves can propagate from a low altitude source, assumed here to be a transmitter on a satellite, towards the Earth-ionosphere boundary and be seen on the satellite as an echo or be seen on the ground. An echo can reach the satellite by single or multiple non-ducted paths (Fig. 6.7d), or propagate down to the Earth-ionosphere boundary in a duct and return back in the same duct (Fig. 6.7e), or be scattered by small scale FAI (Fig. 6.7f). The scattered waves generally show a range of time delay giving echo a diffuse appearance on a spectrogram.

6.4 Observations and Interpretations

The literature describes numerous observations of W-mode waves, from ground stations and from low and high altitude spacecraft [e.g., 1, 48, 51, 54, 56, 74, 115]. We present here a few examples to illustrate the key features of W-mode propagation in the presence of field-aligned irregularities.

W-mode observations are frequently categorized as those found on the ground and those found on spacecraft. The reason for dividing ground and spacecraft observations into separate categories is that somewhat distinct and apparently uncorrelated activity is detected at ground stations and on satellite [Sonwalkar, 115]. However, in this paper, for the reasons discussed in the previous section, we classify observations of W-mode waves of magnetospheric origin into two main categories:

- the observations of W-mode waves with sources on the ground,
- the observations of W-mode waves with sources in the magnetosphere.

6.4.1 W-Mode Observations When the Source is Below the Ionosphere

It is convenient to categorize W-mode observations into two subsections: (1) non-ducted propagation, (2) ducted propagation. Ducted propagation can be seen both on the spacecraft and on the ground, whereas non-ducted propagation can only be seen on the spacecraft. Historically ducted propagation was discovered first (on the ground) and non-ducted later. It is, however, easier to start with observations of non-ducted signals as non-ducted propagation is the natural mode of propagation of W-mode waves in a smooth magnetosphere.

Nonducted Propagation: Effects of Large-scale FAI, Ducts, and Density Drop-offs

We begin with the simplest possible propagation scenario: propagation of a single frequency signal reaching a satellite as a plane wave in a smooth magnetosphere as illustrated by ray B in Fig. 6.4a. Figure 6.8a shows an example of near plane wave received on the DE 1 satellite. A 20-s-long, 4.0 kHz continuous wave Siple transmitter signal was received by the 200-m long electric antenna on the DE 1 satellite. The satellite was located outside the plasmasphere at $L \sim 4.5$ and $\lambda_m \sim 7°N$. The local electron density was about 15 el/cc corresponding to $f_{pe} \approx 35$ kHz and the local magnetic field was about 0.0034 G corresponding to $f_{ce} \approx 9.5$ kHz. The electric field amplitude shows well defined spin fading at half the spin period (6 s), indicating reception of a single plane wave. Reception of plane W-mode waves is relatively rare, but when detected, it can provide testing of some of the most fundamental ideas of W-mode propagation and antenna properties in space plasmas [114, 116]. The depth and the phase of spin fading pattern can be used to obtain the wave normal direction of about 55° with respect to geomagnetic field.

Ground transmitter signals received on the satellite often show fading patterns which cannot be explained as a result of satellite spin motion alone. In fact fading patterns have been observed on satellites that were not spin stabilized. Heyborne [59] and Scarabucci [102] observed amplitude fading at low altitude ($\sim$1000 km) OGO 1, OGO 2, and OGO 4 satellites. Cerisier [32]

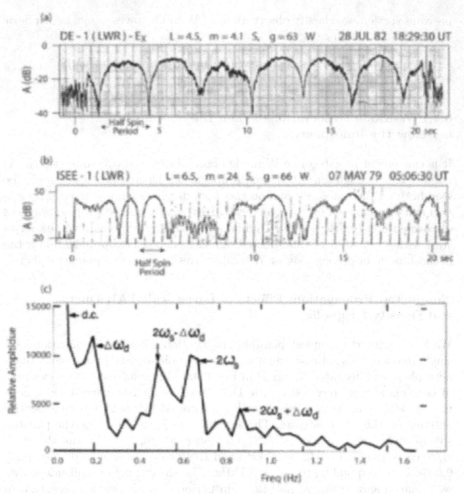

Fig. 6.8. Nonducted whistler mode propagation of Siple transmitter signals: (a) An example of near plane wave received on the DE 1 satellite. A 20-s-long, 4.0 kHz continuous wave Siple transmitter signal was received by the 200-m long electric antenna on the DE 1 satellite. (b) An example of two plane waves received on the ISEE 1 satellite. A 20-s-long, 4.0 kHz continuous wave Siple transmitter signal was received by the 215-m long electric antenna on the ISEE 1. (c) Fourier transform of the wave amplitude shown in (b) [adapted from Sonwalkar et al. 114, 120] (Reprinted with permission of American Geophysical Union)

observed two Doppler shifts in the FUB signal observed on the FR 1 (altitude ~750 km). Neubert et al. [85] and Sonwalkar et al. [120] have observed pulse elongations and amplitude modulations in Omega and Siple transmitter pulses received on the GEOS 1 and ISEE 1 satellites, respectively. These

authors have interpreted their observations as indicating the presence of two
or more plane waves arriving from different directions.

Figure 6.8b shows an example of a 20-s-long, 4.0 kHz continuous wave Siple
transmitter signal received by the 215-m long electric antenna on the ISEE
1 satellite. The fading pattern now contains frequencies other than twice the
spin frequency of the ISEE 1 satellite (spin period = 3 s). This fading pattern
can be explained as that arising from the satellite receiving two plane waves
arriving from two different directions. The Doppler frequency for each plane
wave is different, and thus the satellite receiver measures the beat pattern resulting from this difference. It can be easily shown that the number of spectral
components in the mean square received voltage is given by $(3N^2 - 3N + 4)/2$,
where N is the number of multiple paths [Sonwalkar and Inan, 114]. Figure
6.8c shows the amplitude of the Fourier transform of the electric field amplitude envelope shown in Fig. 6.8b. We clearly notice peaks at five frequencies,
which are identified as the d.c., differential doppler shift $(\Delta\omega_d)$, twice the
spin frequency (ω_s), and the sum and difference of differential Doppler shift
with twice the spin frequency. A detailed analysis of Siple transmitter signals received on the ISEE 1 satellite showed that, in general, at any point
in the magnetosphere the direct signals (before reflections) transmitted from
the ground arrive, within few hundred milliseconds of each other, along two
or more closely spaced multiple paths, as illustrated by rays A and A' in Fig.
6.4b . The multiple paths can be explained by assuming propagation through
1–10 km cross-**B** scale size FAI in the topside ionosphere with a few percent
enhancement or depletion in the plasma density. Such FAI can refract the
wave normal direction by ~5–10°, giving rise to multiple path propagation as
illustrated in Fig. 6.4b [Sonwalkar et al., 120].

The differential Doppler shift Δf resulting from the observation of multi-
path W-mode ground signals on a satellite is of the order of $\Delta n V_s f / c$, where
Δn is the difference in the refractive index values of two closely spaced multiple paths, V_s is the satellite speed and c the velocity of light in vacuum.
Assuming $\Delta n \sim 1$, $f \sim 4$ kHz, $V_s \sim 8$ km, we obtain $\Delta f \sim 0.1$ Hz, a fraction
of Hz. Clearly to resolve such multiple paths we need either long duration fixed
frequency signals, such as the one discussed above, or signals originating in
impulsive (duration ~ ms) lightning discharges which can be resolved in time
domain when they arrive at the satellite by more than one path, or we need
measurements of multiple spacecraft separated by distances of the order of
100 km (e.g. Cluster). It is possible to misinterpret or have more than one interpretation of wave normal direction measurements of a short duration (~1 s)
ground transmitter signals received on a satellite. For example, Lefeuvre et
al. [76] determined wave normal directions of seven Omega transmitter pulses
observed on the GEOS 1 satellite. They found a 0.2- to 0.4-s periodicity in
the observed wave normal directions, which Sonwalkar et al. [120] interpreted
as an indication of two closely spaced multiple paths.

Whistlers propagating in the non-ducted mode also show evidence of propagation along single direct paths in a smooth magnetosphere and that of

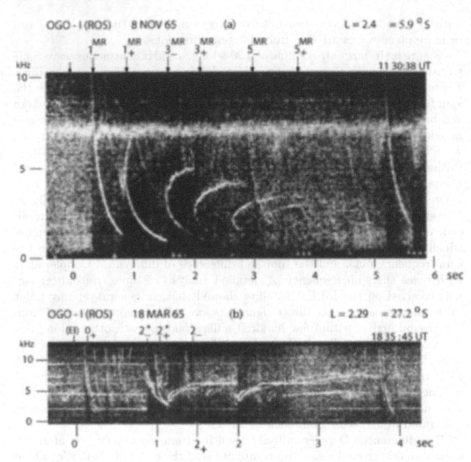

Fig. 6.9. Non-ducted whistler mode propagation of whistlers : (a) Spectrogram of a multi-component MR whistler received on OGO 1 satellite via in a smooth magnetosphere, (b) propagation in the presence of FAI [adapted from Smith and Angerami, 111] (Reprinted with permission of American Geophysical Union)

propagation along multiple paths in a magnetosphere containing large scale FAI. Figure 6.9a shows the spectrogram of a magnetospherically reflected or MR whistler received on OGO 1 satellite [Smith and Angerami, 111]. The MR whistler usually consists of a series of traces or components, each exhibiting a frequency of minimum travel time or nose frequency. The multi-component feature of MR whistlers can be explained in terms of propagation of lightning energy in relatively smooth magnetosphere by non-ducted paths which have undergone multiple MR reflections as illustrated by rays B and B' in Fig. 6.4a [Kimura, 68]. MR whistlers, also called *unducted* or *non-ducted* whistlers, show a wide variety of frequency-time signatures, depending on the location of the receiver (spacecraft) with respect to the causative lightning discharge, and on the density distribution of electrons and ions in the magnetosphere. By the nature of their propagation mode, these unducted whistlers cannot be detected on ground. The spacecraft observations of unducted whistlers have led to a new understanding of the many subtle features of whistler-mode propagation and to the deduction of important plasma parameters in space [17, 35, 36, 37, 38, 111, 118, 129].

Edgar [37] has shown that ray tracing simulations in a smooth magnetosphere can explain the observed spectrogram shown in Fig. 6.9a. In general, MR spectrogram contain features that cannot be explained by propagation in a smooth magnetosphere. For example Fig. 6.9b shows the spectrogram of MR whistler observed on 18 March 1965 on OGO 1 satellite. This whistler has a spectral signature quite different than that of the MR whistler shown in Fig. 6.9a. The traces O_+, 2_-, and 2_+ are expected for an MR whistler propagating in a smooth magnetosphere for the given location of the satellite and the hemisphere of the lightning. Smith and Angerami [111] speculated that extra traces, labelled 2^*_- and 2^*_+, probably resulted from multi-path propagation, possibly due to large-scale irregularities in the magnetosphere. Using ray tracing simulations, Edgar [37] showed that the extra traces 2^*_- and 2^*_+ could be explained by assuming multi-path propagation resulting from a 30% density drop off at $L \sim 1.8$. He also demonstrated that the distortion in the 2_-, and 2_+ components near 6 kHz can be explained by considering a duct with $\sim 20\%$ enhancement in density near $L = 2.4$. In general Edgar [37, 38] and more recently Bortnik et al. [17] have shown that large and small scale FAI in the topside ionosphere and density drop offs and ducts in the magnetosphere can explain specific observed features in MR whistler spectrograms. Some features such as the frequency cutoff of MR whistlers may be explained in terms of their trapping in density gradients, Landau damping, and D-region absorption [17, 38, 131].

It is evident that the same mechanism, viz. bending of rays due to the presence of large scale FAI has been used to explain multi-path propagation of the ground signals (Fig. 6.8b) as well as multi-path propagation of MR whistlers (Fig. 6.9b). This bending permits more than one ray paths of similar type, such as two direct ray paths or two or more MR paths of similar type (e.g. 2_- and 2^*_-), to reach the satellite for a given location in the magnetosphere. It is clear that this type of multi-path propagation (Rays A and A' in Fig. 6.4b), resulting from the presence of large scale FAI in the magnetosphere or topside ionosphere, is different from the one resulting from the multiple magnetospheric reflections in the smooth magnetosphere (Rays B and B' in Fig. 6.4a).

Non-Ducted Propagation: Effects of Small-Scale FAI

Small scale irregularities satisfying condition (2) can scatter W-mode waves into directions which may be drastically different than their initial wave normal directions. For W-mode, with typical wavelengths of the order of a few km, the condition (2) can be easily satisfied by small scale FAI with $L_{FAI} \sim 10$–100 m and $\Delta N_e/N_e \sim$ a few percent. The scattered waves are generally W-mode waves with both small and large wave normal angles. The large wave normal angle waves are quasi-electrostatic and show a lower cutoff at f_{LHR} and are therefore generally call lower hybrid (LH) waves.

Earliest evidence, albeit indirect, of influence of small scale FAI on W-mode propagation came from a strong correlation of whistler activity at mid-latitude with HF backscatter [Carpenter and Colin, 26] and the correlation of auroral hiss activity with HF backscatter [Hower and Gluth, 60]. HF waves are scattered by FAI with scale lengths comparable to HF-wavelength, a few tens of meters.

A dramatic evidence of scattering of W-mode waves by small scale FAI came when the "spectral broadening phenomenon" was discovered for ground transmitter signals received on the low altitude satellites [Bell et al., 11]. Figure 6.10 shows an example of Omega transmitter signals received on the DE 1 satellite. The pulses of ~ 1 s duration from the Omega transmitter in North Dakota are present at 10.2 kHz, 11.05 kHz, and 11.333 kHz. Spectrally broadened Omega transmitter pulses are seen both in Fig. 6.10a, showing a spectrogram of the electric field measured by a 200 m long dipole antenna, and in Fig. 6.10b, showing spectrogram of the electric field measured by a short 8 m long dipole antenna. The spectrally broadened pulses are direct pulses which have propagated from the ground along a ~ 2000 km path directly to the spacecraft. Following the direct pulses are echoes of the direct pulses reflected from the conjugate hemisphere with a delay of ~ 2 s. In both panels the apparent bandwidth of the direct pulses is larger than 100 Hz, ~ 100 times the 1 Hz bandwidth of the transmitted pulses.

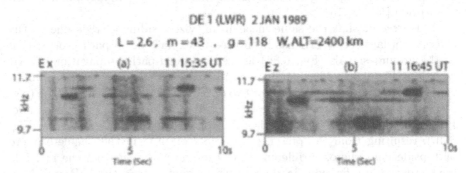

Fig. 6.10. Observations of lower hybrid waves excited by signals from the Omega (ND) transmitter. (**a**) and (**b**) respectively show reception by the E_x and E_z antennas. The E_x antenna consists of two 100 m wires deployed in the spin plane. The E_z antenna consists of two 4 m tubes deployed along spin axis [adapted from Bell et al., 10] (Reprinted with permission of American Geophysical Union)

The LH waves can also be excited by non-ducted whistlers. Figure 11a shows an example of the excitation of LH waves above local f_{LHR} by a series of non-ducted whistlers observed on the ISIS-2 satellite. Note that LH waves are excited for $f \geq f_{LHR}$, consistent with generation by linear mode conversion method (Fig. 6.6).

Observations from many satellites and theoretical analysis show that the excited waves are short wavelength (5 m $< \lambda <$ 100 m) electrostatic lower hybrid (LH) waves excited by electromagnetic whistler mode waves through scattering from magnetic field-aligned irregularities located in the topside ionosphere and magnetosphere [e.g., Bell et al., 12, and references therein]. These LH waves exhibit very large Doppler shifts which cause the observed spectral broadening ($\sim$300–1000 Hz) of the signal as received at the moving satellite [7, 10, 11, 62].

Earlier observations of this phenomenon occurred at altitudes $\leq$7000 km [Bell and Ngo, 6]. Recently, Bell et al. [12] report new observations from the Cluster spacecraft of strong excitation of lower hybrid (LH) waves by electromagnetic (EM) whistler mode waves at altitudes $\approx$20,000 km outside the plasmasphere. These observations provide strong evidence that EM whistler mode waves are continuously transformed into LH waves as the whistler mode waves propagate at high altitudes beyond $L \sim 4$. This process may represent the major propagation loss for EM whistler mode waves in these regions, and may explain the lack of lightning generated whistlers observed outside the plasmapause [Platino et al., 90]. The excited LH waves represent a plasma wave population which can resonate with energetic ring current protons to produce pitch angle scattering on magnetic shells beyond $L\sim$4. Thus linear mode coupling provides a new mechanism by which lightning generated whistler mode waves can affect the lifetimes of energetic ring current protons.

We discuss another interesting manifestation of the effects of FAI, perhaps both small and large scale, on the propagation of whistlers. Figure 6.11b shows an example of whistler-triggered hiss emissions. Using data from the DE 1 satellite, Sonwalkar and Inan [118] showed that lightning generated whistlers

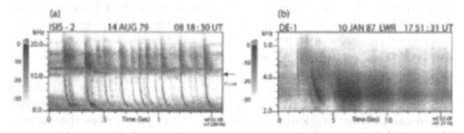

Fig. 6.11. Examples of effects of large and small scale FAI on non-ducted whistler propagation. (a) ISIS-2 satellite observations of lower hybrid waves near 10 kHz excited by whistlers. (b) DE 1 satellite observations of hiss band excited by whistler (courtesy of VLF group, Stanford University)

often trigger hiss emissions that endure up to 10- to 20-s periods and suggested that the lightning may be an embryonic source of plasmaspheric hiss. They recognized that direct multiple path propagation resulting from FAI irregularities may play an important role in the generation of whistler triggered hiss. Subsequently using ray tracing simulations Draganov et al. [35, 36] showed that the combined contribution from whistler rays produced by a single lightning flash entering the magnetosphere at a range of latitudes and with a range of wave normal angles can form a continuous hiss-like signal. It was assumed that at 400 km altitude the initial wave normal angles of the rays are spread around the nominal vertical orientation. The spread in the wave normal angle was presumed to be generated as a result of W-mode propagation through large and small scale FAI in the topside ionosphere. It may be noted that H. C. Koons first suggested the idea in 1984 that plasmaspheric hiss might simply be the accumulated waves of many non-ducted whistlers [See p. 19,486 of 126]. Recently, Green et al. [44] found that the longitudinal distribution of the hiss intensity (excluding the enhancement at the equator) is similar to the distribution of lightning: stronger over continents than over the ocean, stronger in the summer than in the winter, and stronger on the dayside than on the nightside. These observations strongly support lightning as the dominant source for plasmaspheric hiss, which, through particle-wave interactions, maintains the slot region in the radiation belts.

Ducted Propagation: Observations on the Ground

The importance and role of ducts in the propagation of W-mode waves cannot be overemphasized. As discussed in Sect. 6.3, without ducts W-mode waves would not be observable on the ground. The earliest observations of W-mode waves were ducted whistlers observed on the ground. Helliwell [54] in his now classic monograph has given an early history of W-mode wave observations. He gives examples of many types of W-mode waves including lightning-generated whistlers, ground transmitter signals, and VLF emissions of magnetospheric origins. Helliwell et al. [58] were the first to suggest that whistlers might become trapped within field-aligned density irregularities (ducts) and propagate from one hemisphere to another. The duct hypothesis was prompted by the observation that most whistlers observed on the ground show several discrete cases with varying frequency-time dispersion indicative of propagation over several distinct paths.

Figure 6.12 shows examples of ducted propagation of Siple transmitter and lightning-generated whistlers, observed on the ground at Lake Mistissini, Canada and Palmer Station, Antarctica. Figure 6.12a describes the essential idea behind the active experiments performed from Siple Station, Antarctica, during the 1973–1998 period. These experiments provided new results and insight on the effects of cold and hot plasma on W-mode propagation [25, 56, and references therein]. Figure 6.12b shows a 1-s pulse from Siple Station received at Lake Mistassini after propagation along two ducts in which wave

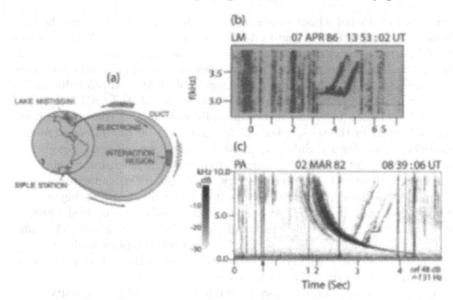

Fig. 6.12. Examples of ducted W-mode propagation observed on the ground: (a) Schematic showing basics of field-aligned whistler mode wave injection experiments between Siple Station, Antarctica (SI), and Lake Mistissini, Quebec, Canada (LM). (b) A 1-s pulse (*lower right*) from SI is received at LM (*upper right*) after propagation along two ducts in which wave growth and triggering of rising emissions occurs. A VLF receiver located at Palmer (PA) provides data on low L shell whistler mode paths and on subionospheric propagation from Siple Station. (c) Spectrogram of a multi-component ducted whistler received at Palmer Station, Antarctica, and the associated triggered emissions [adapted from Helliwell, 56] (Reprinted with permission of American Geophysical Union)

growth and triggering of rising emissions occur. Emission triggering is very common for ducted signals but rarely observed for non-ducted signals. Figure 6.12c shows an example of multi-component whistler and triggered emissions observed at Palmer station, Antarctica. The spheric originating from the same lightning discharge that caused the whistler was also detected and is marked by an arrow.

Spatial and temporal occurrence patterns of ducted and non-ducted whistlers have been a subject of great interest [Helliwell, 54]. Many aspects of the occurrence rates, such as the local time, season, and locations could be related to the occurrence rates of causative lightning, and propagation conditions such as high absorption in the daytime ionosphere and the latitudinal dependance of whistler mode transmission cone. The occurrence of lightning activity reduces at high latitude and the transmission cone is very small at low latitudes. Consequently the whistler activity peaks at nighttime at mid-latitude ($\sim$50°).

Many studies show that large and small scale FAI in the topside ionosphere and plasmapause regions play an important role in the propagation and occurrence of ducted whistlers [see, e.g., 26, 27, 65, and references therein]. Carpenter and Colin [26] found that day-to-day variations of W-mode waves at mid-latitude are strongly correlated with the occurrence of small scale FAI in the F-region as observed by HF backscatter. They interpreted these F-region FAI as being the feet of the large scale FAI – whistler ducts. Using wave normal direction data from FR-1 satellite (altitude = 750 km) and ray tracing, James [65], showed that in addition to factors considered above, large scale FAI in the topside ionosphere influence the rate of occurrence of whistlers. Carpenter and Šulić [27] showed that ionospheric processes such as wave damping, defocusing within mid-latitude trough, focusing within large scale (50–100 km) FAI, and scattering by 10 to 100-m FAI can explain many features of the observed spatial distributions of whistler paths beyond the plasmapause and also the relatively low amplitude of the associated whistlers as compared to that of whistlers found inside the plasmasphere. They also confirmed that ducted whistler propagation outside the plasmasphere occurs at comparatively low rates in comparison to activity within, and tend to occur at certain distances from the plasmapause.

W-mode propagation can also be guided by field aligned density irregularities other than ducts, such as the plasmapause. Carpenter [23] gives examples of whistler and VLF noises propagating just outside the plasmasphere. These whistlers and emissions exhibited features not observed in usual ducted whistlers, including extension of signal frequencies into the range $\simeq 0.5-0.8 f_{Heq}$, where f_{Heq} is the equatorial gyro-frequency of the path, and the echoing or repeated propagation over the path at frequencies above $0.5 f_{Heq}$. The VLF noise bands and bursts tended to occur within the frequency range $0.4-0.8 f_{Heq}$. This kind of propagation is not well understood. Using ray tracing, Inan and Bell [61] have shown that rays can be guided along the plasmapause outer surface, but only for the waves leaving with wave normal angles close to relatively large Gendrin angle. Inan and Bell [61] suggest that large scale FAI with strong density gradients can tilt the vertical wave normal angles into Gendrin angle. Cairo and Cerisier [21] have shown that significant departures of upgoing wave normals from the vertical are observed within the plasmasphere, apparently due to large scale FAI.

There are many aspects of ducted whistler mode propagation that are not well understood. For example, analysis of ground whistler data reveals that a single whistler can contain several hyperfine structures [Hamar et al., 49]. These finer structures in whistlers can be interpreted either in terms of electron density fluctuations of the order of $\sim 1\%$ and spatial scale sizes of the order of 50 km, or in terms of excitation of multiple propagating modes within a duct [Hamar et al., 50]. It is possible that the finer details of wave particle interactions that lead to a variety of triggered emissions are related to the details of the ducted propagation.

Ducted Propagation: Observations on the Spacecraft

Ducted W-mode waves are rarely observed on satellites in the magnetosphere due to the apparently small volume (0.01%) occupied by the ducts [Burgess and Inan, 20], but are most commonly observed on the ground after they have exited from a duct and propagated in the Earth-ionosphere wave guide to a receiving site [Helliwell, 54]. While the duct hypothesis has been extensively used in interpreting ground observations of VLF waves, only a limited set of experimental data are available that provides for the determination of the properties of magnetospheric ducts based on in situ observations [2, 24, 71, 90, 103, 111, 122].

Detection of ducts using spacecraft observations was based in most cases on measurements of whistler dispersion [2, 24, 90, 103, 111]. In one case it was based on enhancements of whistler mode hiss correlated with density enhancements [Koons, 71]. In one case discussed below, wave normal direction measurement was used to establish ducted propagation and electron density measurement was used to determine duct parameters. These authors provided the following measurements of ducts: $\sim$ 400 km near $L = 3$ by Smith and Angerami [111], 223–430 km and 6–22% for $L = 4.1 - 4.7$ by Angerami [2], 68–850 km and 10–40 % for $L = 3.1 - 3.5$ by Scarf and Chappell [103], 630–1260 km and $\leq$30% between $L = 4$ and $L = 5$ by Carpenter et al. [29], and $\sim$500 km and $\sim$40% by Koons [71].

Observations listed above, excepting those by Platino et al. [90], were all made in the plasmasphere. Recently, using data from the Cluster spacecraft, Platino et al. [90] have shown that whistlers observed outside the plasmasphere in the low density regions occur only in the presence of large scale irregularities within which the waves are "ducted". They found that dispersion characteristics of observed whistlers are only matched by ray tracing simulations if the whistlers are ducted. They also propose a possible explanation why whistlers outside the plasmasphere are rarely observed, based on wave conversion from whistler mode to lower hybrid mode [Bell et al., 12].

A direct confirmation of many features of both ducted and non-ducted propagation was obtained in a study involving measurement of the VLF transmitter (15 kHz, 4s ON/4s OFF, 800 kW radiated power) at Khabarovsk, Russia (48°N, 135°E; $\lambda_m \sim 38°, \phi_m \sim -158°$) received on the low altitude COSMOS 1809 satellite Sonwalkar et al. [Sonwalkar et al., 121]. Figure 6.13 shows an example of signals from the VLF transmitter at Khabarovsk, Russia (48°N, 135°E; $\lambda_m \sim 38°, \phi_m \sim -158°$) received on the low altitude COSMOS 1809 satellite (near-circular orbit at an altitude of $\sim$960 km and inclination of $\sim$82.5°). Figure 6.13a top and bottom panel display the DE 1 and COSMOS 1809 satellite orbits in the magnetic meridional and magnetic equatorial planes, respectively, when Khabarovsk signals were observed on the two satellites on 23 Aug 89. Simultaneous observations of ground transmitter signals and electron density on COSMOS 1809 provided unique opportunity to

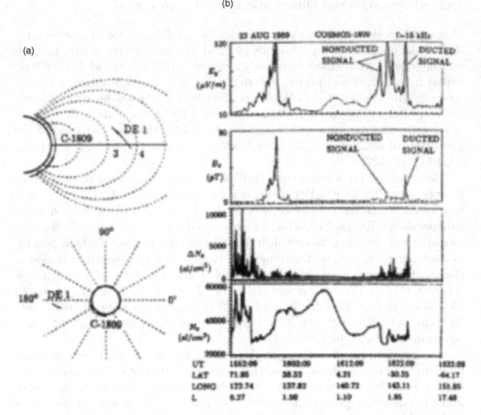

Fig. 6.13. (a) Orbits of DE 1 and COSMOS 1809 satellites in magnetic meridional and equatorial planes on 23 Aug 1989 during the period when the Khabarovsk transmitter was operating. The thickened parts of the orbits indicate the time period during which the transmitter signals were detected on the satellites. (b) Amplitude of electric (*top panel*) and magnetic fields (*second panel*) of Khabarovsk signals on COSMOS 1809 satellites for 23 Aug 89. Also shown are the mean square root variation in the electron density dN_e (*third panel*) and electron density N_e (*fourth panel*) [adapted from Sonwalkar et al., 122] (Reprinted with permission of American Geophysical Union)

measure the parameters of density irregularities responsible for direct multiple path propagation and for ducted propagation.

Figure 6.13b shows COSMOS 1809 observations of Khabarovsk signals in the 15 kHz-channel of the spectrum analyzer when the satellite passed over the transmitter at ~1600 UT and when the satellite passed over a region roughly magnetically conjugate to the transmitter location in the southern hemisphere at ~1622 UT. The figure presents electric field E_y, magnetic field B_x, mean square root variation in the electron density dN_e, and the total electron density N_e along the satellite orbit. Since the satellite altitude during this interval was approximately constant, dN_e is a measure of horizontal gradients in the ionospheric electron density.

The top two panels of Fig. 6.13b show the amplitude fading of Khabarovsk transmitter signals due to multiple path propagation effects and the third panel shows the fluctuations ΔN_e in the electron density believed to be responsible for the multiple path propagation. The typical horizontal extent of these irregularities was ≤100 km and the density fluctuations were of the order of ≤5%, consistent with previous theoretical estimates obtained using wave observations and ray tracing simulations [Sonwalkar et al., 120].

When COSMOS 1809 was in the southern hemisphere, the magnetic field of the transmitter signal showed a large value near 1625 UT. The value of the ratio $(cB_x)/E_y$ gives an indication of the wave normal angle of the wave. Based on ray tracing simulations, for non-ducted wave propagation of ground transmitter signals injected in the northern hemisphere, a large wave normal angle corresponding to a lower value of the ratio $(cB_x)/E_y$ is to be expected in the southern hemisphere. From Fig. 6.14b second and third panel, we find this to be the case in general except for a signal observed between 1624:57 and 1625:14 UT when rather large values of B_x and $(cB_x)/E_y$ were measured indicating low wave normal angles ($< 30°$). Sonwalkar et al. [122] interpreted this low wave normal angle signal to be a case of ducted propagation. To further test this assumption, the electron density data were examined. It was found that between 1625:04 and 1625:12 there was a 3000 to 3800 el/cc enhancement over the background electron density of 29500 el/cc, consistent with the duct hypothesis (other variations in electron density seen in Fig. 6.14b third panel are presumed to be local field aligned irregularities which are commonly present in the ionosphere). Further, the equatorial gyrofrequency corresponding to L =2.85, the duct location, is 34.4 kHz. Thus the 15 kHz Khabarovsk signal is well below the half gyrofrequency cutoff of $f_{Heq}/2 = 17.2$ kHz for ducted propagation [54]. The L shell thickness of the duct was 0.06 and corresponds to a duct cross section of 55 km at 981 km altitude and 367 km at the geomagnetic equator. The density enhancement was 10-13% over the background electron density of 29500 el/cc. Only a lower limit of $\Delta\phi \geq 0.2°$ can be placed on the longitudinal extent of the duct, consistent with the 3-4° of width in longitude estimated by Angerami [2].

The efficiency of signal coupling between the Earth-ionosphere waveguide and magnetospheric ducts is of importance to studies of VLF propagation and wave amplification in the magnetosphere. This coupling is critically dependent on the duct end points in the ionosphere [16]. The above study indicates that duct end points can extend down to at least ∼980 km at ∼0130 local time in the southern hemisphere during austral winter. This result is consistent with theoretical studies of Bernhardt and Park [16] who predict a duct endpoint as low as 300 km at night during winter and equinoxes.

The ducted signals were observed over an L shell range of 0.13, about twice the L shell width of the duct. This could be the result of ducted signals leaking from the duct as previously noted in both experimental and theoretical works

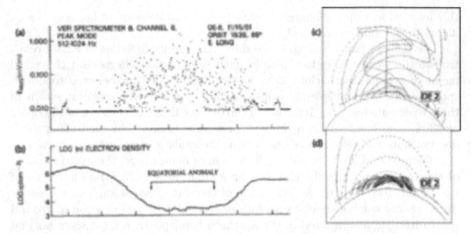

Fig. 6.14. (a) DE 2 observations of impulsive signals near the geomagnetic equator in the 512- to 1024-Hz channel of the VEFI instrument. (b) Langmuir probe observations of electron density illustrating equatorial (Appleton) anomaly. (c) Ray tracings through a horizontally stratified ionosphere. Trajectories of rays (750 Hz frequency) injected at 200-km altitudes with vertical wave normal angles. (d) Ray tracings in the presence of an equatorial anomaly. Trajectories of rays (750 Hz frequency) injected at 200-km altitudes with vertical wave normal angles [adapted from Sonwalkar et al., 121] (Reprinted with permission of American Geophysical Union)

[2, 127, 128]. The peak electric and magnetic field were detected inside the duct near the duct center. Both the peak electric ($>520\,\mu\mathrm{V/m}$) and magnetic field (36 pT) intensities of the ducted signals were comparable to those observed for the non-ducted signals over the transmitter in the northern hemisphere, but were about 20 dB higher than those of the non-ducted signals observed in the southern hemisphere in the vicinity of the duct, consistent with observations of Koons [71]. Inside the duct electric and magnetic field intensities show a fine structure, consistent with recent reports of whistler fine structure by Hamar et al. [49, 50]].

Ducted whistlers are rarely observed below $L = 2$. An unusual observation of lightning energy reaching the low altitude equatorial regions was made by Sonwalkar et al. [Sonwalkar et al., 121]. They show that equatorial anomaly can focus W-mode waves from lightning into a region near the equator. The equatorial anomaly is a tropical ionospheric effect arising from equatorial electrodynamics, which essentially leads to the formation of a ductlike structure with enhanced density along a field line near $\pm 20°$ geomagnetic latitude [66].

The Vector Electric Field Instrument (VEFI) on the Dynamic Explorer 2 (DE 2) satellite observed impulsive ELF/VLF electric field bursts on almost every crossing of the geomagnetic equator in the evening hours (Fig. 6.14a). These signals were interpreted as originating in lightning discharges. These signals that peak in intensity near the magnetic equator were observed within

5-20° latitude of the geomagnetic equator at altitudes of 300–500 km with amplitudes of the order of ∼ millivolts/m in the 512-1024 Hz frequency band of the VEFI instrument. The signals are observed in the same region near the equator where the equatorial anomaly, as detected by Langmuir probe, was found (Fig. 6.14b). Whistler-mode ELF/VLF wave propagation through a horizontally stratified ionosphere predicts strong attenuation of sub-ionospheric signals reaching the equator at low altitudes. However, ray tracing analysis showed (Figs. 6.14c and 6.14d) that the presence of the equatorial density anomaly, commonly observed in the upper ionosphere during evening hours, leads to the focusing of the wave energy from lightning near the geomagnetic equator at low altitudes, thus accounting for all observed aspects of the observed phenomenon. The observations presented here indicate that during certain hours in the evening, almost all the energy input from lightning discharges entering the ionosphere at < 30° latitude remains confined to a small region (in altitude and latitude) near the geomagnetic equator. The net wideband electric field, extrapolated from the observed electric field values in the 512–1024 Hz band, can be ∼10 mV/m or higher. These strong electric fields generated in the ionosphere by lightning at local evening times may be important for the equatorial electrodynamics of the ionosphere.

6.4.2 W-Mode Observations When the Source is in the Magnetosphere

As shown in Fig. 6.1, a variety of W-mode waves, collectively called VLF emissions, are observed in the magnetosphere. All of them, excepting VLF transmitter signals and lightning generated whistlers, have apparent origins in the magnetosphere. A general discussion of these naturally occurring VLF emissions, beyond the scope of this paper, can be found in several review papers on the subject [1, 48, 53, 54, 74, 101, 115]. This paper briefly reviews those aspects of the propagation of VLF emissions which are affected by the presence of FAI of all types, including plasmapause. The generation and propagation of the VLF emissions is closely related to the properties of the cold and hot plasma distributions found in the regions where these emissions are generated.

W-Mode Observations when the Source is Near the Equator: Ducted Propagation

It is believed that plasmaspheric hiss, mid-latitude hiss, chorus, periodic and quasi-periodic emissions, and triggered emissions are generated in the equatorial region of the magnetosphere via some kind of gyroresonance related instability [e.g., 1, 115, and references therein]. The energy sources in most of these cases are the energetic electrons (∼1–100 keV) trapped in the magnetosphere. Figure 6.15 shows dynamic spectra illustrating ground observations

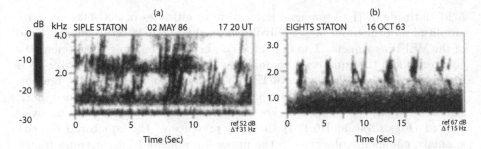

Fig. 6.15. (a) Double banded ELF/VLF chorus observed at Siple, Antarctica ($\lambda_g =$ 76°S, $\phi_g = 84°$W, $\lambda_m = 60°$S). In this case chorus bands are accompanied by weak bands of midlatitude hiss. (b) Periodic emissions observed at Eights, Antarctica ($\lambda_g = 75°$S, $\phi_g = 77°$W, $\lambda_m = 64°$S) (courtesy of VLF group, Stanford University)

of chorus (Fig. 6.15a, discrete rising tones), mid-latitude hiss (Fig. 6.15a, diffuse bands between 500–1500 Hz and 2–3 kHz), and periodic emissions (Fig. 6.15b).

Excepting plasmaspheric hiss, VLF emissions generated near the equator are observed on the ground. As shown in Fig. 6.7a, rays A and B, in general W-mode waves generated near the equator will be reflected back into the magnetosphere by either MR reflection or by total internal reflection. The fact that we routinely observe these emissions on the ground indicates that these emissions are somehow trapped inside one of the nearby ducts and brought down to the ground, as illustrated by rays C and C′ in Fig. 6.7.

As an example of ducted propagation of VLF emissions with the source in the magnetosphere, consider periodic emissions. Figure 6.15b shows a spectrogram of periodic emissions recorded at Eights, Antarctica. A sequence of discrete emissions or clusters of discrete emissions showing regular spacing is called periodic emissions [54, 101]. Usually their period is constant, but on occasion it changes slowly. They generally occur at frequencies below 5–10 kHz with a few kHz bandwidth and a period in the range from two seconds to six seconds. It was established that the period of the emissions is identical to the two-hop whistler transit time [Dowden, 33]. Observations of periodic emissions at conjugate stations Byrd-Hudson Bay showed that the emissions appeared at the two stations with a delay of about 0.8 s [Lokken et al., 77]. Periodic emissions are observed at latitudes that correspond to closed magnetic field lines and are rarely observed on satellites [Sazhin et al., 100]. These characteristics indicate that the periodic emissions, like other VLF emissions observed on the ground, have propagated in field aligned ducts. The fact that they are rarely observed on the satellite hints that probably they are also generated at small wave normal angles within ducts and thus remain trapped inside the ducts. Thus a satellite has to be inside a duct to record these emissions, the probability of which is very small as discussed earlier.

Chorus is routinely observed both on the ground and on spacecraft and many features, though not all, of chorus observed on the ground and spacecraft are similar [e.g., 53, 101, 115, and references therein]. This indicates that chorus may be propagating in both ducted and non-ducted modes. Recent analysis of chorus observed on the POLAR [Lauben et al., 75] and Cluster spacecraft [Santolik et al., 95, 96] support this viewpoint. Lauben et al. [75] find that the upper-band chorus waves ($f \geq 0.5 f_{Heq}$, where f_{Heq} is the equatorial gyrofrequency) are emitted with wave normal $\theta_s \simeq 0$, where θ_s is the wave normal angle at the source region near the equator, while the lower-band chorus waves ($f \leq 0.5 f_{Heq}$) are emitted with $\theta_s \simeq \theta_G$, where θ_G is the Gendrin angle, giving minimum value of refractive index parallel to $\mathbf{B}_0$. For both frequency bands, these respective θ_s values lead to wave propagation paths which remain naturally parallel to the static magnetic field in the source region over a latitude range of typically $3° - 5°$, providing an ample opportunity for cumulative wave/particle interaction and thus rapid wave growth, notably in the absence of field-aligned cold plasma density enhancements (i.e., ducts). Chorus generated at low wave normal angle can easily be trapped in a nearby duct and brought down to the Earth. Chorus at wave normal angles near θ_G can be guided to the Earth by the plasmapause [Inan and Bell, 61].

Mid-latitude hiss is frequently accompanied by whistlers echoing in the same path (duct) and amplified in the same frequency band. Dowden [33] suggested that at least part of the midlatitude hiss may be generated by the superposition of highly dispersed, unresolved, overlapping whistlers continuously echoing from hemisphere to hemisphere along a duct. However to overcome geometric loss of 10–20 dB corresponding to the small fraction of reflected energy being re-ducted, a magnetospheric amplification of 10–20 dB is required to maintain steady hiss intensity over time scales of tens of minutes [Thompson and Dowden, 132]. On the other hand, ground based direction finding measurements suggest that midlatitude hiss is generated just inside the plasmapause [Hayakawa, 51].

There are a few examples in the literature where a more direct correlation between W-mode emissions and FAI is made. Koons [71], using data from VLF/MR swept frequency receiver on AMPTE IRM spacecraft, found strong enhancement of W-mode waves (identified as hiss or chorus) correlated with density enhancement in the outer plasmasphere, in the vicinity of plasmapause, between $L = 4$ and $L = 6$, which he identified as ducts. The ducts consisted of more than 40% density enhancement and had a typical half width of 250 km. The wave intensity inside the duct was an order of magnitude larger than that outside. On Cluster spacecraft Moullard et al. [83] found that electron density fluctuations are regularly observed near the plasmapause together with hiss or chorus emissions at frequencies below the electron cyclotron frequency. Instruments on board Cluster spacecraft often observe two such emission bands with fluctuating wave intensities that suggest wave ducting in density enhancements as well as troughs. During a plasmapause crossing on June 5, 2001 (near the geomagnetic equator, $L = 4 - 6$, afternoon sector), density fluctuations up to hundreds cm^{-3} were found while whistler mode waves were observed in two separate frequency bands, at 100–500 Hz (correlated to the density fluctuations) and 3–6 kHz (anti-correlated). Using the electron density and wave data from Cluster spacecraft, Masson et al. [80] found that in the vicinity of the plasmapause, around the geomagnetic equator, the four Cluster satellites often observe banded midlatitude hiss, typically 1–2 kHz bandwidth and center frequency between 2–10 kHz. They found that the location of occurrence of the hiss was strongly correlated with the position of the plasmapause with no MLT dependence.

Plasmaspheric hiss is a broadband and structureless whistler-mode radiation that is almost always present in the Earth's plasmasphere and is commonly observed by magnetospheric satellites, but not observed on the ground [53, 130]. It is observed in the frequency range extending from a few hundred Hz to 2–3 kHz with a peak below 1 kHz. Plasmaspheric hiss should be distinguished from auroral hiss which is observed at low altitudes in the auroral regions and covers a much wider frequency band ($\sim$100 Hz – 100 kHz) and from mid-latitude hiss that is observed from ground stations and on satellites in 2–10 kHz range. Plasmaspheric hiss is found throughout the plasmasphere and is stronger in the daytime sector compared to the midnight-to-dawn sector, and generally peaks at high ($>$40°) latitudes; it often shows a sharp cutoff at the plasmapause for frequencies below $\sim$1 kHz, though it is also observed outside the plasmasphere at higher frequencies.

W-mode waves generated in the magnetosphere cannot reach the ground (Fig. 6.7a, rays A and B). VLF emissions such as mid-latitude hiss and chorus, discussed above, are observed on the ground, presumably they are generated at low wave normal angles, and get trapped in a duct and thereby propagate down to the Earth. A question naturally arises: why is plasmaspheric hiss not observed on the ground? The answer to this question may come from studies devoted to determining the wave normal directions of plasmaspheric hiss [e.g., 117, 126, and references therein]. The general conclusion from these studies is that plasmapheric hiss most frequently propagates with large wave normal angles. As a result it cannot be trapped inside a duct and be seen on the ground. On the other hand, recent results from POLAR spacecraft show that plasmaspheric hiss often propagates at wave normals that make small angle with the geomagnetic field [Santolik et al., 97]. It thus appears further work is warranted to understand why plasmaspheric hiss is not observed on the ground.

It appears that FAI may play an important role in the generation of plasmaspheric hiss. In one generation mechanism, proposed by Thorne et al. [131], plasmapause plays a central role. In this mechanism, hiss propagates along certain cyclic trajectories, made possible by reflections from plasmapause. This allows hiss to reach the equatorial region repeatedly, each time with a small wave normal angle, thus permitting maximum possible cyclotron resonance interaction with energetic electrons. Alternatively, hiss may be generated by lightning-whistlers as discussed in Sect. 6.1.2 [35, 118]. In this mechanism, topside ionospheric FAI play a central role.

W-Mode Observations when the Source is in the Auroral Region: Non-Ducted Propagation

W-mode emissions commonly observed in the auroral and polar regions include auroral hiss, lower hybrid emissions, and Lion's roar [74, 115, and references therein].

Auroral hiss (AH), shown in Fig. 6.16, is one of the most intense whistler mode plasma wave phenomenon observed both on the ground at high latitudes and on spacecraft in the auroral zone [e.g., 74, and references therein]. The apparent difference in the spectral form of continuous AH (Fig. 6.16a) and impulsive AH (Fig. 6.16b) has been attributed to differences in non-ducted versus ducted propagation [108, 109, 119]. The continuous structure-less spectra of continuous hiss can be explained in terms of non-ducted propagation and the impulsive, or sometimes falling tone spectra, as in Fig. 6.16b, of impulsive hiss in terms of ducted propagation.

There has been some difficulty in understanding the propagation of AH from its source region to the ground and to a high altitude satellite. There is strong experimental evidence as well as theoretical analysis that indicates that AH is generated at large wave normal angle by Cerenkov resonance mechanism [74, 119]. The standard whistler mode propagation in a smooth magnetosphere, discussed in Sect. 6.3, predicts that auroral hiss generated at large wave normal angle along the auroral field lines by Cerenkov resonance cannot penetrate to the ground. Two solutions for this problem have been suggested. Matsuo et al. [81] proposed that large scale FAI refract AH with $\theta \sim \theta_{RES}$ into AH with small wave normal angle which can fall into the transmission cone. Sonwalkar and Harikumar [119] argue that if the waves start with θ within a degree or so of the resonance cone angle, as the theory demands, then the relatively small tilts of horizontal gradients in the auroral ionosphere

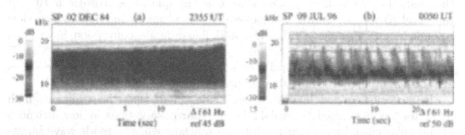

Fig. 6.16. Examples of auroral hiss (AH) observed at South Pole, Antarctica ($\lambda_g = 90°S$, $\lambda_m = 79°S$). (a) Example of continuous auroral hiss (CAH). (b) Example of impulsive auroral hiss (IAH) showing dispersion [adapted from Sonwalkar and Harikumar, 119] (Reprinted with permission of American Geophysical Union)

are insufficient to bend the wave normals enough for them to intercept the transmission cone. They propose that small scale FAI scatter AH propagating close to the resonance cone into small angle W-mode waves with small enough wave normal angles such that they fall into the transmission cone and thus propagate down to the Earth. This mechanism is shown schematically in Fig. 6.7f. There is some experimental evidence supporting this mechanism. Radar observations show a strong correlation between the occurrence of VLF hiss in the 1–10 kHz band and 18 MHz radar echoes from the F-region field-aligned irregularities [However and Gluth, 60]. Thirty two of the thirty three hiss events observed were associated with 18 MHz radar events. On occasion the hiss and radar events showed very close time correlation. Assuming that 18 MHz radar echoes result from scattering of radar signals from FAI with scale sizes equal to half the radar wavelength, the radar observations indicate presence of ∼15 m scale size FAI.

Auroral cavity, a region of reduced electron density, may affect propagation of AH upward from its source region. Snell's law when applied to a sharp density boundary slanted with respect to the geomagnetic field predicts that large wave normal angle W-mode waves can refract into small wave normal angle waves. Morgan et al. [82] suggest that through this mechanism, upward propagating auroral hiss converts from the short wavelengths (m) observed at low altitudes to longer wavelengths (1–3 km) inferred at high altitudes by undergoing multiple reflections from the tilted sides of an auroral cavity which spatially diverges with increasing altitude.

Satellite observations have often shown quasi-electrostatic plasma wave bands with a sharp cutoff near the local lower hybrid resonance (f_{LHR}) frequency [4, 5, 18, 72, 73]. These bands are observed at mid- to high latitudes in the altitude range of ∼1000 km to a few thousand kilometers. These waves are often excited by whistlers, or they could simply be hiss emissions with a cutoff at f_{LHR}. The cold plasma theory predicts a resonance at the lower hybrid frequency for whistler mode waves propagating perpendicular to the static magnetic field. At frequencies higher than the local LHR frequency, the resonance takes place at wave normal angles smaller than 90°. This explains the sharp lower cutoff at f_{LHR} as well as the quasi-electrostatic nature of the band [Brice and Smith, 18]. Several mechanisms including wave-particle resonances, parametric instabilities, and linear mode conversion have been proposed to explain the generation of lower hybrid waves. Bell and Ngo [7] have proposed that lower hybrid waves are excited when whistler mode waves are scattered by 10–100 m scale plasma density irregularities (Fig. 6.6).

Propagation guided by FAI irregularities along open field lines at high latitudes has been used to explain Lion's roar observations at low altitudes in the polar cusp. Lion's roar are intense sporadic whistler mode wave bursts observed in the Earth's magnetosheath [1, 115, and references therein]. They are found in the 100–200 Hz frequency range and typically last for ∼10 s, though at times they can last for ≥5 minutes.

Gurnett and Frank [46, 47], using Injun 5 data, detected lion roars in the high latitude magnetosheath near the polar cusp at 70–80° geomagnetic latitude and at ≤3000 km propagating parallel to the geomagnetic field. They suggested that these waves were generated in the magnetosheath and trapped there in open tubes of force of the geomagnetic field. These waves propagate along $\mathbf{B}_0$, being guided by field-aligned irregularities, and reach polar zone at ∼3000 km altitude.

W-Mode Observations when the Source is at Low Altitude (<5000 km): RPI Signals

The operation of a whistler-mode transmitter in the magnetosphere had been an unrealized goal of space scientists for many years. This goal was achieved, almost surreptitiously, when it was realized that Radio Plasma Imager (RPI) on the IMAGE satellite, designed to operate in free-space mode, could operate as a whistler mode sounder at the low end of its 3-kHz to 3-MHz sounding frequency range [94, 123]. Whistler mode echoes have been regularly observed by RPI when the IMAGE satellite operates in the inner plasmasphere and at moderate to low altitudes over the polar regions. Three types of irregularities – ducts, large scale features (1–10 km), and small scale features (10 m-1 km) – appear to influence the whistler-mode propagation at 10 kHz – 200 kHz. Each type of FAI seems to be associated with certain characteristics of RPI whistler-mode echoes.

Most of the whistler-mode echoes are detected when the satellite is within the plasmasphere or at low altitudes (<5000 km) over the southern polar region. The plasmagrams of Fig. 6.17a-c show examples of three types of echoes, fairly typical of whistler mode echoes observed on IMAGE in the first two years of its operation. Echoes shown in Fig. 6.17a and 6.17b were produced during soundings with 3.2 ms pulses, sequentially transmitted in 60 kHz – 1 MHz range in 144 logarithmic steps. For the case of Fig. 6.17c, a single 3.2 ms pulse was transmitted at each of 78 frequencies logarithmically spaced over the range 40–800 kHz. An RPI plasmagram normally plots the virtual range of echoes in Earth radii (R_E) as a function of sounding frequency [Galkin et al., 42], with virtual range calculated from the measured time delay assuming that the signal has propagated at the velocity of light in free space. Figure 6.17d shows the plot of the low altitude portion of the IMAGE polar orbits for the three cases and the approximate locations of IMAGE, indicated by red dots. Dipole field lines at $L = 4$ are shown as a reference.

The whistler-mode (W-mode) echoes in Figs. 6.17a and 6.17b are the traces below ≈ 300, and 240 kHz, respectively. Echoes such as these with narrowly defined time delay (proportional to virtual range) as a function of frequency are called discrete echoes.

Figure 6.17c shows an example of diffuse whistler mode echoes below ∼ 100 kHz. In contrast to the discrete echo events of Figs. 6.17a and 6.17b, in which virtual range (time-delay) spreading at each frequency was about

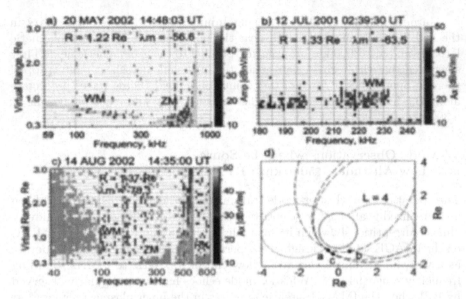

Fig. 6.17. Typical examples of whistler-mode echoes received during the first two years of RPI operations. (**a**) Plasmagram diplaying virtual range versus frequency for whistler mode (WM) echoes received on 20 May 2002 with relatively small spreading in virtual range (**b**) Plasmagram diplaying virtual range versus frequency for whistler mode echoes received on 12 July 2001. The whistler mode echoes are discrete traces in the 180–230 kHz frequency range with noticeable spreading in range (∼15–20 ms spread in time delay). (**c**) Plasmagrams showing example of wide spreading in virtual range (∼100 ms spread in time delays) of whistler mode echoes received outside the plasmapause at low altitudes over the southern polar region on 14 August 2002. Whistler mode echoes are accompanied by Z-mode (ZM) echoes and by a free space R-X mode echo. (**d**) Plot of the low altitude portion of the IMAGE polar orbits for the three cases and the approximate locations of IMAGE, indicated by dots. Dipole field lines at $L = 4$ are shown as a reference. Echoes such as in (a) and (b) with relatively small spreading in time delays (narrow spreading in range on a plasmagram) at each frequency are called "discrete echoes" in order to distinguish them from "diffuse echoes," shown in (c) which have considerable spread in time delays [adapted from Sonwalkar et al., 123] (Reprinted with permission of American Geophysical Union)

a fraction of R_E, in Fig. 6.17c the diffuse echoes are spread over as much as $2R_E$ at frequencies between ≈40 kHz and ≈100 kHz, with the most pronounced spreading at the lower frequencies.

In regions poleward of the plasmasphere, diffuse Z-mode echoes of a kind reported earlier by Carpenter et al. [31], were found to accompany both discrete and diffuse whistler-mode echoes 90% of the time, and were also present during 90% of the soundings when no whistler-mode echoes were detected. In Fig. 6.17a and 6.17c, Z-mode (ZM) echoes occur above 300 kHz and 200 kHz, respectively.

Sonwalkar et al. [123] proposed that the observed discrete whistler-mode echoes are a consequence of RPI signal reflections at the bottom side of the ionosphere, as illustrated in Fig. 6.7d. The larger spread in time delays at each frequency for the echo shown in Fig. 6.17b, relative to that for the echo shown in Fig. 6.17a, may result from echoes returning to the satellite by more than one path, also illustrated in Fig. 6.7d. Diffuse whistler-mode echoes are a result of scattering of RPI signals by FAI located within 2000 km Earthward of the satellite and in directions close to that of the field line passing through IMAGE (Fig. 6.7f). Diffuse Z-mode echoes are believed to be due to the scattering of RPI signals from electron density irregularities within 3000 km of the satellite, particularly those in the generally cross-**B** direction (Fig. 6.7f).

6.5 Discussion and Concluding Remarks

Given the spatial sizes of FAI and typical wavelength of W-mode waves found in the magnetosphere, it is convenient to classify FAI into three broad categories: large scale FAI, duct-type FAI, and small scale FAI. Observations indicate that FAI, large or small scale, influence the propagation of every kind of W-mode wave originating on the ground or in space.

There are two ways FAI can influence W-mode propagation:

- They provide W-mode waves accessibility to regions otherwise not reachable. This has made it possible for W-mode waves to probe remote regions of the magnetosphere, rendering them as a powerful remote sensing tool.
- They modify the wave structure which may have important consequences for radiation belt dynamics via wave-particle interactions.

We can illustrate this two-fold influence of FAI on W-mode propagation with specific examples. Ducts provide W-mode waves accessibility to the ground and allow long distance propagation along the duct [Helliwell, 54]. On the other hand, by restricting wave normal directions of W-mode waves close to B_0, ducts help maximize cyclotron resonance interaction between waves and energetic electrons [Helliwell, 55]. Large scale FAI in the topside ionosphere provide W-mode waves, entering ionosphere from the ground, accessibility to the ducts, and those exiting from the ducts to the ground [James, 65]. On the other hand, whistler propagation through these same FAI may lead to the generation of hiss emissions [35, 118] that were identified early on as a dominant contributor to the loss of radiation belt particles [67, 79]. Small scale FAI can help bring auroral hiss, propagating with wave normal angles close to the resonance cone, down to the ground [Sonwalkar and Harikumar, 119]; these same FAI can convert W-mode waves into quasi-electrostatic lower hybrid waves that can lead to the heating of ions in the auroral and subauroral regions [Bell et al., 9].

The influence of ducts on whistler propagation was recognized as early as 1956 by Helliwell et al. [58], three years after Storey's publication of the classic paper explaining the origins of whistlers [Storey, 125]. Remote sensing of magnetospheric plasma using ground observations of whistlers (whistler-technique) depends upon ducted propagation and major discoveries such as plasmasphere and plasmapause were based on this technique. It is interesting that though the whistler technique has been useful and has led to numerous new results, very little is known about the properties of ducts, their origins, and their distribution in space and time. Only in recent years has it been found that a single ducted whistler can contain multiple hyperfine elements, which may result because of structures within the duct or because of multi-mode propagation within a duct [Hamar et al., 50]. Carpenter [25] has discussed some of the outstanding problems in the area of ducted whistler propagation.

Importance of other large and small scale FAI was also recognized early [e.g., 26, 60, 65, and references therein]. Here also our knowledge of the distribution of these irregularities and their origins remains qualitative. For a long time, it was believed that small scale FAI may scatter W-mode waves below ~7000 km [Bell and Ngo, 6]; but recently Bell et al. [12] found that W-mode waves can excite LH waves at altitudes >20,000 km, implying the existence of small scale FAI in those regions of the magnetosphere. In the literature there is some discussion about the formation and/or maintenance of ducts and small scale FAI by precipitation of energetic particles [6, 16, 87]. Such mechanisms need to be further investigated. One direction for future work is to gain a better understanding of the structure, occurrence patterns, and generation mechanisms of large and small scale FAI.

It is clear that FAI alter or determine the properties of W-mode waves which in turn determine the nature of wave particle-interactions that occur in the magnetosphere.

Important open questions are the following:

- Are the effects of FAI on W-mode waves geophysically important?
- What is the role of ducts – or ducted whistlers – in the equilibrium of radiation belt electrons?
- What role do FAI in the topside ionosphere play in the generation of whistler triggered hiss?
- Is the generated hiss of sufficient intensity to maintain the slot region in the radiation belts?
- What role is played by lower hybrid waves that are continuously excited when all kinds of W-mode waves are scattered by small scale FAI?

These outstanding questions require answers in order for us to properly evaluate the importance of FAI in the propagation of W-mode waves.

Acknowledgements

The work at University of Alaska Fairbanks was supported by the Los Alamos National Laboratory under contracts 389AZ0017-97 and 65520-001-03-97, by

NASA under contract NNG04GI67G, and by NSF under contract NSF ATM-0244171. Author acknowledges discussions with many colleagues with whom he has worked on the problems related to whistler mode propagation. Author would like to thank Bodo Reinisch and the RPI instrument team of the IMAGE project for developing a wonderful new tool to study the effects of plasma density irregularities on plasma wave propagation. The author thanks Radha Krisha Proddaturi and Arun Venkatasubramanian for their assistance in preparing figures and editing early drafts of the paper.

References

[1] Anderson, R. R. and W. S. Kurth: Discrete electromagneticemissions in planetary magnetospheres, *Plasma Waves and Instabilities at Comets and in Magnetospheres*, Geophys. Monograph Vol. 53, (B. T. Tsurutani and H. Oya, Eds.), p. 81, 1989.

[2] Angerami, J. J.: Whistler duct properties deduced from VLF observations-made with the OGO-3 satellite near the magnetic equator, J. Geophys. Res. 75, 6115, 1970.

[3] Angerami, J. J., and J. O. Thomas, Studies of planetary atmosphere 1,the distribution of electrons and ions in the Earth's exosphere, J. Geophys. Res. 69, 4537, 1964.

[4] Barrington, R. E., and Belrose, J. S., (1963). Preliminary results from the very-low-frequency receiver aboard Canada's Alouette satellite, Nature 198, 651, 1963.

[5] Barrington, R. E., J. S. Belrose, D. A. and Keeley: Very low frequency noise bands observed by the Alouette 1 satellite, J. Geophys. Res. 68, 6539, 1963.

[6] Bell, T. F. and H. D. Ngo: Electrostatic waves stimulated by coherent VLF signals propagating in and near the inner radiation belt, J. Geophys. Res. 93, 2599, 1988.

[7] Bell, T. F. and H. D. Ngo: Electrostatic lower hybrid waves excited by electromagnetic whistler mode waves scatteringfrom planar magnetic-field-aligned plasma density irregularities, J. Geophys. Res.95, 149, 1990.

[8] Bell, T. F., James, H. G., Inan, U. S., and Katsufrakis, J. P.: The apparent spectral broadening of VLF transmitter signals during transionospheric propagation, J. Geophys. Res., 88, 4813, 1983.
Bell, T. F., R

[9] Bell, T. F., R. A. Helliwell, and M. K. Hudson: Lower-hybrid waves excited through linear mode coupling and the heating ofions in the auroral and subauroral magnetosphere, J. Geophys.Res. 96, 11379, 1991a.

[10] Bell, T. F., U. S. Inan, V. S. Sonwalkar, and R. A. Helliwell: DE-1 Observations of lower hybrid waves excited by VLF whistler mode waves, Geophys. Res. Lett., 18, 393, 1991b.

[11] Bell, T. F., R. A. Helliwell, U. S. Inan, and D. S. Lauben: The heating of suprathermal ions above thunderstorm cells, Geophys. Res. Lett. 20, 94, 1993.

[12] Bell, T. F., U. S. Inan, M. Platino, J. S. Pickett, P. A. Kossey, and E. J. Kennedy: CLUSTER observations of lower hybrid waves excited at high altitudes by electromagnetic whistler mode signals from the HAARP facility, Geophys. Res. Lett. 31, L06811, 2004.

[13] Benson, R. F. and W. Calvert: ISIS 1 observations at the source of auroral kilometric radiation, Geophys. Res. Lett. 6, 479, 1979.

[14] Benson, R. F., V. A. Osherovich, J. Fainberg, and B. W. Reinisch: Classification of IMAGE/RPI-stimulated plasma resonances for the accurate determination of magnetospheric electron-density and magnetic field values, J. Geophys. Res. 108, 1207, 2003.

[15] Benson, R. F., P. A. Webb, J. L. Green, D. L. Carpenter, V. S. Sonwalkar, H. G. James, and B. W. Reinisch: Active wave experiments in space plasmas: The Z mode, these Proceedings, Ringberg Workshop, 2005.

[16] Bernhardt, P. A. and C. G. Park: Protonospheric-ionospheric modeling of VLF ducts, J. Geophys. Res. 82, 5222, 1977.

[17] Bortnik, J., U. S. Inan, and T. F. Bell: Frequency-time spectra of magnetospherically reflecting whistlers in the plasmasphere, J. Geophys. Res. 108, 1030, 2003.

[18] Brice, N. M. and R. L. Smith: Lower hybrid resonance emissions, J. Geophys. Res. 70, 71, 1965.

[19] Budden, K. G.: *The propagation of radio waves*, Cambridge University Press, Cambridge, 1985.

[20] Burgess, W. C. and U. S. Inan: The role of ducted whistlers in the precipitation loss and equilibrium flux of radiation belt electrons, J. Geophys. Res. 98, 15,643, 1993.

[21] Cairo, L. and J. C. Cerisier: Experimental study of ionospheric electron density gradients and thier effect on VLF propagation, J. Atmos. Terr. Phys. 38, 27, 1976.

[22] Calvert, W.: Wave ducting in different wave modes, J. Geophys. Res. 100, 17,491, 1995.

[23] Carpenter D. L.: Whistlers and VLF Noises Propagating Just Outside the Plasmasphere, J. Geophys. Res. 83, 44, 1978.

[24] Carpenter, D.L.: A study of the outer limits of ducted whistler propagation in the magnetosphere, J. Geophys. Res.86, 839-45, 1981.

[25] Carpenter, D. L.: Remote Sensing the Earth's Plasmasphere, Radio Science Bull. (URSI), 308 (March), 13, 2004.

[26] Carpenter, G. B. and L. Colin: On a remarkable correlation between whistler mode propagation and high frequency northscatter, J. Geophys. Res. 68, 5649, 1963.

[27] Carpenter, D. L. and D. M. Šulić: Ducted whistler propagation outside the plasmapause, J. Geophys. Res.93, 9731, 1988.

[28] Carpenter, D. L., F. Walter, R. E. Barrington, and D. J. McEwen: Alouette 1 and2 observations of abrupt changes in whistler rate and of VLF noise variationsat the plasmapause-a satellite-ground study, J. Geophys. Res. 73, 2929, 1968.

[29] Carpenter, D. L., R. R. Anderson, T. F. Bell and T. R. Miller: A comparison of equatorial electron densities measured by whistlers and by a satellite radio technique, Geophys. Res. Lett. 8, 1107, 1981.

[30] Carpenter, D. L., M. A. Spasojevic´, T. F. Bell, U. S. Inan, B. W. Reinisch,I. A. Galkin, R. F. Benson, J. L. Green, S. F. Fung, and S. A. Boardsen: Small-scale field-aligned plasmaspheric density structures inferred fromthe Radio Plasma Imager on IMAGE, J. Geophys. Res. 107, 1258, doi:10.1029/2001JA009199, 2002.

[31] Carpenter, D. L., T. F. Bell, U. S. Inan, R. F. Benson, V. S. Sonwalkar, B. W. Reinisch, and D. L. Gallagher: Z-mode sounding within propagation "cavities" and other inner magnetospheric regions by the RPI instrument on the IMAGE satellite, J. Geophys. Res. 108, 1421, 2003.

[32] Cerisier, J. C.: A theoretical and experimental study of non-ducted VLF waves after propagation through the magnetosphere, J. Atmos. Terr. Phys. 35, 77, 1973.

[33] Dowden, R. L.: Distinction between mid latitude VLF hiss and discrete emissions, Planet. Space Sci. 19, 374, 1971.

[34] Dowden, R. L., and Heliwell, R. A., Very-low-frequency discrete emissions received at conjugate points, Nature, 195, 64, 1962.

[35] Draganov, A. B., U. S. Inan, V. S. Sonwalkar, and T. F.Bell: Magnetospherically reflected whistlers as a source ofplasmaspheric hiss, Geophys. Res. Lett. 19, 233, 1992.

[36] Draganov, A. B., U. S. Inan, V. S. Sonwalkar, and T. F. Bell: Whistlers and plasmaspheric hiss: Wave directions and three- dimensional propagation, J. Geophys. Res. 98, 11,401, 1993.

[37] Edgar, B. C.: *The structure of the magnetosphere as deduced from magnetosphericallyreflected whistlers*, Ph.D. Thesis, Stanford Univ., Dept. Electr. Engng., 1972.

[38] Edgar, B. C.: The upper and lower frequency cutoffs of magnetospherically reflected whistlers, J. Geophys. Res. 81, 205, 1976.

[39] Fejer, B. G. and M. C. Kelley: Ionospheric irregularities, Rev. Geophys. 18, 401, 1980.

[40] Ferencz, C. et al. (Eds.): *Whistler phenomena : short impulse propagation*, Kluwer Academic Publishers, Dordrecht and Boston, 2001.

[41] Fung, S. F., R. F. Benson, D. L. Carpenter, B. W. Reinisch, and D. L. Gallagher: Investigations of plasma irregularities in remote plasma regions by radio sounding: Applications of the Radio Plasma Imager on IMAGE, Space Sci. Rev. 91, 391, 2000.

[42] Galkin, I, G. Khmyrov, A. Kozlov, B. Reinisch, X. Huang, and G. Sales: New tools for analysis of space-borne sounding data, in: *Proc. 2001 USNC/URSI National Radio Science Meeting, July 8-13, 2001,* p. 304, 2001.

[43] Gendrin, R.: Le guidage des whistlers par le champ magnétique, Planet, Space Sci. 5, 274, 1961.

[44] Green, J. L., S. Boardsen, L. Garcia, W. W. L. Taylor, S. F. Fung, and B. W. Reinisch: On the origin of whistler mode radiation in the plasmasphere, J. Geophys. Res. 110, A03201, 2005.

[45] Groves, K. M., M. C. Lee, and S. P. Kuo: Spectral broadening of VLF radio signals traversing the ionosphere, J. Geophys. Res. 93, 14,683, 1988.

[46] Gurnett, D. A. and L. A. Frank: VLF hiss and related plasma observations in the polar magnetosphere, J. Geophys. Res. 77, 172, 1972a.

[47] Gurnett, D. A. and L. A. Frank: ELF noise bands associated with auroral electron precipitation, J. Geophys. Res. 77, 3411, 1972b.

[48] Gurnett, D. A. and U. S.Inan: Plasma wave observations with the Dynamics Explorer 1 spacecraft, Rev. Geophys. Space Phys. 26, 285, 1988.

[49] Hamar, D., G. Tarcsai, J. Lichtenberger, A. J. Smith, and K. H. Yearby: Fine structure of whistlers recorded digitally at Halley, Antarctica, J. Atmos. Terr. Phys. 52, 801, 1990.

[50] Hamar, D., C. Ferencz, J. Lichtenberger, G. Tarcsai, A. J. Smith, and K. H. Yearby: Trace splitting of whistlers: a signature of fine structureor mode splitting Radio Sci. 27, 341, 1992.

[51] Hayakawa, M.: Whistlers, in: *Handbook of Atmospheric Electrodynamics*, H. Volland (Ed.), CRC Press, Boca Raton, Florida, USA, 1995.

[52] Hayakawa, M., Parrot, M., and Lefeuvre, F.: The wave normals of ELF hiss emissions observed on board GEOS 1 at the equatorial and off-equatorial regions of the plasmasphere, J. Geophys. Res., 91, 7989, 1986.

[53] Hayakawa, M. and S. S. Sazhin: Mid-latitude and plasmaspheric hiss: a review, Planet Space Sci. 40, 1325, 1992.

[54] Helliwell, R. A.: *Whistlers and Related IonosphericPhenomena*, Stanford Univ. Press, 1965.

[55] Helliwell, R. A.: A theory of discrete VLF emissions from the magnetosphere, J. Geophys. Res. 72, 4773, 1967.

[56] Helliwell, R. A.: VLF wave stimulation experiments in the magnetosphere from Simple Station, Antarctica, Rev. Geophys. Space Phys. 26, 551, 1988.

[57] Helliwell, R.A.: The role of the Gendrin mode of VLF propagation in the generation of magnetospheric emissions, Geophys. Res. Lett. 16, 2095, 1995.

[58] Helliwell, R. A., J. H. Crary, J. H. Pope, and R. L. Smith: The "nose" whistler- a new high latitude phenomenon, J. Geophys. Res. 61, 139, 20, 1956.

[59] Heyborne, R. L.: Observations of whistler-mode signals in the OGO satellites from VLF ground station transmitters, Tech. Rep. No. 3415/3418-1, Stanford Electron. Lab., Stanford, Calif., 1966.

[60] Hower, G. L. and W. I. Gluth: Associations between VLF hiss and HF radar echoes from field-aligned ionization, J. Geophys. Res. 70, 649, 1965.

[61] Inan, U. S. and T. F. Bell: The plasmapause as a VLF waveguide, J. Geophys. Res. 82, 2819, 1977.

[62] Inan, U.S. and T.F. Bell: Spectral broadening of VLF transmitter signals observed on DE1: a quasi-electrostatic phenomenon?, J. Geophys. Res. 90, 1771, 1985.

[63] Jacobson, A. B. and W. C. Erickson: Observations of electron-density iregularities in the plasmasphere using the VLA radio-interferometer, Ann. Geophys. 11, 869, 1993.

[64] Jacobson, A. B., R. C. Carlos, R. S. Massey, G. Wu, and G. Hoogeveen: Total-electron-content signatures of plasmasphere motions, Geophys. Res. Lett., 22, 2461, 1995.

[65] James, H.G.: Refraction of whistler-mode waves by large-scale gradients inthe middle latitude ionosphere, Ann. Geophys. 28, 301, 1972.

[66] Kelley, M.: *The Earth's Ionosphere*, Academic Press, San Diego, California, 1989.

[67] Kennel, C. F. and H. E. Petschek: Limit on stably trapped particle fluxes, J. Geophys. Res. 71, 1, 1966.

[68] Kimura, I.: Effects of ions on whistler-mode ray tracing, Radio Science 1, 269, 1966.

[69] Kivelson, M. G. and C. T. Russell (Eds.): *Introduction to Space Physics*, Cambridge University Press, Cambridge 1995.

[70] Kletzing, C. A., F. S. Mozer, and R. B. Torbert: Electron temperature and density at high latitude, J. Geophys. Res. 103, 14,837, 1998.

[71] Koons, H.C.: Observations of large-amplitude, whistler mode wave ducts inthe outer plasmasphere, J. Geophys. Res. 94, 15,393, 1989.

[72] Laaspere, T. and H. A. Taylor: Comparison of certain VLF noise phenomena with the lower hybrid resonance frequency calculated from simultaneous ion-composition measurements, J. Geophys. Res. 75, 97, 1970.

[73] Laaspere, T. and W. C. Johnson: Additional results from an OGO-6 experiment concerning ionospheric electric and electromagnetic fields in the range 20 Hz to 540 kHz, J. Geophys. Res. 78, 2926, 1973.

[74] LaBelle, J. W. and R. A. Treumann: Auroral radio emissions, Space Sci. Rev. 101, 295, 2002.

[75] Lauben, D. S., U. S. Inan, and T. F. Bell: Source characteristics of ELF/VLF chorus, J. Geophys. Res. 107, 1429, 2002.

[76] Lefeuvre, F., T. Neubert, M. Parrot, and C. Delannoy: Wave normal angles and wave distribution functions for ground transmitter signals, J. Geophys. Res. 87, 6203, 1982.

[77] Lokken, J. E., J. A. Shand, C. S. Sir, K. C. B. Wright, L. H. Martin, N. M. Brice, and R. A. Helliwell: Stanford-Pacific Naval Laboratory conjugate point experiment, Nature 192, 319, 1961.

[78] Lyons, L. R. and D. J. Williams: *Quantitative Aspects of Magnetosphere Physcs*, D. Reidel Publishing Company, Dordrecht, Holland 1984.

[79] Lyons, L. R., R. M. Thorne, and C. F. Kennel: Pitch-angle diffusion of radiation belt electrons with theplasmasphere, J. Geophys. Res. 77, 3455, 1972.

[80] Masson, A., U. S. Inan, H. Laakso, O. Santolík, and P. Décréau: Cluster observations of mid-latitude hiss near the plasmapause Ann. Geophys. 22, 2565, 2004.

[81] Matsuo, T., T. Nishiyama, and D. Matuhara: Propagation of a quasi electrostatic whistler mode auroral hiss to the ground, Proc. NIPR Symp. Upper Atmos. Phys. 12, 12, 1998.

[82] Morgan, D. D., D. A. Gurnett, J. D. Menietti, J. D. Winningham, and J. L. Burch: Landau damping of auroral hiss, J. Geophys. Res. 99, 2471, 1994.

[83] Moullard, O., A. Masson, H. Laakso, M. Parrot, P. Décréau, O. Santolík, and M. André: Density modulated whistler mode emissions observed near the plasmapause, Geophys. Res. Lett. 29, 1975, 2002.

[84] Muldrew, D. G.: Nonvertical propagation and delayed-echo generation observed by the topside sounders, Proc. IEEE 57, 1097, 1969.

[85] Neubert, T., E. Ungstrup, and A. Bahnsen: Observations on the GEOS 1 satellite of whistler mode signals transmitted by the omega navigation system transmitter in northen Norway, J. Geophys. Res. 88, 4015, 1983.

[86] Ngo, H. D. and T. F. Bell: Quasi-electrostatic waves excited during the scattering of coherent whistler-mode waves excited from magnetic-field-aligned plasma density irregularities, Eos Trans. AGU 66, 1038, 1985.

[87] Park, C. G. and R. A. Helliwell: The formation by electric fields of field-aligned irregularities in the magnetosphere, Radio Sci. 6, 299, 1971.

[88] Persoon, A.M.: Electron density distributions, Adv. Space Res. 8, 79, 1988.

[89] Persoon, A. M., D. A. Gurnett, and S. D. Shawhan: Polar cap electron densities from DE 1 plasma wave observations, J. Geophys. Res. 88, 10,123.

[90] Platino, M., U. S. Inan, T. F. Bell, D. A. Gurnett, J. S. Pickett, P. Canu, and P. M. E. Décréau: Whistlers observed by the Cluster spacecraft outside the plasmasphere, J. Geophys. Res. 110, A03212, 2005.

[91] Poeverlein, H.: Strahlwege von Radiowellen in der Ionosphäre, Sitz. Ber. Bayerische Akad. Wiss. 1, 175, 1948.

[92] Raghuram, R.: A new interpretation of subprotonospheric whistler characteristics, J. Geophys. Res. 80, 4729, 1975.

[93] Rastani, K., U. S. Inan, and R. A. Helliwell: DE 1 observations of Siple transmitter signals and associated sidebands, J. Geophys. Res. 90, 4128, 1985.

[94] Reinisch, B. W. et al.: The Radio Plasma Imager investigation on the IMAGE spacecraft, Space Sci. Rev. 91, 319, 2000.

[95] Santolík, O., D. A. Gurnett, J. S. Pickett, M. Parrot, N. Cornilleau-Wehrlin: Spatio-temporal structure of storm-time chorus, J. Geophys. Res. 108, 1278, 2003.

[96] Santolík, O., D. A. Gurnett, J. S. Pickett, M. Parrot, and N. Cornilleau-Wehrlin: A microscopic and nanoscopic view of storm-time chorus on 31 March 2001, Geophys. Res. Lett. 31, L02801, 2004.

[97] Santolík, O., M. Parrot, L. R. O. Storey, J. Pickett, and D. A. Gurnett: Propagation analysis of plasmaspheric hiss using Polar PWI measurements, Geophys. Res. Lett. 28, 1127, 2001.

[98] Santolík, O., A. M. Persoon, D. A. Gurnett, P. M. E. Decreau, J. S. Pickett, O. Marsalek, M. Maksimovic, and N. Cornilleau-Wehrlin: Drifting field-aligned density structures in the night-side polar cap, Geophys. Res. Lett. 32, L021696, 2005.

[99] Sazhin, S. S. and M. Hayakawa: Magnetospheric chorus emissions: a review, Planet Space Sci., 40, 681, 1992.

[100] Sazhin, S. S. and M. Hayakawa: Periodic and quasiperiodic VLF emissions, J. Atmos. Terr. Phys. 56, 735, 1994.

[101] Sazhin, S. S., K. Bullough, and M. Hayakawa: Auroral hiss: A review, Planet. Space Sci. 41, 153, 1993.

[102] Scarabucci, R. R.: Analytic and numerical treatment of wave propagation in the lower ionosphere, Technical Report No. 3412-11, Stanford Electronics Laboratories, Stanford Univ., Stanford, Calif., 1969.

[103] Scarf, F. L. and C. R. Chappell: An association of magnetosphericwhistler dispersion characteristics with changes in local plasma density, J. Geophys. Res. 78, 1597, 1973.

[104] Schultz, M. and L.J. Lanzerotti, Particle diffusion in the radiation belts, Springer-Verlag, Berlin – Heidelberg – New York, 1974.

[105] Schunk, R. W. and A. F. Nagy: Ionospheres: Physics, Plasma Physics, and Chemistry, Cambridge Univ. Press, New York, 2000.

[106] Scudder, J. et al.: Hydra-A 3-dimensional electron and ion hot plasma instrument for the POLAR spacecraft of the GGS mission, in: The Global Geospace Mission, C. T. Russell (Ed.), Kluwer Acad. Publ., Norwell, Mass., 1995.

[107] Shawhan, S. D.: Magnetospheric plasma wave research 1975-1978, Rev. Geophys. Space Phys. 17, 705, 1979.

[108] Siren, J. C.: Dispersive auroral hiss, Nature 238, 118, 1972.

[109] Siren, J. C.: Fast hisslers in substorms, J. Geophys. Res. 80, 93, 1975.

[110] Smith, R. L.: An explanation of subprotonospheric whistlers, J. Geophys.Res. 69, 5019, 1964.

[111] Smith, R. L. and J. J. Angerami: Magnetospheric properties deduced from OGO 1 observations of ducted and non-ductedwhistlers, J. Geophys. Res. 73, 1, 1968.

[112] Smith, R. L., R. A. Helliwell, and I. Yabroff: A theory of trapping of whistlers in field-aligned columns of enhanced ionization, J. Geophys. Res. 65, 815, 1960.

[113] Sojka, J. J., L. Zhu, M. David, and R. W. Schunk: Modeling the evolution of meso-scale ionospheric irregularities at high latitudes, Geophys. Res. Lett. 27, 3595, 2000.

[114] Sonwalkar, V. S.: *New Signal Analysis Techniques and Their Applications to Space Physics*, Ph.D. Thesis, Stanford University, June 1986.

[115] Sonwalkar, V. S.: Magnetospheric LF- VLF-, and ELF-waves, in: *Handbook of Atmospheric Electrodynamics*, H. Volland (Ed.), CRC Press, BocaRaton, Florida, USA, 1995.

[116] Sonwalkar, V. S. and U. S. Inan: Measurements of Siple transmitter on the DE 1 satellite: Wave normal direction and antenna effective length, J. Geophys. Res. 91, 154, 1986.

[117] Sonwalkar, V. S. and U. S. Inan: Wave normal direction and spectral properties of whistler-mode hiss observed on the DE-1 satellite, J. of Geophys. Res. 93, 7493, 1988.

[118] Sonwalkar, V. S., and Inan, U. S., Lightning as an embryonic source of VLF hiss, J. Geophys. Res. 94, 6986, 1989.

[119] Sonwalkar, V.S. and J. Harikumar: An explanation of ground obesrvations of auroral hiss: Role of density depletions and meter-scale irregularities J.Geophys. Res. 105, 18,867, 2000.

[120] Sonwalkar, V. S., T. F. Bell, R. A. Helliwell, and U. S.Inan: Direct multiple path propagation: A fundamental property of non-ducted VLF waves in the magnetosphere, J. of Geophys. Res. 89, 2823, 1984.

[121] Sonwalker, V. S., U. S. Inan, T. L. Aggson, W. M. Farrell, and R. Pfaff: Focusing of non-ducted whistlers by the equatorial anomaly, J. Geophys. Res. 100, 7783, 1995.

[122] Sonwalkar, V. S., U. S. Inan, T. F. Bell, R. A. Helliwell, V. M. Chmyrev, Ya. P. Sobolev, O. Ya. Ovcharenko, and V. Selegej: Simultaneous observations of VLF ground transmitter signals on the DE 1 and COSMOS 1809 satellites: Detection of a magnetospheric caustic and a duct, J. Geophys. Res. 99, 17,511, 1994.

[123] Sonwalkar V. S. et al.: Diagnostics of magnetospheric electron density and irregularities at altitudes <5000 km using whistler and Z mode echoes from radio sounding on the IMAGE satellite, J. Geophys. Res. 109, A11212, 2004.

[124] Stix, T. H.: *Waves in Plasmas*, American Inst. Physics, New York, 1992.

[125] Storey, L. R. O.: An investigation of whistling atmospherics, Philos. Trans. R. Soc. London, Ser. A. 246, 113, 1953.

[126] Storey, L. R. O., F. Lefeuvre, M. Parrot, L. Cairo, and R. R. Anderson: Initial survey of the wave distribution functions for plasmaspheric hiss observed by ISEE 1, J. Geophys. Res. 96, 19,469, 1991.

[127] Strangeways, H. J. and M. J. Rycroft: Systematic errors in VLF direction-finding of whistler ducts, II, J. Atmos. Terr. Phys. 42, 1009, 1980.

[128] Strangeways, H. J., M. J. Rycroft, and M. J. Jarvis: Multi-station VLF direction-finding measurements in eastern Canada, J. Atmos. Terr. Phys. 44, 509, 1982.

[129] Thorne, R. M. and R. B. Horne: Landau damping of magnetospherically reflected whistlers, J. Geophys. Res., 99, 17, 249, 1994.

[130] Thorne, R. M., E. J. Smith, R. K. Burton, and R. E. Holzer: Plasmaspheric hiss, J. Geophys. Res., 78, 1581, 1973.

[131] Thorne, R. M., S. R. Church, and D. J. Gorney: On the origin of plasmaspheric hiss: The importance of wave propagation and plasmapause, J. Geophys. Res. 84, 5241, 1979.

[132] Thompson, R. J. and R. L. Dowden: Ionospheric whistler propagation, J. Atmos. Terr. Phys. 40, 215, 1978.

[133] Titova, E. E., V. I. Di, V. E. Yurov, O. M. Raspopov, V. YuTrakhtengertz, F. Jiricek, and P. Triska: Interaction between VLF waves and turbulent ionosphere, Geophys. Res. Lett. 11, 323, 1984.

[134] Voss, H. D., W. L. Imhof, J. Mobilia, E. E. Gaines, M. Walt, U. S. Inan, R. A. Helliwell, D. L. Carpenter, J. P. Katsufrakis, H. C. Chang: Lightning induced electron precipitation, Nature 312, 740, 1984.

7

Dipole Measurements of Waves in the Ionosphere

H.G. James

Communications Research Centre Canada, Ottawa, Ontario K2H 8S2 Canada
gordon.james@crc.ca

Abstract. The theory of distributed dipole antennas in magnetoplasmas has been tested through the analysis of the data from the two-point propagation experiment OEDIPUS C (OC). The transmission of electromagnetic signals over a 1-km distance in the ionosphere has been used to substantiate the theory of emission, propagation and detection of waves in a cold magnetoplasma. Confirmations and insights about the dipole theory arising from the OC research results have occasioned a return to some older data that can be profitably interpreted with them. The concept of dipole effective length L_{eff} has been re-examined quantitatively with the help of the reciprocity principle. It is found that previous measurements of electromagnetic whistler-mode propagation give L_{eff} values similar to those predicted using a classic reciprocity definition brought over from the vacuum dipole theory. In contrast, OC investigations found that L_{eff} can be many times the dipole physical length for propagation near the whistler-mode resonance cone. The finding has important consequences for the interpretation of the measured strength of the radio emission auroral hiss and, by extension, the nature of its source. Since intra-ionospheric experiments on propagation near either the lower- or the upper-oblique-resonance cone produced extremely strong transmission, the L_{eff} applied to the interpretation of the strength of plasma-wave phenomena in the corresponding frequency domains must be chosen with care.

Key words: Dipole antenna, effective length, magnetoplasma, electromagnetic, active, passive, emissions

7.1 Introduction

Understanding of the physics of dipole antennas has been important because dipoles of different sorts have had a variety of applications to the measurement of electric fields **E**, from dc to high frequency (HF) and beyond. This paper discusses distributed dipoles. Since wire and tubular dipoles have had extensive use and have gained a reputation for dependable deployment, they

H.G. James: *Dipole Measurements of Waves in the Ionosphere*, Lect. Notes Phys. **687**, 191–210 (2006)
www.springerlink.com

have been widely applied to spinning and stabilized space vehicles. Distributed dipoles have been employed: as communications transmit-receive antennas; for the reception of spontaneous emissions originating in the earth's atmosphere, at the sun and beyond; and for the diagnoses of plasma parameters at and near space payloads. This has happened on orbital, rocket and balloon payloads. In the context of the data to be discussed here, an important subtype of the distributed dipole is the stiff tubular distributed dipole. It has provided much interesting data in active radio experiments like the ionospheric topside sounder spacecraft missions.

Some features of published research on space radio data indicate that the current methodology of space dipoles is not yet mature. One feature is the disinclination of authors to analyze absolute magnitudes of received signals, even when the observation of a radio emission is made at its source. Spectra of radio emissions and their presence/absence in different regions of the magnetospheric topology are regularly discussed, their absolute field strengths much less often. Another feature is that, when absolute wave electric-field strengths are discussed, there is often no concern about the understanding of the measurement technique. As an example, the assumption that the effective length L_{eff} of a dipole of physical length L is

$$L_{eff} = \frac{L}{2} \cos \Theta' \tag{7.1}$$

is often used without discussion of its relevance (Θ' is the angle between $\mathbf{E}$ of the incident wave and the dipole-axis direction). A third feature is that statements about measurement precision or accuracy are typically not made; published reports of electric field measurements in space rarely contain error estimates.

Accurate measurements of wave $\mathbf{E}$-field parameters, such as absolute strength, wave mode, polarization and direction of propagation, all depend on the knowledge of L_{eff}. Thus, accurate L_{eff} values are fundamentally important in such areas as the determination of the location and extent of natural noise sources, imaging techniques for ionospheric density structure, and the investigation of the microphysics of radio-wave sources, whether spontaneous or manmade.

Notwithstanding the widespread use of dipoles in experimental space science, evidence from space experiments about the emitting and receiving properties of dipoles is limited. The analysis of data from the OEDIPUS-C (OC) two-point rocket experiment has expanded our understanding of dipoles. Research on OC confirmed some expected properties of dipoles and also one area where conventional methods do not predict observations. In Sect. 7.2, a summary of OC results from different domains of the diagram of Clemmow, Mullaly and Allis (CMA) [28] serves as a framework for a review of the contributions to date from the OC two-point propagation experiment. Observed radiation and propagation features of cold-plasma modes are compared with theoretical expectations. A bibliography of the busy period for the cold-plasma antenna theory in the 1960s is the references cited by Wang et al. [30]. The representative theoretical work of Kuehl [18] is relevant to the observations discussed in this paper, in which it is cited several times.

Results from OC suggest that some old subjects in ionospheric radio science involving dipoles could be profitably re-examined in the area of L_{eff} of a dipole. First, the results about radiation fields for the electromagnetic whistler and fast Z modes have shown that a classic cold-plasma treatment correctly predicts the radiated E field. In Sect. 7.3, a conventional concept of antenna L_{eff} and the reciprocity principle are invoked to find the L_{eff} of some dipoles from the same cold-plasma theory, for comparison with specific measurements of L_{eff} in space experiments reported by other authors.

Second, remarkably strong transmissions of quasi-electrostatic whistler-mode waves at 25 kHz near the oblique resonance cone were explained using a novel formalism to derive the dipole L_{eff}. The success of that investigation prompted the return in Sect. 7.3 to observations on the ISIS satellites of strong signals carried by waves near the lower- and upper-oblique-resonance cones. The resonance cone, an important feature of the whistler mode often invoked in space and laboratory plasma wave research, is a mathematical singularity of continuing interest in plasma wave theory [James, 12, and references therein].

Electric-field antennas and probes continue to be studied and used in laboratory plasmas. Insight thus obtained can be relevant to space plasmas. However the behaviour of dipoles in the unbounded space plasma is not easy to deduce from laboratory plasma results. Laboratory magnetoplasmas with physical dimensions very large compared with wavelengths of interest and having the magnetospherically relevant condition that the plasma frequency f_p is less than the gyrofrequency f_c are apparently yet to be used to study dipole radiation effects, if such facilities exist at all. The need has persisted for active space experiments on the physics of dipoles.

7.2 Summary of OEDIPUS-C Results

7.2.1 Overview

Figure 7.1 serves as a summary of two-point propagation results from the OC experiment [Horita and James 7]. OC was carried out using a double sounding rocket payload, with a transmitter driving crossed dipoles on one subpayload and crossed dipoles connected to a synchronized receiver on the other subpayload. In the payload mode that produced the data of Fig. 2.1, a 25–8000 kHz sweep started with a 600-μs rectangular pulse emitted at a carrier frequency of 25 kHz and received over a 6-ms period. This was followed by a series of 300-μs rectangular pulses with a 3-ms listening time, for carrier frequencies 100, 150, 200...8000 kHz. This linear sweep lasted 0.5 s.

Figure 7.1 is a concatenation of all ionogram data recorded with the aforementioned frequency sweep of the synchronized transmitter-receiver pair

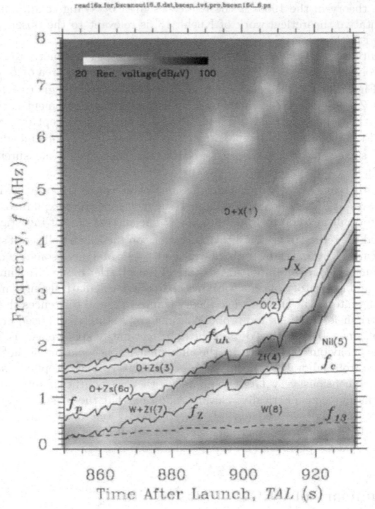

Fig. 7.1. Summary of transmitted signal strengths in the OC experiment on dipoles, from the last part of the down-leg trajectory [Horita and James, 7]. The data display combines all data recorded with the 0.1 to 8.0 MHz sweep. The darkness of the pixels is proportional to the received signal voltage, which can be read from the inserted gray-scale. The CMA regions 1–8 are labelled in the frequency domains in this figure. The regions designated are also preceded by the cold-plasma wave modes present in that region. The O mode is present in CMA1, 2, 3 and 6a; the X mode in CMA1; the slow Z mode, Z_s, in CMA3 and CMA6a; the fast Z mode, Z_f, in CMA4 and 7; the W mode in CMA7 and 8; and no mode is present in CMA5 (Reprinted with permission of American Geophysical Union)

starting at a time after launch (TAL) of 850 s. The blackness of any pixel is proportional to the signal strength in the prompt pulse arriving at the receiver with less than 10-μs delay, the typical delay for most observable modes with refractive indices near 1. The overlaid lines on the figure are relevant electron characteristic frequencies. Values of f_p were scaled from observable characteristic frequencies in the associated OC ionograms [James and Calvert, 15]. Values of f_c were obtained from an IGRF1995-based field model. The overlaid lines from the lower to the higher frequencies are: the cutoff frequency of the Z mode $f_Z = [-f_c + (f_c^2 + 4f_p^2)^{1/2}]/2$, f_p, f_c, the upper hybrid resonance frequency $f_{uh} = (f_c^2 + f_p^2)^{1/2}$, and the cutoff frequency of the X mode $f_X = [f_c + (f_c^2 + 4f_p^2)^{1/2}]/2$.

Various cold-plasma wave modes are evidently transmitted across the HEX-REX bistatic link in the direct pulse. Figure 7.1 includes the labelling of modes and of the CMA domains in which they occur. For instance, in the "O+X(1)" area above f_X, the O and X modes propagate, and this is the CMA1 domain. The piecewise-discontinuous light loci in CMA1 are Faraday interferences explained by James and Calvert [15]. These Faraday interference "fringes" occur in CMA regions where two cold-plasma wave modes propagate and therefore give rise to destructive interference of transmitted signals on account of the resulting linear electric-field polarization. When the two modes are O and X in CMA1, interference possibilities give rise to various orders m of the null condition

$$(n_O - n_X)\frac{\pi f_Z}{c} - \alpha \approx \frac{m\pi}{2}, \tag{7.2}$$

where n_O and n_X are respectively the O- and X-mode refractive indices at frequency f, z is the separation of HEX and REX, α is the azimuthal angular separation of the axes of the effective dipoles of the HEX and REX and m is an odd integer. The two uppermost O-X fringes in Fig. 7.1, rising to the top right of the diagram, were found to be orders $m = -1$ and $+1$, respectively. James and Calvert [15] demonstrated that the fringes can be used for density magnitude and gradient measurements. The stepwise discontinuous nature of the $m = -1$ (highest) fringe in $870 <$ TAL < 890 indicates that the payload is descending through some periodic density structure in the corresponding height range, 348 to 291 km. The OC Faraday observations thus substantiated the cold-plasma theory predictions about electric-field polarization and wave dispersion.

Strong transmissions in the fast Z mode (CMA4) [Horita and James, 7] and in electromagnetic (CMA8) whistler-mode [James, 13] have been compared with cold-plasma theory for short dipoles. Analysis of these particular domains was deemed worthwhile for two reasons. Both show strong levels in Fig. 7.1. Also, there is only one cold-plasma mode in each CMA domain, avoiding the complication in analysis of the contributions of two modes to the total signal.

Z-mode signal enhancement is evident in Fig. 7.1 in the CMA4 region, in which the fast Z-mode propagation occurs at frequencies below f_p and above the greater of f_c and f_Z. The Z-mode signal is the relatively dark swath between f_Z and f_p. These signal levels, seen to be strikingly greater in CMA4 than in neighbouring domains, are predicted by the theory. The computed transmitting dipole impedance in the $f_Z - f_p$ frequency range provides a better match to the output of the transmitter than in the bands of other wave modes examined. In addition, the observed shapes of individual intensity-versus-frequency spectra are reproduced by the theory in the OC sweep across the domain.

In the CMA 8 region of Fig. 7.1, the grey-scaled whistler-mode signal strengths are observed to be uniform at frequencies between about 0.5 and 1.3 MHz, but to have a checkered pattern at lower frequencies. For the set of plasma parameters and transmitter-receiver geometry at every instant, there is a frequency f_{13} above which there is only one dispersion-relation solution with the required group direction linking the transmitter to the receiver. This relates to the fact that the whistler-mode dispersion surface is inflected for frequencies below about $f_c/2$. f_{13} is very close to the frequency that makes the Gendrin angle zero. The theoretical value of f_{13}, based on plasma parameters and geometry, is plotted with a broken line. In the flight survey of Fig. 7.1, the demarcation between the frequency domains of three saddle-point solutions and one solution is clearly seen. Below the f_{13} line, the grey-scale is checkered because in certain pixels the three wave components interfere constructively and in others, destructively. The theory of Kuehl [18] when adapted to the present OC conditions was found to predict absolute transmitted signal strengths to within a few decibels, working best at $f > f_{13}$.

Other features of the cold-plasma propagation theory were confirmed. For example, in Fig. 7.1, the region to the right and below f_Z and above f_c shows no signals present. This is CMA5 where, indeed, no cold-plasma wave propagation is predicted. The gap in CMA8 between the f_c curve and the grey shading arises because the prompt-pulse sampling upon which the figure is based automatically excludes signals with group speeds less than c, the case for signals near f_c in the OC ionograms.

All of the description in Sect. 7.2.1 above is about direct-propagation pulses with refractive indices close to 1, i.e., prompt pulses. In certain other cases, the received OC pulses were dispersed and delayed for times of the order of the 3-ms receiver listening time. This occurred in the very strong transmission in the whistler-mode around 25 kHz and at the lowest frequency in Fig. 7.1, 100 kHz. That pixel bandwidth, 100 ± 25 kHz, was found to correspond to group resonance-cone directions lying about 5° away from the axis of the earth's magnetic field $\mathbf{B}$. That is, the transmitter-receiver separation direction lay close to the resonance cone at these frequencies. Expected enhancement in resonance-cone signals indeed materialized, as shown by the heavy horizontal line running through 100 kHz. This subject is taken up in the next section. Dispersed-delayed pulses were also strongly received close to f_p when $f_c > f_p$, presumably Langmuir waves, and in CMA3 in the form of slow-Z-mode waves [James, 14]. The explanation of the slow-Z-mode observations called upon hot-plasma theory and an intermediate process of radiation by sounder-accelerated electrons to explain the characteristics of the waves at the OC receiver subpayload.

7.2.2 Whistler-Mode Propagation Near the Resonance Cone

The two lowest pulse carrier frequencies transmitted in the OC two-point propagation experiment were 25 and 100 kHz, each in a receiver bandwidth of 50 kHz. Figure 7.2 is a detailed spectral display of the whistler-mode waves received in these two bands on the flight downleg [James, 13]. These two grey-scaled displays considerably expand the lowest part of Fig. 7.1, and show the spectra of received signals for selected portions of the history. The lower figure is for the bandwidth centred at 25 kHz and contains evidence of strong signals at that band centre. The same is true for the upper panel showing some of the 100-kHz band history. The two panels have been positioned in the diagram to have both the time after launch (TAL) and frequency scales maintained between the two. It is therefore easy to see that in addition to the spectral strength at the 25- and 100-kHz carriers, there is a broad swath of signal enhancement centred first near the carrier in the bottom frame, then running diagonally toward its upper right corner. This same swath is seen to continue diagonally up through the top frame.

It was suspected that the enhancement swath was a signature of lower-oblique-resonance effects [James, 12]. The "resonance cone frequency" f_{rc} is the whistler-mode frequency that puts the group velocity direction, at angle θ_{gr} with respect to B, along the transmitter-receiver direction, which is at angle δ with respect to B. Applying an expression for $\tan \theta_{gr}$ as a function of f, f_p and f_c [Stix, 28] and equating it to $\tan \delta$ leads to the relation

$$\tan^2 \delta = \tan^2 \theta_{gr} = \frac{-1 + f_p^2/(f_{rc}^2 - f_c^2)}{1 - f_p^2/f_{rc}^2} \tag{7.3}$$

Inverting this equation for f_{rc} as a function of all the known variables throughout the total period in Fig. 7.2 produced the thin continuous line in that diagram labelled "f_{rc}". The f_{rc} locus is found to track the centre of the swath, confirming the suspicion about the resonance-cone nature of the waves. This conclusion was supported by the observation of spin modulation of the signal in Fig. 7.2. Seen in both panels, spin modulation is especially clear in the lower frame in $700 < TAL < 750$ as a series of pixels at 25 kHz that are alternatively strong and weak. In theory, the electric field is polarized linearly along the wave vector direction for waves near the resonance condition. Fortuitously on OC, the major-frame duty cycle of HEX-REX is 3 seconds and the spin period of the receiving subpayload is very close to four times greater, 12 s. During each spin half rotation, HEX-REX executes two duty cycles; on one cycle, the receiving dipole is aligned close to the transmitted wave **E**; 3 s later, the alignment is perpendicular. Hence the clear 1-on /1-off pattern emerges.

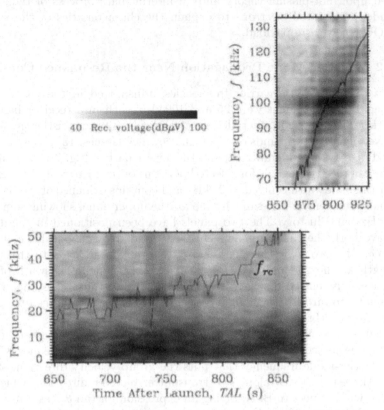

Fig. 7.2. Summary plot of the detailed signal spectra of the 25- and 100-kHz pulses, for those portions of the downleg flight when the resonance-cone frequency f_{rc}, as defined in (7.2) and traced with the overlaid line, lies within either of the bandwidths centred on those two frequencies. The two panels have been positioned in the diagram to have both the time and frequency scales maintained between the two

The resonance cone signals at the carrier frequencies in Fig. 7.2 were much stronger than the electromagnetic whistler-mode signals at higher carrier frequencies in Fig. 7.1. These absolute signal strengths have been evaluated in a link calculation by Chugunov et al. [4]. Their detailed computation for the 25-kHz case took into consideration the geometry of the double-V dipoles on both ends of the link. A radiating theory for the transmitting dipoles that is pertinent to propagation directions near the cone was used [Mareev and Chugunov, 22]. For the receiving dipoles, a new application of the reciprocity principle [Chugunov et al. 3] was employed to estimate the receiving dipole's L_{eff}, which was found to be 30 times its physical length. Good agreement between observed and theoretical resonance-cone signals was obtained.

The good agreement obtained has implications for the interpretation of whistler-mode waves in space plasma. Sects. 7.3.2 through 7.3.4 following will illustrate this with a retrospective examination of some past observations of oblique-resonance-cone waves in the ionosphere.

7.3 Retrospective on Past Observations

Section 7.1 reviewed analyses of OC data that confirmed the theory for EM waves. The success of the classic short-dipole theory, represented here by the Kuehl [18] CW theory, has motivated a re-examination of reports about dipole L_{eff} for EM cases; this is in Subsect. 7.3.1. The success of the novel Chugunov [3] theory of dipole L_{eff} for quasi-electrostatic resonance-cone waves has prompted a return to published analyses of the amplitudes of such waves. Discussion along this line for both lower-oblique resonance waves (whistler mode) and upper-oblique resonance-cone waves (slow-Z mode) follows in Sects. 7.3.2 through 7.3.4.

7.3.1 Dipole Effective Length for EM Propagation

Jordan's [16, pp. 336] definition of a generalized L_{eff} is taken as a point of departure. His (10-81) when applied to a cold plasma theory implies that the complex effective-length vector can be written

$$\mathbf{L}'_{eff} = \mathbf{e}_k L^k_{eff} + \mathbf{e}_\theta L^\theta_{eff} + \mathbf{e}_{-\phi} L^\phi_{eff} \tag{7.4}$$

This refers to a wave-vector space specified in spherical coordinates with the directions corresponding to radial direction k, polar-angle direction θ and azimuthal direction ϕ. The **E** polarization in magnetoionic theory in general has components in all three directions, as opposed to the vacuum case of Jordan's (10–81) where the polarization ellipse is perpendicular to the wave vector. Under the reciprocity principle, it is assumed that an expression of the same form as Jordan's (10 82) can be used to represent the vector field **E** of an active dipole in a magnetoplasma as

$$\mathbf{E} = \frac{60\pi\, I(0)}{r\lambda_0} \mathbf{L}'_{eff}, \tag{7.5}$$

where $I(0)$ is the current injected at the antenna terminal, r is the distance and 0 is the vacuum wavelength. An expression for **E** was obtained from the Kuehl [18] theory used in the above-described tests of the OC radiation field. His (14) can be rewritten as

$$\mathbf{E} = \frac{k_0^2}{4\pi\epsilon_0 r} A_s e^{-i\phi} \left(\mathbf{e}_{ks} B_s + \mathbf{e}_{\theta s} C_s + \mathbf{e}_\phi D_s \right), \tag{7.6}$$

in which k_0 is the vacuum wave number, ϕ is the phase, and we concentrate on a specific mode and solution of the dispersion relation, labelled with s. Tests of theory for radiated fields or L_{eff} must be carried out with due regard for two complications: First, the possibility that in some modes like the whistler, a single group-velocity direction can correspond to between one and three different propagation directions; second, in space measurements, waves can take different paths from a source to a receiver [Sonwalkar et al. 25].

The coefficients A'_s, B'_s, C'_s, D'_s depend on the plasma parameters, the angle θ between the direction of propagation and $\mathbf{B}$, and the dipole moment of the antenna current. When a triangular current distribution is assumed on the antenna, the dipole moment has a magnitude $I(0)L/(4\pi f)$. With this and the dipole-moment direction cosines inserted into the coefficients, (7.6) can be rearranged to read

$$\mathbf{E} = \frac{60\pi I(0)}{r\lambda_0}\frac{L}{2}\left(\mathbf{e}_{ks}A'_sB'_s + \mathbf{e}_{\theta s}A'_sC'_s + \mathbf{e}_{\phi}A'_sD'_s\right) . \qquad (7.7)$$

The Kuehl [18] expression is seen to be of the same form as Jordan's (10–82) in (7.5). The resulting coefficients of the unit vectors are identified with the corresponding L_{eff} values in (7.4). It thus appears that the cold-plasma theory is consistent with the Jordan concept of general L_{eff} involving a vector description. For conformity with the principle of reciprocity, the open-circuit voltage induced on a receiving dipole is presumed to be

$$V_{oc} = \mathbf{E} \cdot \frac{L}{2}\left(\mathbf{e}_{ks}A'_sB'_s + \mathbf{e}_{\theta s}A'_sC'_s + \mathbf{e}_{\phi}A'_sD'_s\right)$$

$$= \frac{L}{2}(A'_sB'_sE_k + A'_sC'_sE_\theta + A'_sD'_sE_\phi) . \qquad (7.8)$$

The coefficients B'_s, C'_s, D'_s are consistent with the polarization ratios given by the cold-plasma theory for the propagation direction. The coefficients also depend on dipole orientation. Let us now see where the hypothesis of (7.8) leads when its predictions are compared with observations.

Sonwalkar and Inan [26] scaled electric and magnetic fields measured on the DE-1 spacecraft to determine L_{eff} of the 200-m wire dipoles flown on that satellite. Two measurement sets were found to give $L_{eff} = 219.8$ and 224.6 m. These evaluations were based on their expression (1) for L_{eff}, an assumption that has had wide use in space radio science. Often, L_{eff} as used in Sonwalkar and Inan [26] has been assigned a value of $L/2$, half the physical length of the dipole, appropriate for vacuum propagation detected by dipoles with $L \ll \lambda$, the wavelength.

The specific plasma parameters and geometry reported in Sonwalkar and Inan [26] allow L_{eff} to be evaluated from present (7.8) as $L/2$ times the multiplier factors $A'_sB'_s, A'_sC'_s, A'_sD'_s$ for the three respective components. In the case of the whistler mode, the k and θ electric field components are in phase with each other and in quadrature with the ϕ component. One axis of the $\mathbf{E}$ ellipse is the vector addition of the E_k and E_θ components and the other is the E_ϕ. All three components are generally needed to determine V_{oc}. The magnitude of the effective length of the dipole is

$$|L_{eff}| = (L/2) \left[(A'_s B'_s)^2 + (A'_s C'_s)^2 + (A'_s D'_s)^2 \right]^{1/2} \equiv (L/2)M . \qquad (7.9)$$

The multiplier M thus defined for the Sonwalkar and Inan [26] cases is plotted as a function of the spin angle ξ in Fig. 7.3 with continuous and broken line. The inset figure shows that the spin axis of the DE-1 spacecraft is approximated to lie along the y axis. The wave vector direction and the plasma parameters take the values reported by the authors. The results are M values that oscillate between about four and zero. The theory thus confirms that a short dipole receiving EM whistler-mode propagation can have $|L_{eff}| \approx L$ as reported by SI. Further comparison is not possible since Sonwalkar and Inan [26] assume that L_{eff} is isotropic whereas it is a vector here, depending on the orientations of both the wave vector and the dipole axis.

Imachi et al. [8] determined L_{eff} of wire dipoles on the Geotail spacecraft by analyzing electric- and magnetic-field amplitudes of chorus emissions. This result has been checked against the present method assuming that these whistler-mode waves propagate along **B**. The stated plasma and working frequencies and a reasonable electron gyrofrequency, 10 kHz, provide the variation of $M(\xi)$ in the dot-dash line in Fig. 7.3. Although the details of the method of determining L_{eff} are not given by Imachi et al. [8], it is found that the range of M for all ξ brackets 1, the value deduced by those authors.

Finally it is noted that (7.5) above simplifies to the familiar expression

$$E_\theta = \frac{60\pi I(0)}{r\lambda_0} \frac{L}{2} \cos \theta' \qquad (7.10)$$

for short dipoles and for working frequencies much higher than the electron characteristic frequencies, i.e., for the vacuum propagation conditions of (7.1). To check this, the multipliers of (7.8) were calculated for the same plasma characteristic frequencies as for Imachi et al. [8] but for the O mode, at a working frequency of 30 MHz. The result in Fig. 7.3 is the dot-dot-dot-dash curve, labelled "HF", showing M when the θ term is the only term. The k has been put along the z axis. For the particular case of $\xi = 0°, \theta = 0°$, for which the propagation direction is at right angles to this vacuum dipole axis, the r, θ, and ϕ multipliers have values of 0, $-0.5i$ and -0.5, respectively. This same set becomes 0, $-0.5i$ and 0.5 when run for the X-mode. The total L_{eff} magnitudes found from the sum of the two solutions are then $(L/2)(0., 1., 0)$, corresponding to the expected linear polarized field at $\theta' = 0°$ from a simple dipole in vacuum, which radiates and detects the θ component only.

7.3.2 Intensity of Auroral Hiss

Correct methodology for the dipole measurement of the absolute magnitudes of EM radio emissions in space includes a correct knowledge of the dipole L_{eff}.

kuehlef6.for, kuehl2.pro, Kuehl2.PS

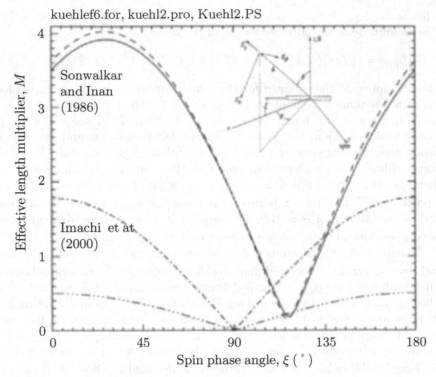

Fig. 7.3. The multiplier M in (7.9) giving the magnitude of L_{eff} as a function of the dipole spin angle ξ, for comparison for three different situations. The continuous and broken lines are the computed M for the two cases reported by Sonwalkar and Inan [26], the dot-dash curve for the Imachi et al. [8] conditions, and the lowest curve (HF) for vacuum conditions

This is illustrated in the history of auroral hiss. In the 1970s, there was interest in understanding the generation of hiss. Earlier work in space and astrophysical contexts had drawn attention to the idea that charged particles moving through magnetoplasma may produce EM radiation through the Cherenkov effect – a particle radiates when its velocity matches the phase speed of EM plasma waves. An important question to be answered was whether a single-particle theory sufficed, wherein the total emission of a hiss source region could be obtained by considering all the particles therein as single incoherent radiators. Or were the absolute hiss levels high enough to require a more powerful process, e.g., a coherent beam-plasma interaction?

Tests were carried out on different satellite data sets. One example was ISIS-I observations of hiss in the dayside cusp at LF and MF [James, 9]. As illustrated in Fig. 7.4, that investigation dealt with passes of the spacecraft through L shells with soft electron precipitation. The observed evolution of hiss spectra was quantitatively consistent with spreading out of radiation

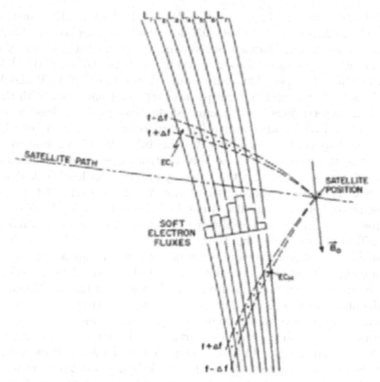

Fig. 7.4. The trajectory of the ISIS-I satellite takes it through magnetic field lines where its soft-particle spectrometer measures soft electron fluxes [James, 9]. Low-frequency hiss is also measured by the sounder receiver as the spacecraft approaches, passes through and recedes from the precipitation. The calculation of total noise power flux spectral density is computed within a bandwidth $f \pm \Delta f$ by tracing resonance-cone group rays from the spacecraft location to establish 2 three-dimensional group cones that intersect the flux sheet and so determine the total volume of energetic particles irradiating the spacecraft for that location and bandwidth (Reprinted with permission of the American Geophysical Union)

from the source flux tubes on resonance-cone group paths. In one case study applying ISIS-I energetic-particle, ambient-density and wave measurements, incoherent-radiation theory was used to calculate theoretical electric-field magnitudes. These were compared with values scaled from receiving-dipole voltages when the dipole was assumed to have an L_{eff} equal to its physical length, L. The ratio of the observed to theoretical field strengths was about 20 dB, which led to the conclusion that the test-particle theory was not applicable.

Similar discrepancies in the same sense were reported in other work, mostly at very low frequencies in the nightside ionosphere. Jorgensen [17] modelled the hiss generation region in the lower ionosphere and computed resulting power flux densities as high as 10^{-14} Wm^{-2} Hz^{-1} at 10 kHz. Gurnett and Frank [6] and Mosier and Gurnett [23] described the morphology of hiss with respect to other auroral-latitude phenomena observed with the Injun-V spacecraft. Power flux densities for strong events were of the order 10^{-11} Wm^{-2} Hz^{-1}. Lim and Laaspere [20] also computed hiss levels at VLF and LF for typical energetic-flux and ambient parameters. For comparison with the foregoing VLF observations, their computed flux intensity was less than 10^{-13} Wm^{-2} Hz^{-1}. Taylor and Shawhan [29] carried out a case study of VLF hiss and energetic-particle spectra observed on Injun V. They came to similar conclusions, that the shape of the hiss spectrum is as expected based on ray paths and geometry but that the intensity of VLF hiss levels calculated was 2 orders of magnitude below observations. Maeda [21] refined the model of the magnetosphere and calculated dayside hiss levels of 10^{-14} Wm^{-2} Hz^{-1} for strong electron fluxes. For a wider summary of observed hiss intensities, see the Table 1 in LaBelle and Treumann [19].

Table 7.1 summarizes the preceding paragraph. The upshot of several parallel investigations of hiss power flux was that observed values exceeded by at least 100 times values obtained from calculation using the incoherent theory [LaBelle and Treumann, 19]. This held true for a variety of different observational circumstances. The power-flux finding was typically followed by the conclusion that the incoherent test-particle theory for radiation was not applicable and that one should look to more powerful wave-particle interaction theory. However, the recent OC work at 25 kHz has now found that for resonance cone waves $L_{eff} = 30\,L$. The ISIS-I data originally were interpreted assuming $L_{eff} = L$. This implies that electric field strengths deduced were 30 times too large, or that the power flux values were $20\log_{10} 30 = 29.5$ dB too high. Since this figure is similar to the disparity between calculated and observed hiss intensities, the rejection of the incoherent theory in the 1970s should now be reconsidered.

Table 7.1. Auroral Hiss Field Strength Measurements and Calculations

Observation	Incoherent Radiation Theory
VLF (0–100 kHz)	
Gurnett and Frank (1972) 0.5–1 $\times 10^{-11}$	Jorgensen (1968) 10^{-14}
Mosier and Gurnett (1972)	
$>1,5 \times 10^{-11}$	Lim and Laaspere (1972) 10^{-13}
	Taylor and Shawhan (1972)
	0.5 $- 3.2 \times 10^{-13}$
	Maeda (1975) 10^{-14}
	Ratio 11–32 dB
LF-MF (100–500 kHz)	
James (1973) 0.81–4.6 $\times 10^{-17}$	James (1973) $0.21 - 8.8 \times 10^{-19}$
	Ratio: 17–26 dB

That auroral hiss generation in the low-altitude auroral oval is correctly described with an incoherent theory may gain some credence when recent work on particle acceleration is consulted. Auroral electron acceleration is thought to result from the action of the parallel electric field of shear Alfvén waves and to occur at altitudes between about 3500 km and a few earth radii [Paschmann et al., 24, , pp. 187-189]. The low-earth-orbit-satellite observations cited above were all made below 3100 km, that is, below the acceleration region. Electron fluxes in that region are expected to have lost their Alfvén-imparted coherence through beam-plasma interactions. Various instability mechanisms like those producing hiss or auroral kilometric radiation may have the dominant role. In the former case, wave growth and saturation will happen over vertical scales of the order of ten plasma periods multiplied by particle velocities, say, several kilometers. What would then emerge from the bottom of the acceleration region would be turbulent electron fluxes with spatial coherence scales corresponding to the saturation spectrum of the created waves. Paschmann et al. [24] state that the beam-plasma process causes quasilinear plateauing of the beam at the intermediate altitudes studied by James [9] and Taylor and Shawhan [29]. The effect of plateauing is presumed to be one of shifting beam electrons to lower energies. However, at LF and MF, the hiss ray directions at frequencies not far below f_p are not far from horizontal. Hence it can be argued that the particles measured by satellite are indeed responsible for the hiss observed on the same spacecraft, especially inside the precipitation L shells.

The presence of turbulence would mean that overall the descending flux would radiate incoherently but that the individual radiating entities would not be individual electrons but a three-dimensional ensemble of micro-structures. What might be the dimensions of a hiss-radiating microstructure? First of all, it is noted that an upper-limit on the cross-field dimension of the emission source is given by the perpendicular scale size of the Alfvén wave, which is of the order of the electron inertial length, 1 km. Therefore already the acceleration concept implies incoherence because the observed source regions in the auroral oval must contain a large number of such sources. Further, if thermal damping of whistler-mode waves affects resonance-cone waves at refractive indices >1000, then saturated MF hiss at 300 kHz will have wavelengths of around 1 m. 1 m is probably a lower limit on structure sizes because the Debye length has approximately this order of magnitude in the acceleration region. Phase-space electron holes as small as a Debye length have been observed [Ergun et al., 5]. Assume that the hiss source is a flux of 1-m structures. With a differential flux density of 10^7 cm^{-2} s^{-1} ster^{-1} eV^{-1} in the dayside cusp, the density of particles in the beam is about 10^7 m^{-3}. This implies that the electric charge q in the incoherent radiated power expression should be 10^7 times a single-electron charge and that the number of the structures irradiating any observer must be reduced by 10^7. In incoherent emission theory, the spectral power density varies as the square of the charge of the radiating "particle". Hence to suppose that one observes radiating microstructure leads to a total volume emission of $(10^7)^2/10^7$, a $+70\,$dB correction in the absolute emission. The hypothesis of microstructure is discarded because the observation-theory disparity remains as large as before, albeit in the opposite sense. Hence the view is maintained that hiss in the topside ionosphere is incoherent radiation from a downward electron flux below the acceleration region in the auroral magnetosphere.

7.3.3 Intersatellite Whistler-Mode Propagation Near a Resonance Cone

Results from ISIS intersatellite propagation studies resemble the strong transmissions in the OC two-point experiment shown in Fig. 7.2. Whistler-mode transmission at 480 kHz was achieved between ISIS-I and ISIS-II 73-m sounder dipoles during a rendezvous campaign [James, 10]. ISIS I was in a 570 × 3520-km polar orbit, while ISIS II had a nearly circular polar orbit around 1400 km altitude. The satellites were commanded into the "Alternate Mode" wherein the sounders cycled between conventional swept-frequency sounding and periods of about 20 s when the sounder pulse carrier frequency was held fixed at 480 kHz. The rendezvous occurred at high latitude where the two orbital planes intersected. Transmissions in both directions were obtained on several rendezvous.

In Fig. 7.5, a 4-s data excerpt from a rendezvous on 4 April 1974 displays representative pulses received. On the left are pulses emitted by ISIS II and received on ISIS I. The roles are reversed on the right side. Both satellites had 400-W emitters and produced input voltages at the other receiver of about 70 µV. Pulses about 100 µs long upon emission are seen to be dispersed spectacularly by 100 times and more at reception. All pulses in the 20-s interval from which this set is taken exhibit components with delays of the order of 10 ms. The pulses in 28 to 30 s also have sharp rising edges, on account of favourable antenna orientation at that time. Pulse stretching of this order was clearly maintained when the satellite separation lay well inside the group resonance cone for 480 kHz.

Ray-optics studies were carried out based on three-dimensional electron density distributions obtained from the the real-height analysis of the swept-frequency ionograms recorded during the rendezvous. Iterative techniques were used to find the delays associated with direct inter-satellite propagation. Good agreement was found for the time of the sharp rising edges of pulses that had such, like those in Fig. 7.5. It was demonstrated that the pulse stretching was not caused by magnetoionic dispersion of the different frequency components of the pulse. The signal delays required propagation near the resonance cone. Hence, the conclusion was that wave packets could take a great variety of scattering paths between the transmitter and the receiver, because the stretching was seen preferentially in the high-latitude irregular ionosphere. It was concluded that highly dispersed pulses arose from

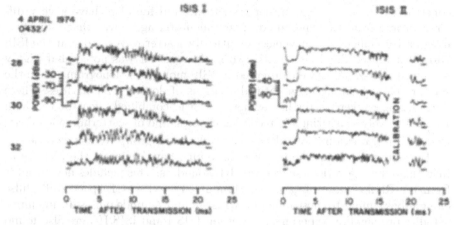

Fig. 7.5. Amplitude scans of whistler-mode pulses received reciprocally during a rendezvous of the ISIS-I and -II satellites [James, 10]. The pulse carrier RF was 480 kHz and the spacecraft separation was about 200 km (Reprinted with permission of the American Geophysical Union)

scattering by density irregularities along and inside the 480-kHz resonance cones through both spacecraft.

An additional feature that supported the existence of quasi-electrostatic short-wavelength waves was the appearance of "primary" and "secondary" components of pulses. It was argued that this structure appeared because both the emitting and receiving dipoles had current distributions that favoured the emission or reception of certain wavelengths at a given frequency. These were only observed when each satellite lay near the 480-kHz group resonance cone of the other. The range of refractive indices, up to about 30, needed to explain the observed group delays correspond to wavelengths down to 20 m. This pointed to resonant- and anti-resonant-length relations between the now-finite-length dipoles and the waves, favouring the transmission of certain wave numbers and blocking others, hence the pulse envelope.

7.3.4 Intersatellite Slow-Z-Mode Propagation
Near a Resonance Cone

There have been observations of apparently intense waves propagating near the upper-oblique resonance cone in CMA 3. Another two-point experiment during the ISIS rendezvous campaign produced observations of strong, highly dispersed Z-mode pulses that appeared to carry an unusually large total energy [James, 11]. In fact the Z pulses resembled the inter-satellite whistler-mode pulses just described, with comparable length and strength. Three-dimensional ray-optics was applied to find the phase paths of many pulse spectral components at the receiver. The received pulse envelope was then constructed by an inverse Fourier transform, and found to have a very different shape from that measured. The conclusion again was that ionization irregularities efficiently quasi-electrostatically scatter waves and that the ISIS 73-m distributed dipole is sensitive to such waves. It was noted that highly dispersed slow-Z pulses were observed in ISIS monostatic ionograms when the satellite passed through high-latitude regions of density irregularities. These Z results are taken as further evidence of the combined effect of the high L_{eff}, effectively sensitizing the receiving dipoles, and of scattering ionospheric irregularities, providing long-delay paths along the group resonance cone.

Benson et al. [2] in Chap. 1 of this volume have reviewed the Z mode as investigated using various spaceborne RF sounders. This includes new insights from the Radio Plasma Imager instrument on the IMAGE spacecraft. Pulse elongations similar to those found for whistler and Z- mode pulses during inter-satellite propagation experiments between ISIS I and ISIS II were also found in recent whistler and Z-mode sounding experiments on IMAGE [Sonwalkar et al., 27]. These authors have put forward an interpretation similar to that given by [James, 10, 11], i.e. scattering by irregularities.

7.4 Conclusion

The success of the cold-plasma CW theory for explaining dipole results from OEDIPUS commends wider use of that theory for supplying the effective lengths of distributed dipoles for space-borne reception. The application of the principle of reciprocity brought over from the vacuum short-dipole theory appears useful for EM propagation. However the particular results about resonance-cone quasi-electrostatic propagation indicate a need for caution in the interpretation of signal levels.

The importance in space physics of the correct dipole L_{eff} concept remains. It appears possible that hiss in the topside ionosphere is correctly described, after all, as incoherent radiation from a thermalized flux left over from acceleration in the auroral magnetosphere. However, the interpretation of whistler-mode absolute intensities today is complicated by the clear evidence for scattered resonance-cone propagation. In the computation of power flux densities from sources of emissions like hiss, which can have horizontal dimensions of hundreds of kilometers, there is now the added geometrical complication of dealing with scatter by irregularities. The scattering of EM VLF whistler-mode waves by field-aligned small-scale irregularities into waves propagating near the cone has been explained as a consequence of either linear mode coupling or nonlinear mechanisms [Bell et al., 1, and references therein]. Whichever the case, the complete account of radiative transfer between source and observer brings a new challenge in understanding RF emissions of the atmosphere. This may be true not only for whistler-mode hiss but for other phenomena like hiss in the slow-Z mode.

Given the state of knowledge about dipoles in 1978, the sustained pulses observed with topside sounders seemed to imply that great power was broadcast by the sounders along the oblique resonance cones. This inference is now tempered by the notion of large sensitivity of the receiving dipole to quasi-electrostatic resonance-cone waves, as represented by a big L_{eff} in the distributed dipole theory.

References

[1] Bell, T.F., R.A. Helliwell and M.K. Hudson: Lower hybrid waves excited through linear mode coupling and the heating of ions in the auroral and sub-auroral magnetosphere, *J. Geophys. Res.* 96, 11,379–11,388 (1991).

[2] Benson, R.F., P.A. Webb, J.L. Green, D.L. Carpenter, V.S. Sonwalkar, H.G. James and B.W. Reinisch: Active wave experiments in space plasmas: the Z mode. In: Proceedings Volume for the Ringberg Workshop, edited by J. LaBelle and R. A. Treumann, Springer Lecture Notes in Physics, Springer New York-Heidelberg (2005).

[3] Chugunov, Yu.V.: Receiving antenna in a magnetoplasma in the resonance frequency band, *Radiophys. Quantum Electron.* 44, 151–160 (2001).

[4] Chugunov, Yu.V., E.A. Mareev, V. Fiala and H.G. James: Transmission of waves near the lower oblique resonance using dipoles in the ionosphere, *Radio Sci.* 38, 1022, doi:10.1029/2001RS002531 (2003).

[5] Ergun, R.E., C.W. Carlson, J.P. McFadden, F.S. Mozer, L. Muschietti, I. Roth and R.J. Strangeway: Debye-scale plasma structures associated with magnetic-field-aligned fields, *Phys. Rev. Lett.* 81, 826–829 (1998).

[6] Gurnett, D.A. and L.A. Frank: VLF hiss and related plasma observations in the polar magnetosphere, *J. Geophys. Res.* 77, 172–190 (1972).

[7] Horita, R.E. and H.G. James: Two point studies of fast Z mode waves with dipoles in the ionosphere, *Radio Sci.* 39, doi:10.1029/2003RS002994 (2004).

[8] Imachi, T., I. Nagano, S. Yagitani, M. Tsutsui and H. Matsumoto: Effective lengths of the dipole antennas aboard Geotail spacecraft. In: Proc. 2000 Int. Sympos. Antennas and Propagat, (ISAP2000), IEICE of Japan, Tokyo 819–822 (2000).

[9] James, H.G.: Whistler-mode hiss at low and medium frequencies in the dayside-cusp ionosphere, *J. Geophys. Res.* 78, 4578–4599 (1973).

[10] James, H.G.: Wave propagation experiments at medium frequencies between two ionospheric satellites, 2, Whistler-Mode pulses, *Radio Sci.* 13, 543–558 (1978).

[11] James, H.G.: Wave propagation experiments at medium frequencies between two ionospheric satellites, 3, Z mode pulses, *J. Geophys. Res.* 84, 499–506 (1979).

[12] James, H.G.: Electrostatic resonance-cone waves emitted by a dipole in the ionosphere, *IEEE Trans. Antennas Propagat.* 48, 1340–1348 (2000).

[13] James, H.G.: Electromagnetic whistler-mode radiation from a dipole in the ionosphere, *Radio Sci.* 38, 1009, doi:10.1029/2002RS002609 (2003).

[14] James, H.G.: Slow Z-mode radiation from sounder-accelerated electrons, *J. Atmos. Solar-Terr. Phys.* 66, 1755–1765 (2004).

[15] James, H.G. and W. Calvert: Interference fringes detected by OEDIPUS C, *Radio Sci.* 33, 617–629 (1998).

[16] Jordan, E.C.: Electromagnetic Waves and Radiating Systems, Prentice-Hall, Englewood Cliffs NJ (1950).

[17] Jørgensen, T.S.: Interpretation of auroral hiss measured on POGO-2 and at Byrd Station in terms of incoherent Cerenkov radiation, *J. Geophys. Res.* 73, 1055–1069 (1968).

[18] Kuehl, H.H.: Electromagnetic radiation from an electric dipole in a cold anisotropic plasma, *Phys. Fluids* 5, 1095–1103 (1962).

[19] LaBelle, J. and R.A. Treumann: Auroral radio emissions, 1. Hisses, roars and bursts, *Space Sci. Rev.* 101, 295–440 (2002).

[20] Lim, T.L. and T. Laaspere: An evaluation of the intensity of Cerenkov radiation from auroral electrons with energies down to 100 eV, *J, Geophys. Res.* 77, 4145–4157 (1972).

[21] Maeda, K.: A calculation of auroral hiss with improved models for geoplasma and magnetic field, *Planet. Space. Sci.* 23:843–865 (1975).

[22] Mareev, E.A. and Yu.V. Chugunov: Excitation of plasma resonance in a magnetoactive plasma by external source, 1, A source in a homogeneous plasma, *Radiophys. Quantum Electron.* 30, 713–718 (1987).

[23] Mosier, S.R. and D.A. Gurnett: Observed correlations betwen auroral and VLF emissions, *J. Geophys. Res.* 77, 1137–1145 (1972).

[24] Paschmann, G., S. Haaland and R.A. Treumann: Auroral plasma physics, *Space Sci. Rev.* 103, No. 1–4 (2002).

[25] Sonwalkar, V.S.: The influence of plasma density irregularities on whistler mode wave propagation. In: Proceedings Volume for the Ringberg Workshop, edited by J. LaBelle and R. A. Treumann, Springer Lecture Notes in Physics, Springer New York-Heidelberg (2005).

[26] Sonwalkar, V.S. and U.S. Inan: Measurements of Siple transmitter signals on the DE 1 satellite: wave normal direction and antenna effective length, *J. Geophys. Res.* 91, 154–164 (1986).

[27] Sonwalkar, V.S., D.L. Carpenter, T.F. Bell, M. Spasojevic, U.S. Inan, J. Li, X. Chen, A. Venkatasubramanian, J. Harikumar, R.F. Benson, W.W.L. Taylor and B.W. Reinisch: Diagnostics of magnetospheric electron density and irregularities at altitudes <5000 km using whistler and Z mode echoes from radio sounding on the IMAGE satellite, *J. Geophys. Res.* 109, A11212, doi:10.1029/2004JA010471 (2004).

[28] Stix, T.H.: Waves in Plasmas, American Institute of Physics, New York (1992).

[29] Taylor, W.W.L. and S.D. Shawhan: A test of incoherent Cerenkov radiation for VLF hiss and other magnetospheric emissions, *J. Geophys. Res.* 79, 105–117 (1974).

[30] Wang, T.N.C. and T.F. Bell: VLF/ELF radiation patterns of arbitrarily oriented electric and magnetic dipoles in a cold lossless multicomponent magnetosplasma, *J. Geophys. Res.* 77, 1174–1189 (1972).

8

Mode Conversion Radiation
in the Terrestrial Ionosphere and
Magnetosphere

P.H. Yoon[1], J. LaBelle[2], A.T. Weatherwax[3], and M. Samara[2]

[1] Inst. for Physical Sci. and Tech., Univ. of Maryland, College Park, Maryland
   yoonp@ipst.umd.edu
[2] Dartmouth College, Dept. of Physics and Astronomy, Hanover, New Hampshire
   jlabelle@einstein.dartmouth.edu
   marilia@aristotle.dartmouth.edu
[3] Siena College, Department of Physics, Loudonville, New York
   aweatherwax@siena.edu

Abstract. A significant fraction of the radiation types observed in the Earth's ionosphere and magnetosphere can be classified as mode-conversion radiation, in that they result from generation of electrostatic waves by unstable particle populations followed by conversion of some fraction of the wave energy to electromagnetic modes which then propagate relatively long distances. In particular, we address the complex frequency structure observed in terrestrial mode conversion radiation. Theory suggests that electrostatic eigenmodes trapped within source-region density structures, analogous to the quantum energy levels of hydrogen atom potential well, may account for the observed fine frequency structure. We review observational results, provide a synthesized, generalized version of the appropriate theory extending existing theoretical work, and assess the current state of comparison between the theoretical predictions and the observations. Understanding the mode-conversion radiation processes in near-Earth geospace may significantly enhance interpretations of observations of similar radiations from more remote space plasma environments such as distant magnetospheres, the solar atmosphere, and astrophysical plasmas.

Key words: Mode conversion radiation, Langmuir waves, wave trapping, non-thermal radiation generation, plasma radiation

8.1 Introduction

Wave phenomena in the Earth's environment or in those of other planets are often characterized by discrete frequency structures. For example, recent satellite and rocket observations show that electric fields directly measured in the auroral ionosphere and magnetosphere are composed of coherent structures,

P.H. Yoon et al.: *Mode Conversion Radiation in the Terrestrial Ionosphere and Magnetosphere*,
Lect. Notes Phys. **687**, 211–234 (2006)
www.springerlink.com

such as solitary waves, wave packets, or eigenmodes imposed by preexisting density structures. Satellite, rocket, and ground-based observations show that remotely-sensed electromagnetic (EM) radiations, such as Earth's auroral kilometric radiation (AKR), auroral roar emissions from the ionosphere, or terrestrial continuum radiation are characterized by frequency fine structure.

Observations of EM radiation in high frequency wave modes can contribute much to understanding Earth's environment, even though the energy density of this radiation is generally lower than that of the particles or lower-frequency and static electromagnetic fields. Through wave-particle interactions high frequency EM waves mediate the exchange of energy between larger energy reservoirs, such as different populations of particles. Radiation also provides a mechanism to transport energy over long distances and can play a cumulative role in energy balance in a system even if the instantaneous energy content in the waves is relatively small. Finally, and perhaps most significantly, EM radiation carries information about the plasma in its source region and the plasma through which it propagates, allowing for passive remote sensing of densities, temperatures, distribution functions, and irregularities, in some cases using inexpensive ground-based measurements. High frequency waves can be particularly effective for this purpose, because of their large information content. For example, early magnetospheric studies used ground-based measurements of whistler waves at 1–10 kHz to determine electron density in the outer magnetosphere. For a more recent example, Weatherwax et al. [62] explored the parameter space for excitation of auroral radio emissions at different electron cyclotron harmonics and described how coincident observations of these two types of emissions can provide a means of remote sensing the F-region electron density scale height.

Radiation mechanisms can be divided into two general types: direct and indirect. Direct emission occurs when the particle distribution function is unstable to electromagnetic wave modes, resulting in direct excitation of these wave modes. Examples in the Earth's auroral ionosphere are Cherenkov amplification of whistler mode waves to generate auroral hiss and cyclotron maser emission excited by "horseshoe" electron distribution functions to generate intense X-mode AKR. These direct radiation mechanisms produce the most intense high frequency EM waves in the Earth's environment.

In contrast, indirect emission occurs when the particle distribution function is unstable to electrostatic modes, which, after excitation to some level, convert to electromagnetic radiation via either linear or nonlinear mechanisms. Therefore these EM waves are often referred to as "mode conversion radiation." Examples in the Earth's environment are auroral roar emissions near electron cyclotron harmonics in the auroral ionosphere and the several types of terrestrial continuum radiation in the outer magnetosphere. These emissions tend to be weaker in intensity than the direct emissions, because of the extra step required in producing them from the unstable particle distribution function, and the resulting inefficiency.

Much previous literature has concentrated on the direct emissions, auroral hiss and AKR in particular. However, despite their relatively weaker intensity, mode conversion radiation effectively conveys information about the plasma in its source region and in the regions through which it propagates. In fact, mode conversion radiation is often highly structured in frequency and time, and understanding this complex fine structure is an outstanding problem in space physics. This paper explores the hypothesis that frequency structure in mode conversion radiation is related to plasma density irregularities in the source region of the radiation. Section 8.2 reviews the observational evidence for fine structure in several example types of mode conversion radiation in the Earth's environment. Section 8.3 describes a theory for the excitation of high frequency electrostatic waves in an inhomogeneous source plasma. The predicted frequency structure of the electrostatic waves is then compared to that of the observed radiation.

8.2 Observations

8.2.1 Auroral-Zone Mode-Conversion Radiation

The Earth's auroral zone is characterized by highly nonthermal electron distribution functions which excite a variety of different wave modes in different altitude ranges. At altitudes of about 5000 km, parallel electric fields accelerate cold plasma sheet electrons into a beam of 1–10 keV; the beam evolves into a ring distribution which upon reflection in the converging magnetic field geometry has a loss cone in the upward going side, as described quantitatively by many authors and models going back to Chiu and Schulz [8]. This energetic electron population overlaps with a cold ionospheric electron population which is roughly Maxwellian with temperature of a few tenths of an eV. As mentioned above, this distribution function gives rise, in various altitude ranges, to direct emission of EM waves such as auroral hiss and AKR.

However, a zeroth-order effect of these beam and loss-cone type electron distribution functions is excitation of electrostatic waves at the plasma frequency resonance parallel to the background magnetic field ($k_{||} >> k_{\perp}$), or at the upper-hybrid resonance perpendicular to the background magnetic field ($k_{\perp} >> k_{||}$), or at the oblique high-frequency resonances with $f_{pe} < f < f_{uh}$. When these excitations are followed by conversion to one of the electromagnetic modes (X or O), the resulting mode conversion radiation can propagate long distances to distant space-borne or ground-based receivers.

Rocket Observations

Indeed, almost every suitably instrumented experiment penetrating the auroral region records plasma waves in the band between the electron plasma frequency and the upper hybrid frequency [1, 15, 58] as reviewed by LaBelle

[34]. Early and recent experiments indicated the bursty nature of these auroral Langmuir waves [see, e.g., Ergun, 9]. Recent experiments include detailed comparisons with electron distribution function measurements. For example, McFadden et al. [43] report on waves near f_{pe} and show evidence that these originate from a Landau resonance with the simultaneously measured electron beams. Similarly, Samara et al. [53] show evidence that waves near f_{uh} result from cyclotron resonance with the measured electrons in which electrons transfer energy into the waves. Ergun et al. [10] and Kletzing et al. [31] use wave-particle correlators to measure directly the interaction giving rise to the auroral Langmuir waves.

Recent experiments also show the frequency structure of these high frequency waves and their conversion to electromagnetic modes. Beghin et al. [2] show high frequency waves near the plasma frequency observed in the auroral zone under a variety of conditions using wave receivers on the Aureol-3 satellite. For underdense conditions ($f_{pe} < f_{ce}$), the wave structure is predominantly at the plasma frequency and below, while for overdense conditions ($f_{ce} < f_{pe}$) the waves are at and above the plasma frequency. The waves below f_{pe} in the underdense case are attributed to mode conversion of Langmuir waves into whistler waves.

McAdams et al. [40] observed similar phenomena during an auroral rocket flight, PHAZE-II in 1997, which penetrated both underdense and overdense plasma conditions. Their wave receiver had unprecedented resolution achieved by full sampling of the waveforms. With this resolution, McAdams et al. [40] observed that in the underdense case, the waves just below the plasma frequency form multiple constant-frequency "bands". These bands are of long-duration, up to tens of seconds (tens of km distance). Sometimes individual band structures are punctuated by an intense burst of emission at the plasma frequency, when the rocket crosses the location where the band frequency matches the plasma frequency. McAdams et al. [40] attribute these bands to mode conversion to whistler mode, as suggested earlier by Beghin et al. [2]. The banded structure is a natural consequence of the bursty intermittent nature of the causative Langmuir waves.

In the overdense case, McAdams et al. [41] report completely different and more complex wave structure. As reported by Beghin et al. [2] the waves occur at and above f_{pe} in this case. However, the high resolution shows that they consist of multiple discrete features, with bandwidths less than a few hundred Hz, separated by the order of one kilohertz. Sometimes up to four or five multiplets are observed, but more often doublets. Figure 8.1a shows a spectrogram of the high frequency electric fields observed during a 3 s time interval of the PHAZE-II experiment. The distinct, rapidly varying wave cutoff corresponds to the plasma frequency as argued by McAdams et al. [41]. Just above the plasma frequency, structured waves with amplitudes of $\sim 0.7\,\mathrm{mV/m}$ occur. Where their frequency nearly matches f_{pe}, their amplitude maximizes, and they are clearly associated and appear trapped in electron density depletions. Further away from these electron density depletions, where their frequency

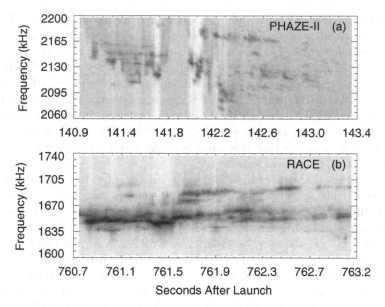

Fig. 8.1. (a) High frequency electric fields observed during the PHAZE-II experiment show multiple discrete frequency features, with bandwidths less than a few hundred Hz, separated by the order of one kHz. (b) High frequency electric fields observed by RACE show similar structures but with different frequency-time variations (Reprinted with permission of the American Geophysical Union)

exceeds the plasma frequency by ten percent or so, they appear as descending tones with short durations of order 100 ms. Hundreds of descending tones occur during a 10 s interval. By the time the rocket travelled 10 s (or about 10 km) away from the region with the electron beam and the electron density irregularities, the waves died out to an undetectable level.

Since 1997, several further auroral rocket experiments have included full-waveform measurements of high frequency plasma waves. Figure 8.1b shows a spectrogram of the high frequency electric fields observed during 10 s interval on one of these experiments, RACE, launched in 2002. The Langmuir waves in the overdense auroral ionospheric plasma show many features in common with those previously observed by McAdams et al. [41]. They are associated with the combination of auroral electron beam and electron density structure evidenced, as before, by variations in the wave cutoff associated with the plasma frequency. Again the waves are most intense where their frequency nearly matches f_{pe}, and there is evidence for trapping of the waves in density depletions in this region. Though not shown in Fig. 8.1b, observations from many kilometers away from this source region, where the wave frequency exceeds the local plasma frequency by ten percent or so, the wave intensity is lower, and again by the time the rocket travels about 10 km from the source region, the waves are undetectable [Samara et al., 52]. The detailed wave structure consists of discrete features separated by about 1–2 kHz. The discrete features often occur in pairs. However, unlike the previous case, features with decreasing frequency are not observed exclusively. Several examples of features with increasing frequency are shown in Fig. 8.1b. In addition, some examples last far more than 100 ms. In one case not shown in Fig. 8.1b, a discrete feature is intermittent but continuously detectable for many seconds. McAdams et al. [41] labelled their observed features "chirps" based on the consistent decreasing frequency and short 100 ms lifetime. However, the recent observations show that this name is overly restrictive, since a far broader variety of frequency variations occurs.

As mentioned above, auroral HF waves are detected throughout the band between the plasma and upper hybrid frequencies, and recent experiments include full waveform measurements of the waves near the upper-hybrid frequency, complementing the observations near f_{pe} cited above. Figure 8.2, from Samara et al. [53], shows a spectrogram of waves at and just below f_{uh} in an active aurora penetrated by the HIBAR rocket, launched in 2003. Two bursts of highly structured waves occur where the upper hybrid frequency matches the electron cyclotron harmonic ($f_{uh} \approx 2f_{ce}$). The waves lie at and just below f_{uh}, which is indicated by the weak wave cutoff and confirmed by a second wave cutoff at the plasma frequency (not shown). In the first wave burst, of duration 0.5 s and total amplitude up to 10 mV/m, the waves occur in 2–4 visible bands with bandwidth of about 1–5 kHz separated by about 8 kHz. Each of these bands is composed of several sub-bands with separation of order 1–2 kHz. The entire band structure decreases in frequency with increasing time and altitude, at a rate of 21 kHz/s, approximately ten times too high to be explained by the decrease of the electron cyclotron harmonic frequency with increasing altitude. The second wave burst is of shorter duration. In the second burst, the emission is also banded, but the separations of the bands are of order 2–4 kHz. It is not clear whether the band separation in the second event should be compared to the larger separation of the principal bands in

Fig. 8.2. A spectrogram highlighting waves just below f_{uh} in an active aurora penetrated by the HIBAR rocket [from 53]. The HIBAR observations suggest trapping of the upper hybrid waves in electron density enhancements, as discussed in Sect. 8.3 (Reprinted with permission of the American Geophysical Union)

the first event, or whether it should be identified with the sub-band structure in the first event, which is about the same size. The electron density observations from HIBAR suggest that these upper hybrid waves are trapped in electron density enhancements: the second wave burst clearly corresponds to an enhancement, whereas the first wave burst corresponds to a "shoulder" in the electron density profile, a signature which may result from the one-dimensional sampling of a density enhancement by the rocket trajectory. Earlier lower-resolution rocket observations of auroral upper hybrid waves near f_{uh} also suggested this trapping, supported by careful consideration of the wave dispersion relation by Carlson et al. [6].

Ground-Based Observations

Strong evidence suggests that the structured waves excited near the upper hybrid frequency under the matching conditions $f_{uh} \approx 2f_{ce}$ or $f_{uh} \approx 3f_{ce}$ have converted to O mode EM radiation and been detected remotely with satellites overflying the auroral zone [e.g., James et al., 24] and with ground-based observatories at northern and southern auroral zone sites [e.g., Kellogg and Monson, 29, 30] and [Weatherwax et al., 59, 61].

Using a topside ionospheric sounder in a passive configuration, James et al. [24] observe LO radiation emerging from sources in the topside ionosphere, and by ray-tracing the observed LO radiation from the satellite to its source, into the topside ionospheric density profile measured with periodic active soundings, they conclude that the radiation is generated near the upper hybrid frequency at the matching conditions.

Similarly, Hughes and LaBelle [22] ray-trace the observed radiation from a ground-based direction-finding antenna array into the bottom-side ionospheric density profile measured with an incoherent scatter radar, and they likewise conclude that the radiation is generated near the upper hybrid frequency at the matching condition. Theoretical work shows that Z-modes at the $f_{uh} = Nf_{ce}$ matching condition can be excited in auroral conditions [see, Kaufman, 28] and [Yoon et al., 64]. Furthermore, based on such density structures observed at times of auroral roar emissions, Yoon et al. [65, 66] show that for a narrow range of frequencies and initial wave phase angles, trapped Z mode can be converted to O mode via the "Ellis" radio window. Weatherwax et al. [62] also present a detailed maser instability analysis of upper-hybrid, Z, and X mode wave excited along an auroral field line, discussing the generation of topside and bottomside radiation within the context of the observations of James et al. [24], Benson and Wong [3], and Hughes and LaBelle [8].

In the case of the ground-level observations, called auroral roar, deployment of high-bandwidth receivers with large data rates reveals that the mode conversion radiation is composed of complex patterns of discrete frequency features as shown by LaBelle et al. [35] and Shepherd et al. [55]. Figure 8.3 shows an example of this fine structure, recorded April 21, 2003, at South Pole Station. The most intense of these emissions have amplitudes up to $500\,\mu\mathrm{V/m}$.

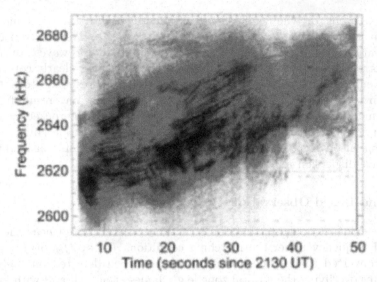

Fig. 8.3. 2595–2685 kHz spectrogram recorded at South Pole Station at 2130 UT on April 21, 2003, showing auroral roar with complex fine structure

As known from previous studies [35, 55] auroral roar consists of multiple fine structures, with width as narrow as <10 Hz, which shift upward and/or downward in frequency with time in complex patterns, as seen in Fig. 8.3. Shepherd et al. [55] report that among features with significant frequency drifts, downward drifts are more common than upward drifts. The phenomenon is highly nonstationary, with periods of coherent multiplet structures lasting typically only a second or so; this characteristic is supported by direction finding data suggesting that multiple sources contribute to auroral roar events [Hughes and LaBelle, 22].

In addition to auroral roar emissions, a second type of HF auroral emission is observable at ground-level. Auroral MF-burst is a broadband impulsive emission occurring for a few minutes at substorm onsets and correlated with VLF/LF impulsive auroral hiss [36, 61], as reviewed by LaBelle and Treumann [37]. MF-burst has a typical bandwidth of about 1 MHz, extending for example from 1.5–2.5 MHz or from 3–4.5 MHz, or sometimes both. The polarization of MF-burst appears consistent with LO-mode ionospheric propagation [56], which suggests that it may be generated by a mode conversion mechanism. However, as yet there is no convincing mechanism explaining the generation of auroral MF-burst.

8.2.2 Magnetospheric Mode-Conversion Radiation

The Earth's magnetosphere also contains ample sources of energy for generation of HF EM radiations, most notably the trapped electrons of the radiation belts. These energetic particles with energies of kcVs to MeVs have distribution functions characterized by a loss cone and are superposed on a background plasmaspheric electron population with eV energies. This situation leads readily, for example, to excitation of oblique electrostatic waves near the upper hybrid frequency. Another region of the Earth's environment favorable for HF wave generation is the electron foreshock region upstream of the Earth's bow shock, where reflected electrons form a counterstreaming beam superposed on the background solar wind plasma. In this region and in the analogous region upstream of the Jovian bow shock, intense waves near the plasma frequency are observed by Filbert and Kellogg [11] and Gurnett et al. [18]. Regions of magnetic reconnection on the dayside magnetopause and in the magnetotail produce highly nonthermal electron distribution functions with complex beam and ring features, potentially unstable to high frequency waves. In another article in this issue, Pottelette and Treumann [50] explore how the electron distribution functions resulting from these regions generate EM radiation through a direct mechanism at some distance away from the reconnection sites.

Satellites in the magnetosphere regularly detect electrostatic waves near the plasma or upper hybrid frequency, and these are in fact extensively used for passive plasma diagnostics. The emission frequency determines the electron density, as exploited for example by Carpenter and Anderson [7] to perform a statistical study of plasmaspheric densities. In cases where the emission can be assumed in equilibrium with a quasi-stable nonthermal electron distribution, the line width and line amplitude may be related to other moments of the electron distribution function such as temperature [Lund et al., 38, 39] in analogy to passive wave diagnostic techniques [23, 48] developed and extensively applied to quasi-thermal plasmas in the solar wind [see the review by Meyer-Vernet and Perche, 47].

However, a significant consequence of these emissions is the generation via mode conversion of terrestrial continuum radiation, an electromagnetic emission which lies in the same frequency range, roughly 1 kHz to 1 MHz. Originally reported by Gurnett [17], this radiation is commonly divided into two components: at low frequencies, up to tens of kHz, the emissions are generally featureless, exhibit a lower cutoff near the local plasma frequency, and often exhibit a fairly distinct upper cutoff which is attributed to the maximum plasma frequency in the magnetosheath. This component of the continuum radiation is believed to be trapped in between the relatively high density plasmasphere on the inside and the relatively high density magnetosheath on the outside [Gurnett and Shaw, 16], and the waves in reflecting back and forth lose whatever fine structure might be indicative of their source and appear as a broadband continuum. At high frequencies, up to hundreds of kHz or even 1000 kHz, the emission is referred to as escaping continuum, and it is observed to be more time variable, frequency and time structured [Kurth et al., 32], and is observed outside the magnetopause and deep in the magnetotail [see, e.g., Steinberg et al., 57] as well as inside the magnetosphere. This component has frequency too high for containment within the magnetosheath and simply escapes. Both components of the continuum are believed to result from mode conversion of the upper-hybrid waves generated in the inner magnetosphere in the presence of the trapped energetic electrons and the cold plasmasphere, as proposed by Jones [25]. The mode conversion mechanism is not known, although the strong density gradient on the plasmapause makes linear conversion on the density gradient, via refraction into the "Ellis window," a plausible mechanism, as explored analytically by Jones [25, 26] and later using the HOTRAY code by Horne [20], although many other mode conversion mechanisms have been put forth, both linear and nonlinear [e.g., Rönnmark, 51] and [Melrose, 44].

The early observations of the escaping continuum showed frequency structure, including bands separated by something close to the electron gyrofrequency in the source region [e.g., Kurth et al., 32, 33]. These observations generated interest in explaining this structure in terms of, for example, enhancement of upper hybrid wave growth at certain matching conditions relating to the electron gyrofrequency, mode conversion or interaction of the so-called $n + 1/2$ electrostatic waves lying between electron gyroharmonics in the outer magnetosphere, or dependence of the mode conversion process on wave frequency [Horne, 21]. Recently, identification of other components of continuum radiation have been reported: for example, enhanced continuum, a type occurring in the midnight to dawn sector associated with substorms [12, 27], and kilometric continuum, a distinctive subset of the escaping continuum at 100s of kHz frequency which has special beaming characteristics restricting it to low latitudes [Hashimoto et al., 19]. Recently, Green et al. [13, 14] report evidence from the IMAGE satellite showing that kilometric continuum is generated in density "bite-out" indentations in the plasmasphere which co-rotate with it.

Early observations also suggested the presence of fine structure of the continuum emission, based on for example the wideband downconverting receiver on ISEE-1; see for example Figs. 1–2 of Kurth et al. [32]. Recent satellite observations from Geotail, Polar, Cluster, and Image provide similar measurements but with vastly improved coverage and time- and frequency-resolution. Most recently, Menietti et al. [45], in a study of continuum radiation simultaneously observed with combinations of the Cluster, Geotail, and Polar satellites, show several wideband examples of kilometric continuum emission composed of large numbers of narrow-band structures separated by 2–3 kHz in one case (their Fig. 9) and a few hundred Hz or less in another case (their Fig. 10); in either case the separation is far less than the probable electron gyrofrequency in the source region, so this type of structure cannot be explained by the mechanisms reviewed above which explain banding on the order of f_{ce}. Menietti et al. [46] show an even more striking example of low frequency continuum (14–17 kHz) detected with the Cluster satellite, their Fig. 6, which bears a strong superficial resemblance to ground level observations of auroral roar fine structure shown by Shepherd et al. (1998a), their Fig. 2c or 2h. Figure 8.4 of this paper compares the satellite and ground-based observations reported

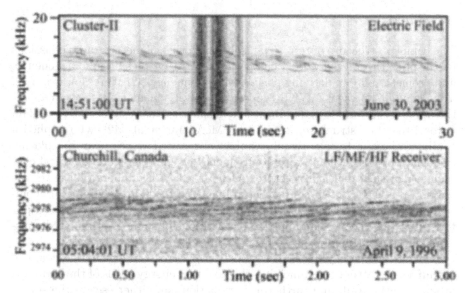

Fig. 8.4. Several bands of upper-hybrid emissions observed by the Cluster-II space-craft are depicted in the top panel. These emissions are at least superficially related to ground-level observations of auroral roar fine structure shown in the bottom panel; note that the frequency and time scales are different for each panel. Both type of emissions result from mode conversion of upper hybrid waves generated in regions with electron density irregularities. See Menietti et al. [45, 46] and Shepherd et al. [55] for further details about the above observations

by Menietti et al. [46] and Shepherd et al. [55]. This resemblance suggests a commonality between the source of the fine structure of these emissions, both of which result from mode conversion of upper hybrid waves generated in a region with electron density inhomogeneity, as pointed out by Menietti et al. [46].

Other magnetospheric mode-converted EM waves bear similar resemblance to auroral ionosphere counterparts, despite significant differences in the plasma parameters in the source region. For example, the structure of Langmuir waves in the electron foreshock is highly time variable, as is that of Langmuir waves along auroral field lines. Cairns et al. [4] show that the electric fields in the former case have a Gaussian distribution, if care is taken to sort the observations according to distance from the foreshock boundary. Recently Samara et al. [54] explore the statistics of the electric fields in the auroral case and find that they are also Gaussian over a range of electric field values but with a power-law tail. Studies to compare the fine structure of these emissions may be warranted, for example finding whether the plasma waves in the foreshock exhibit features analogous to the "chirp" features observed by McAdams and LaBelle [41] in the overdense auroral zone plasma.

8.3 Theory

The complex structure of terrestrial, solar and other planetary radio emissions has inspired theoretical work focused on the role that plasma inhomogeneities and geometry play in deriving plasma wave frequency structure. For example, Yoon et al. [67] present a WKB-type wave analysis calculation associated with the upper-hybrid mode in a horizontal ionospheric density structure to explain auroral roar fine structure, inspired by McAdams et al. [42], who applied a similar technique to successfully explain structured waves near the plasma frequency. In both studies, discrete frequency eigenmodes were a natural consequence of such density structures. Osherovich and Fainberg [49] have also developed a theory of force-free electromagnetic fields in a cylindrical geometry to predict eigenmodes with frequency spacings varying as $n^{1/2}$ and put forth that this theory may apply to observations of sounder-stimulated emissions in the ionosphere.

Complex frequency structure results from electrostatic eigenmodes trapped within density structures, much like the quantum energy levels of the hydrogen atom potential well, and furthermore radiation can either escape the source region and be detected by remote means or be observed in situ. To describe these waves, it is appropriate to derive effective Schrödinger equations for the trapped electrostatic wave eigenfunctions. In the following sections, the nonlocal linear theory of electrostatic waves is briefly outlined and reviewed, and then discussed within the context of trapped Langmuir and upper-hybrid waves.

8.3.1 Nonlocal Theory of Electrostatic Waves in Density Structures

For the sake of simplicity, we employ the electrostatic approximation and ignore ion dynamics. Nonlocal kinetic equations for linear electrostatic waves are given by the following:

$$\nabla^2 \phi(\mathbf{r}, t) = 4\pi e n \int d\mathbf{v} \, f(\mathbf{r}, \mathbf{v}, t) \,,$$

$$f(\mathbf{r}, \mathbf{v}, t) = -\frac{e}{m} \int_{-\infty}^{t} ds \, \frac{\partial}{\partial \mathbf{r}(s)} \, \phi[\mathbf{r}(s), s] \cdot \frac{\partial F[\mathbf{v}(s)]}{\partial \mathbf{v}(s)} \,, \tag{8.1}$$

where n, F, f, and ϕ denote, respectively, the average electron density, velocity-space distribution function, perturbed distribution and perturbed electrostatic potential. Other quantities are standard; e and m being the unit electric charge and mass, respectively. In what follows, we shall further simplify the situation by restricting ourselves to a uniform magnetic field and one-dimensional density gradient. In (8.1), the quantity $\mathbf{r}(s)$ is the unperturbed orbit of an electron.

In the above nonlocal theory the orbit integral cannot be carried out in closed form, so that certain expansions have to be employed. It turns out that there are two convenient limiting cases: one is when the magnetic field is sufficiently strong such that the electron gyro-radius can be considered small, and the integrand can thus be expanded for small gyro-radius; and the other is when the magnetic field is sufficiently weak such that the electron can be thought of as executing a straight orbit, i.e., unmagnetized plasmas. Nonlocal formalism carried out under the unmagnetized assumption will turn out to correspond to discrete Langmuir wave theory, while the expansion in small gyro-radius will lead to discrete upper-hybrid wave theory.

Unmagnetized Case

Let us first consider the straight "unmagnetized" orbit,

$$\mathbf{r}(s) = \mathbf{r} - \mathbf{v}(t - s), \qquad \mathbf{v} = \text{const}. \tag{8.2}$$

Assuming a Gaussian distribution for F and allowing the density gradient to lie along the x axis, we may Fourier analyze (8.1) along uniform spatial coordinates and employ Laplace transformation in time. That is, we assume that perturbed quantities are given by the form $a(\mathbf{r}, t) = a(x) e^{-i\omega t + i\mathbf{k} \cdot \mathbf{r}}$, where $\mathbf{k} = k_y \hat{y} + k_z \hat{z}$. This leads to

$$\phi''(x) - k^2 \phi(x) = \frac{2\omega_p^2(x)}{v_T^2} \int d\mathbf{v} \, F(v) \int_0^\infty d\tau \, e^{i(\omega - \mathbf{k} \cdot \mathbf{v})\tau}$$

$$\times \left[v_x \, \phi'(x - v_x \tau) + i\mathbf{k} \cdot \mathbf{v} \, \phi(x - v_x \tau) \right], \tag{8.3}$$

where $\omega_p^2 = 4\pi e^2 n(x)/m$ is the square of the nonlocal plasma frequency, $v_T^2 = 2T/m$ is the square of the electron thermal speed, T being the electron temperature, and we have redefined the integration dummy variable to $\tau = t - s$.

The above equation is a complicated integro-differential equation. We therefore simplify the situation by adopting the customary approximation,

$$\phi(x - v_x \tau) \approx \phi(x) - v_x \tau \, \phi'(x) + \cdots, \tag{8.4}$$

and retain the lowest-order term. This leads to

$$\phi'' - k^2 \left(1 - \frac{\omega_p^2}{\omega^2} \zeta^2 \, Z'(\zeta) \right) \phi = 0, \qquad \zeta = \frac{\omega}{k v_T}. \tag{8.5}$$

where Z is the usual plasma dispersion function:

$$Z(z) = \frac{1}{\pi^{1/2}} \int_{-\infty}^{\infty} \frac{e^{-y^2} \, dy}{y - z} \tag{8.6}$$

Equation (8.5) is in the form of a Schrödinger equation, and as such, it may support discrete frequency eigenvalues. Note that (8.5) is appropriate for either sufficiently weak ambient magnetic field, or when the electrons execute predominantly straight orbits.

In the limit of $\omega/kv_T \gg 1$ ("fluid" limit), we may expand the plasma Z function in asymptotic series, and the resulting effective Schrödinger equation takes on the form

$$\phi''(x) - k^2 \left[1 - \frac{\omega_p^2(x)}{\omega^2} \left(1 + \frac{3k^2 v_T^2}{2\omega^2} \right) \right] \phi(x) = 0 \ . \tag{8.7}$$

This limiting form makes clear why this "unmagnetized" case is relevant for Langmuir waves in density irregularities, since the uniform plasma limit of the above equation reduces to the familiar Langmuir wave dispersion relation.

Strongly Magnetized Case

Next we consider the case of magnetized electron orbit. For uniform magnetic field directed along the z axis, the unperturbed electron orbit is given by the standard form,

$$\mathbf{v}(\tau) = \left(v_\perp \cos(\varphi - \Omega\tau), \ v_\perp \sin(\varphi - \Omega\tau), \ v_\parallel \right) ,$$

$$\mathbf{r} - \mathbf{r}(\tau) = \left(\frac{v_\perp}{\Omega} [\sin\varphi - \sin(\varphi - \Omega\tau)], \ -\frac{v_\perp}{\Omega} [\cos\varphi - \cos(\varphi - \Omega\tau)], \ v_\parallel \tau \right) . \tag{8.8}$$

where $\tau = t - s$, the same integration dummy variable introduced in (8.3); $v_\perp^2 = v_x^2 + v_y^2$; $v_\parallel = v_z$; $\varphi = \arctan(v_y/v_x)$ is the phase angle; $\Omega = eB_0/mc$ is the electron cyclotron frequency, with B_0 representing the intensity of the constant magnetic field; and $\rho = v_\perp/\Omega$ is the electron gyro-radius.

The kinetic nonlocal wave equation in the magnetized case is expressed as

$$\phi''(x) - k^2\phi(x) = \frac{2\omega_p^2(x)}{v_T^2} \int d\mathbf{v} \, F(v) \int_0^\infty d\tau \, e^{i\omega\tau - i\mathbf{k}\cdot[\mathbf{r}-\mathbf{r}(\tau)]}$$

$$\times \left(v_x(\tau) \, \phi'[x(\tau)] + i\mathbf{k} \cdot \mathbf{v}(\tau) \, \phi[x(\tau)] \right) . \tag{8.9}$$

As with the unmagnetized case, the exact orbit integration is not possible. The expansion parameter in this case is the electron gyro-radius under the assumption that

$$\rho = v_\perp/\Omega \ll 1 \ . \tag{8.10}$$

The above assumption is equivalent to taking the electrons to be strongly magnetized. One could employ the small gyro-radius expansion in the integrand of (8.9) and integrate over τ. However, more straightforward manipulation is to treat the right-hand side of (8.9), which represents the perturbed charge density, as if it were locally uniform. Thus, we introduce an effective wave number along the direction of inhomogeneity, $k_x \rightarrow -id/dx$, and express the perturbed electrostatic field as $\phi(x) \to \phi\, e^{ik_x\, x}$. With this treatment, the results of uniform plasma theory are applicable, and the right-hand side of (8.9) can be expressed as

$$\text{RHS of (8.9)} = \frac{2\omega_p^2}{v_T^2}\left(1 + \frac{\omega}{k_z v_T} \sum_{n=-\infty}^{\infty} I_n(\lambda)\, e^{-\lambda}\, Z(\xi_n)\right)\phi\,,$$

$$\lambda = \frac{k_\perp^2 v_T^2}{2\Omega^2}\,, \qquad \xi_n = \frac{\omega - n\Omega}{k_z v_T}\,, \tag{8.11}$$

where $k_\perp^2 = k_x^2 + k_y^2$.

The above expression, superficially identical to the that of locally uniform plasma theory, contains k_x which is a differential operator. Thus, the above result cannot be used directly in conjunction with (8.9) but rather, the above-mentioned expansion in terms of small gyro-radius must be employed first. The assumption of $\rho = v_\perp/\Omega \ll 1$ is equivalent to $\lambda \ll 1$ in (8.11). After employing the expansion in λ and converting k_x back to its original operator form, we arrive at

$$\frac{v_T^2}{2\Omega^2}\frac{\omega_p^2(x)}{\Omega^2}\frac{\omega^2}{\omega^2 - \Omega^2}\frac{d^4\phi(x)}{dx^4} - \frac{\omega^2 - \omega_p^2(x) - \Omega^2}{\omega^2 - \Omega^2}\frac{d^2\phi}{dx^2}$$

$$= -k^2\left(\frac{k_z^2}{k^2}\frac{\omega^2 - \omega_p^2(x)}{\omega^2} + \frac{k_y^2}{k^2}\frac{\omega^2 - \omega_p^2(x) - \Omega^2}{\omega^2 - \Omega^2}\right)\phi(x)\,. \tag{8.12}$$

In the cold uniform plasma limit, this equation predicts upper-hybrid oscillation for waves propagating perpendicular to the ambient magnetic field. For nonuniform density, the above equation can be used for the discussion of trapped upper-hybrid waves with fine frequency structure. A final simplification is to assume that the trapped eigenmodes are characterized by $d/dx \gg k$. Under this condition, we may ignore the right-hand side of (8.12).

8.3.2 Trapped Langmuir Waves with Discrete Frequency Spectrum

The close association between electron density irregularities and structured plasma waves in the overdense auroral ionosphere, as described above for the PHAZE-II rocket observations, led McAdams et al. [42] to put forth a model of eigenmode trapping in density depletions. Using cold plasma theory and assuming a parabolic density depletion, they calculated a discrete spectrum of electrostatic plasma waves evenly spaced in frequency as observed, and with spacing of order $1\,\text{kHz}$ for a 800-m one-percent depletion, also in agreement with the observations. With this idea as a starting point, we consider a model of the density structure whose number density profile is given by a Lorentzian form embedded in a constant background,

$$n(x) = n_0\left\{1 + \delta\left[1 + (x/L)^2\right]^{-1}\right\}\,, \tag{8.13}$$

where the parameter δ is a positive real number. The quantity δ is a measure of the density increase at the center $x = 0$. For instance, $\delta = 0.1$ represents 10% density enhancement at $x = 0$, and $\delta = 0.9$ implies 90% density increase, etc. The quantity n_0 represents the asymptotic number density outside the structure. Inserting the model (8.13) into the wave equation (8.7), we have

$$\frac{d^2\phi}{dx^2} + V(x)\,\phi = 0, \quad V(x) = V_0 \frac{x_0^2 - x^2}{L^2 + x^2}, \quad V_0 = k^2 \left(1 - \frac{\omega_{p0}^2}{\omega^2} - \frac{3}{2}\frac{k^2 v_T^2}{\omega^2}\right),$$

$$\frac{x_0^2}{L^2} = \frac{(1+\delta)(\omega_{p0}^2 + 3k^2 v_T^2/2) - \omega^2}{\omega^2 - \omega_{p0}^2 - 3k^2 v_T^2/2}. \tag{8.14}$$

In the above $\omega_{p0}^2 = 4\pi n_0 e^2/m$ is the square of the asymptotic plasma frequency. In order for thermal Langmuir waves to be trapped within the density structure an oscillatory solution must exist within the cutoffs, $-x_0 < x < x_0$, while outside the cutoffs the solutions must decay away exponentially. Careful consideration reveals that in the underdense case ($\omega_{p0} < \Omega$), the density cavity ($\delta < 0$) cannot support such a solution, and that only the enhanced density structure ($\delta > 0$) can satisfy this criterion. The reverse is true in the overdense case ($\omega_{p0} > \Omega$). Below, we restrict our analysis to the underdense case. Yoon and LaBelle [68] treat the overdense case.

The discrete frequency spectrum associated with eigenmodes trapped within the density structure can be derived on the basis of the continuity of the eikonal solutions of (8.14) across the cutoff points,

$$\int_{-x_0}^{x_0} [V(x)]^{1/2} \, dx = (n + 1/2)\,\pi, \tag{8.15}$$

where $n = 0, 1, 2, \ldots$ Note that the cutoffs $x \pm x_0$ are real and positive if the condition

$$\omega_{p0}^2 + 3k^2 v_T^2/2 < \omega^2 < (1+\delta)(\omega_{p0}^2 + 3k^2 v_T^2/2) \tag{8.16}$$

is satisfied.

Inserting the specific expression for $V(x)$ defined in (8.14) to the matching condition (8.15) and carrying out closed-form analytical integration over x, we arrive at

$$4\delta^{1/2}\, kL\, \frac{\omega_{p0}}{\omega} \left(1 + \frac{3k^2 v_T^2}{\omega_{p0}^2}\right)^{1/2} [K(\eta) - E(\eta)] = (2n+1)\,\pi,$$

$$\eta = \frac{\left[(1+\delta)(\omega_{p0}^2 + 3k^2 v_T^2/2) - \omega^2\right]^{1/2}}{\delta^{1/2}\left(\omega_{p0}^2 + 3k^2 v_T^2/2\right)^{1/2}}. \tag{8.17}$$

In the above $K(\eta)$ and $E(\eta)$ are complete elliptic integrals. Equation (8.17) constitutes a transcendental dispersion equation which supports a discrete

frequency spectrum as its solution. The simplest approximation is to re-
place the full elliptic integrals with their small-argument series approximation, $K(\eta) - E(\eta) \approx \pi\eta^2/4$, which is valid if the normalized cutoffs $\pm x_0/L$
are sufficiently small, $x_0^2/L^2 \ll 1$. Adopting such a simplification, we obtain
an analytical expression for the eigenvalues,

$$\omega = \frac{\omega_{p0}}{2} \left(1 + \frac{3k^2 v_T^2}{2\omega_{p0}^2}\right)^{1/2} \left\{ \left[4(1+\delta) + a_n^2\right]^{1/2} - a_n \right\} , \qquad (8.18)$$

$$a_n = \frac{(2n+1)\,\delta^{1/2}}{kL} .$$

Applying the above approximate solution for the discrete spectrum requires
that the solution ω^2 be confined to the range specified by (8.16). For typical
ionospheric conditions, the condition (8.16) is easily satisfied.

In the numerical computation of the discrete wave spectrum (8.16), the
width of the density structure plays a key role. Taking an example from the
auroral ionosphere, if the plasma frequency is $f_{p0} = \omega_{p0}/2\pi \sim 2$ MHz, and
the background electron thermal energy is $T \sim 0.1$ eV, then we calculate

$$\alpha \equiv L\omega_{p0}/v_T \approx 10^5 L[\text{km}] .$$

$L = 100$m implies $\alpha \approx 10^4$, $L = 1$ km implies $\alpha \sim 10^5$, and $L = 1$ m implies
$\alpha = 10^2$. The characteristic parallel wave number associated with Langmuir
waves is determined by the condition for maximum wave growth rate. For
$V_0/v_{Tb} \approx 10$ and $T \approx T_b$, we have $kv_T/\omega_{p0} \approx 0.1$.

In Figure 8.5, we plot the discrete eigenvalue f versus n, for $f_{p0} = 2$ MHz,
$kv_T/\omega_{p0} = 0.1$, density enhancement factor $\delta = 0.05$, and for two different
structure widths, $L = 100$ m and $L = 500$ m. Note that the wider structure
leads to smaller frequency spacing between individual discrete modes. The

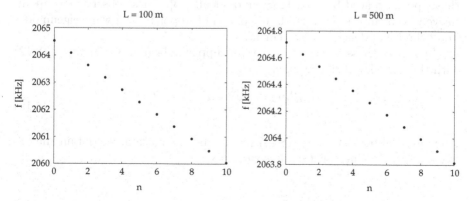

Fig. 8.5. Discrete Langmuir wave spectrum f [kHz] versus the mode number n, for
two different density structure widths, L, corresponding to 100 m and 500 m

predicted eigenmode spacings are of order 0.5 kHz and 0.1 kHz, respectively. Yoon and LaBelle [68] find similar eigenmode spacings in the overdense case, which is appropriate to the observations of McAdams and LaBelle [41]. These spacings are comparable to those observed by McAdams and LaBelle [41] and calculated by McAdams et al. [42].

8.3.3 Discrete Upper-Hybrid Waves in Density Structures

The frequency spacing associated with auroral roar fine structure was first addressed by Shepherd et al. [55], who proposed an explanation based upon geometrical considerations in which the excitation of multiplet discrete modes resulted from standing waves within planar field-aligned density cavities with vertically converging walls. This approach was similar to an idea originally proposed by Calvert [5] to account for AKR fine structure. This explanation faces several difficulties, however, which Shepherd et al. [55] discuss. The principal problem is that it requires the wave-vector to vanish at the reflection points, which means that the wave cannot continuously grow as it stands in the cavity. Combined with the relatively low growth rate, it is difficult to envision how sufficient gain can be achieved to allow the discrete eigenmodes to grow.

Yoon et al. [67] present a model which corrects the primary difficulty of the Shepherd et al. [55] approach by considering circulating eigenmodes in a cylindrical geometry. This geometry is similar to that suggested by Carlson et al. [6] to explain the enhancement of upper hybrid waves in an artificially enhanced density structure, although they did not consider eigenmode structure.

As discussed above, recent observations in the auroral ionosphere suggest that waves near the upper hybrid frequency, polarized with wave-number primarily perpendicular rather than parallel to the background magnetic field, exhibit ordered frequency spectra with peaks spaced by ~ 10 kHz, and within those, peaks spaced by ~ 1 kHz [Samara et al., 53]. These observations qualitatively confirm the Yoon et al. [67] model of upper hybrid wave eigenmodes in a cylindrical density enhancement.

To model these waves, consider the upper-hybrid wave equation (8.12) with the right-hand side ignored,

$$
\frac{v_T^2}{2\Omega^2} \frac{\omega_p^2(x)}{\Omega^2} \frac{\omega^2}{\omega^2 - \Omega^2} \frac{d^4\phi(x)}{dx^4} - \frac{\omega^2 - \omega_p^2(x) - \Omega^2}{\omega^2 - \Omega^2} \frac{d^2\phi}{dx^2} \approx 0 . \tag{8.19}
$$

Applying the density model (8.13) to the above equation, we obtain the following equations in dimensionless variables:

$$\frac{d^2\phi}{dX^2} + k_{uh}^2(X)\,\phi = 0, \qquad k_{uh}^2(X) = Q\,\frac{X_0^2 - X^2}{1 + \delta + X^2}\,,$$

$$Q = \frac{2}{\rho^2\,\alpha^2}\,\frac{w^2 - w_n^2}{w^2}, \qquad X_0^2 = \frac{w_x^2 - w^2}{w^2 - w_n^2}\,,$$

$$w_n^2 = 1 + \alpha^2, \qquad w_x^2 = 1 + (1+\delta)\,\alpha^2\,, \tag{8.20}$$

where

$$X = \frac{x}{L}, \quad \rho = \frac{v_T}{L\Omega}, \quad \alpha^2 = \frac{\omega_{p0}^2}{\Omega^2}, \quad w = \frac{\omega}{\Omega}\,. \tag{8.21}$$

If $w^2 < 1+\alpha^2$, then $k_{uh}^2(X) > 0$ everywhere, implying that only continuous eigenmodes exist. If $w^2 > (1+\delta)\,\alpha^2$, on the other hand, then $k_{uh}^2(X) < 0$, and only damped modes exist. Thus, the possibility of trapped eigenmodes exists only in the intermediate range $w_n^2 < w^2 < w_x^2$. One can show that discrete modes cannot exist if $\delta < 0$ (i.e., density depletion), since in this case the internal region $(X^2 < X_0^2)$ is characterized by damped solutions while the outer region supports oscillatory solutions. Therefore, the situation is similar to that of discrete Langmuir waves, except that the discrete modes occur for the case of density enhancements rather than density depletions.

Asserting the matching condition (8.15) across the cutoff points results in an equation for the eigenfrequencies similar to (8.17),

$$\frac{(2\,\delta)^{1/2}}{\rho\,\alpha}\,\frac{(w^2 - 1)^{1/2}}{w}\,[K(\zeta) - E(\zeta)] = \left(n + \frac{1}{2}\right)\frac{\pi}{2}\,,$$

$$\zeta = \frac{[1 + (1+\delta)\alpha^2 - w^2]^{1/2}}{\delta^{1/2}\,(w^2 - 1)^{1/2}}\,, \tag{8.22}$$

which can be approximately solved for w^2,

$$w^2 = \frac{1}{1 - a_n^2}\left[w_x^2 - \frac{a_n^2}{2} - a_n\left(w_x^2\,(w_x^2 - 1) + \frac{a_n^2}{4}\right)^{1/2}\right]\,,$$

$$a_n = (2n+1)\left(\frac{\delta}{2}\right)^{1/2}\rho\,\alpha\,. \tag{8.23}$$

In Fig. 8.6 we plot the solution of (8.23) for $\alpha = 0.7746$, $\delta = 4$, $\rho = 10^{-2}/L$, and for two choices for the size of the density structures: $L = 100\text{m}$ and $L = 500\,\text{m}$. In addition to the above normalized input parameters, we take the electron cyclotron frequency to be 1 MHz, as is approximately the case in the auroral ionosphere.

The resulting eigenmode spacings are of order 0.2 kHz for $L = 100\,\text{m}$ and 0.1 kHz for $L = 500\,\text{m}$; both cases fall within the range of frequency spacings observed in fine structures of auroral roar emissions [see Shepherd et al., 55]. These predictions are narrower than the principal band spacing, but not much narrower than the substructure spacing, observed in rocket data of structured upper hybrid waves [Samara et al., 53].

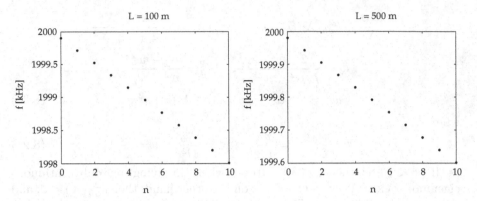

Fig. 8.6. Discrete upper-hybrid wave spectrum f [kHz] versus the mode number n, for two different density structure widths, L, corresponding to 100 m and 500 m

8.4 Conclusions

As reviewed above, the terrestrial ionosphere and magnetosphere include numerous examples of mode-conversion radiation, which occurs when nonthermal particle distributions excite electrostatic waves which convert to EM waves. In the theory section above, we have presented linear analysis of the normal modes of high-frequency plasma waves in inhomogeneous plasma, where the inhomogeneity, either a density enhancement or depletion, can be described by a Lorentzian form. This generalized treatment contains both the Langmuir wave eigenmodes proposed by McAdams et al. [42] and the upper hybrid wave eigenmodes proposed by Yoon et al. [67]. The frequency spacings derived from these models, for inhomogeneities of scale size 100–1000 m, match to the observed spacings of frequency fine structure observed in auroral Langmuir wave "chirps" [McAdams and LaBelle, 41] and auroral roar fine structure [35, 55]; upper hybrid wave frequency fine structure recently observed directly with a rocket experiment shows somewhat wider frequency spacings [Samara et al., 53]. The theoretical treatment outlined above is independent but does not exclude the possibility that other mechanisms may act to produce wave structure, such as for example the discretization upon wave conversion in inhomogeneous plasma discussed by Willes and Cairns [63].

Perhaps the most significant aspect of these studies lies in the potential use of observations of mode conversion radiation to remotely sense the plasma environment of the source region, gaining information about electron densities, magnetic fields, and depth and spatial scale of irregularities. The solar system is replete with examples of the mode-conversion radiation exhibiting fine frequency structures, ranging from nonthermal emissions of planetary ionospheres and magnetospheres to some types of radio emissions originating in the solar wind. To remotely sense terrestrial ionospheric and magnetospheric plasma using mode conversion radiation has intriguing implications for interpretation of the emissions originating in more remote plasmas, where direct measurements are not available and parameters are less well-known.

Acknowledgments

This research was supported by the following National Science Foundation grants: ATM-0223764 to the University of Maryland; OPP-0338105, OPP-0341470 and ATM-0243645 to Siena College; and OPP-0090545 and ATM-0243595 to Dartmouth College. Dartmouth College was further supported by NASA grant NNG04WC27G. We thank Doug Menietti for the Cluster-II data depicted in Fig. 8.4 and Shengyi Ye for the South Pole data shown in Fig. 8.3.

References

[1] Bauer, S.J., and R.G. Stone: Satellite observations of radio noise in the magnetosphere, Nature 128, 1145, 1968.

[2] Beghin, C., J.L. Rauch, and J.M. Bosqued: Electrostatic plasma waves and HF auroral hiss generated at low altitudes, J. Geophys. Res. 94, 1359, 1989.

[3] Benson, R.F., and H.K. Wong: Low-altitude ISIS 1 observations of auroral radio emissions and their significance to the cyclotron maser instability, J. Geophys. Res. 92, 1218, 1987.

[4] Cairns, I. H. and J.D. Menietti: Radiation near $2f_p$ and intensified emissions near f_p in the dayside and nightside auroral region and polar cap, J. Geophys. Res. 102, 4787, 1997

[5] Calvert, W.: A feedback model for the source of auroral kilometric radiation, J. Geophys. Res. 87, 8199, 1982.

[6] Carlson, C.W. et al.: unpublished manuscript, 1987.

[7] Carpenter, D.L., and R.R. Anderson: An ISEE/whistler model of equatorial electron density in the magnetosphere, J. Geophys. Res. 97, 1097, 1992.

[8] Chiu, Y.T., and M. Schulz: Self-consistent particle and parallel electrostatic field distributions in the magnetospheric – ionospheric auroral region, J. Geophys. Res. 83, 629, 1978.

[9] Ergun, R.E., C.W. Carlson, J.P. McFadden, J.H. Clemmons, and M.H. Boehm: Evidence of transverse Langmuir modulational instability in a space plasma, Geophys. Res. Lett. 18, 1177, 1991a.

[10] Ergun, R.E., C.W. Carlson, J.P. McFadden, J.H. Clemmons, and M.H. Boehm: Langmuir wave growth and electron bunching: Results from a wave-particle correlator, J. Geophys. Res. 96, 225, 1991b.

[11] Filbert, P.C. and P.J. Kellogg: Electrostatic noise at the plasma frequency beyond the earth's bow shock, J. Geophys. Res. 84, 1369, 1979.

[12] Gough, M.P.: Nonthermal continuum emissions associated with electron injections: Remote plasmapause sounding, Planet. Space Sci. 30, 657, 1982.

[13] Green, J.L., B.R. Sandel, S.F. Fung, D.L. Gallagher, and B.W. Reinisch: On the origin of kilometric continuum, J. Geophys. Res. 107, 1105,10.1029/2001JA000193, 2002.

[14] Green, J.L., Scott Boardsen, Shing F. Fung, H. Matsumoto, K. Hashimoto, R.R. Anderson, B.R. Sandel, and B.W. Reinisch: Association of kilometric continuum with plasmapsheric structures, J. Geophys. Res. 109, A03203, doi:10.1029/2003JA010093, 2004.

[15] Gregory, P.C.: Radio emissions from auroral electron, Nature 221, 350, 1969.

[16] Gurnett, D.A. and R.R. Shaw: Electromagnetic radiation trapped in the magnetosphere above the plasma frequency, J. Geophys. Res. 78, 8136, 1973.

[17] Gurnett, D.A.: The Earth as a radio source: The nonthermal continuum, J. Geophys. Res. 80, 2751, 1975.

[18] Gurnett, D.A., F.L. Scarf, W.S. Kurth, R.R. Shaw, and R.L. Poynter: Determination of Jupiter's Electron Density Profile from Plasma Wave Observations, J. Geophys. Res. 86, 8199-8212, 1981.

[19] Hashimoto, K., W. Calvert, and H. Matsumoto: Kilometric continuum detected by Geotail, J. Geophys. Res. 104, 28645, 1999.

[20] Horne, R.B.: Path-integrated growth of electrostatic waves: The generation of terrestrial myriametric radiation, J. Geophys. Res. 94, 8895, 1989.

[21] Horne, R.B.: Narrowband structure and amplitude of terrestrial myriametric radiation, J. Geophys. Res. 95, 3925, 1990.

[22] Hughes, J.M. and J. LaBelle: Plasma conditions in auroral roar source regions inferred from radio and radar observations, J. Geophys. Res. 106, 21157, 2001.

[23] Issautier, K., N. Meyer-Vernet, M. Moncuquet, S. Hoang, and D.J. McComas: Quasi-thermal noise in a drifting plasma: theory and application to solar wind diagnostic on Ulysses, J. Geophys. Res. 104, 6691, 1999.

[24] James, H.G., E.L. Hagg, and L.P. Strange: Narrowband radio noise in the topside ionosphere, AGARD Conf. Proc., AGARD-CP-138, 24, 1974.

[25] Jones, D.: Source of terrestrial nonthermal radiation, Nature 260, 385, 1976.

[26] Jones, D.: Latitudinal beaming of planetary radio emissions, Nature 288, 225, 1980.

[27] Kasaba, Y., H. Matsumoto, K. Hashimoto, R.R. Anderson, J.-L. Bougeret, M.L. Kaiser, X.Y. Wu, and I. Nagano: Remote sensing of the plasmapause during substorms: Geotail observation of nonthermal continuum enhancement, J. Geophys. Res. 103, 20389, 1998.

[28] Kaufmann, R. L.: Electrostatic wave growth: Secondary peaks in measured auroral electron distribution function, J. Geophys. Res. 85, 1713, 1980.

[29] Kellogg, P.J. and S.J. Monson: Radio emissions from aurora, Geophys. Res. Lett. 6, 297, 1979.

[30] Kellogg, P.J. and S.J. Monson: Further studies of auroral roar, Radio Sci. 19, 551, 1984.

[31] Kletzing, C.A., S.R. Bounds, J. LaBelle and M. Samara: Observation of the reactive component of Langmuir wave phase-bunched electrons, Geophys. Res. Lett., in press, 2005.

[32] Kurth, W.S., D.A. Gurnett, and R.R. Anderson: Escaping nonthermal continuum radiation, J. Geophys. Res. 86, 5519, 1981.

[33] Kurth, W.S.: Detailed observations of the source of terrestrial narrowband electromagnetic radiation, Geophys. Res. Lett. 9, 1341, 1982.

[34] LaBelle, J.: Radio Noise of Auroral Origin: 1968–1988, J. Atmos. Terr. Phys. 51, 197, 1989.

[35] LaBelle, J., M.L. Trimpi, R. Brittain, and A.T. Weatherwax: Fine structure of auroral roar emissions, J. Geophys. Res. 100, 21953, 1995.

[36] LaBelle, J., S.G. Shepherd, and M.L. Trimpi: Observations of auroral medium frequency bursts, J. Geophys. Res. 102, 22221, 1997.

[37] LaBelle, J. and R.A. Treumann: Auroral Radio Emissions, 1. Hisses, Roars, and Bursts, Space Sci. Rev. 99, 295, 2002.

[38] Lund, E.J., J. LaBelle, and R.A. Treumann: On quasi-thermal fluctuations near the plasma frequency in the outer plasmasphere: A case study, J. Geophys. Res. 99, 23651, 1994.

[39] Lund, E.J., R.A. Treumann, and J. LaBelle: Quasi-thermal fluctuations in a beam-plasma system, Phys. Plasmas 3, 1234, 1996.

[40] McAdams, K.L., J. LaBelle, M.L. Trimpi, P.M. Kintner, and R.A. Arnoldy: Rocket observations of banded structure in waves near the Langmuir frequency in the auroral ionosphere, J. Geophys. Res. 104, 28109, 1999.

[41] McAdams, K.L. and J. LaBelle: Narrowband structure in HF waves above the plasma frequency in the auroral ionosphere, Geophys. Res. Lett., 26, 1825, 1999.

[42] McAdams, K.L., R.E. Ergun, and J. LaBelle: HF Chirps: Eigenmode trapping in density deletions, Geophys. Res. Lett. 27, 321, 2000.

[43] McFadden, J.P., C.W. Carlson, and M.H. Boehm: High-frequency waves generated by auroral electrons, J. Geophys. Res. 91, 12079, 1986.

[44] Melrose, D.B.: A theory for the nonthermal radio continua in the terrestrial and Jovian magnetospheres, J. Geophys. Res. 86, 30, 1981.

[45] Menietti, J.D., R.R. Anderson, J.S. Pickett, and D.A. Gurnett: Near-source and remote observations of kilometric continuum radiation from multispacecraft observations, J. Geophys. Res. 108, 1393, doi:10.1029/2003JA009826, 2003.

[46] Menietti, J.D., O. Santolik, J.S. Pickett, and D.A. Gurnett: High resolution observations of continuum radiation, Planet. Space Sci., submitted 2004.

[47] Meyer-Vernet, N., and C. Perche: Tool kit for antennae and thermal noise near the plasma frequency, J. Geophys. Res. 94, 2405, 1989.

[48] Meyer-Vernet, N.: On the thermal noise in an anisotropic plasma, J. Geophys. Res. 21, 397, 1994.

[49] Osherovich, V. and J. Fainberg: Dependence of frequency of nonlinear cold plasma cylindrical oscillations, Phys. Plasmas 11, 2314, 2004.

[50] Pottelette, R. and R.A. Treumann: Auroral acceleration and radiation, this volume, 2005.

[51] Rönnmark, K.: Emission of myriametric radiation by coalescence of upper hybrid waves with low frequency waves, Anal. Geophys. 1, 187, 1983.

[52] Samara, M., J. LaBelle, C.A. Kletzing, and S.R. Bounds: Rocket Measurements of Polarization of Auroral HF Waves, EOS Trans. Am. Geophys. Union, Fall Meeting, 2002.

[53] Samara, M., J. LaBelle, C.A. Kletzing, and S.R. Bounds: Rocket observations of structured upper hybrid waves at $f_{uh} = 2f_{ce}$, Geophys. Res. Lett. 31, L22804, 10.1029/2004GL021043, 2004.

[54] Samara, M., J. LaBelle, I.H. Cairns and R.A. Treumann, Statistics of Auroral Langmuir Waves, J. Geophys. Res., submitted, 2005.

[55] Shepherd, S.J., J. LaBelle, and M.L. Trimpi: Further investigation of auroral roar fine structure, J. Geophys. Res. 103, 2219, 1998a.

[56] Shepherd, S.J., J. LaBelle, and M.L. Trimpi: The polarization of auroral radio emissions, Geophys. Res. Lett. 24, 3161, 1998b.

[57] Steinberg, J.-L., S. Hoang, and M.F. Thomsen: Observations of the Earth's continuum radiation in the distant magnetotail with ISEE-3, J. Geophys. Res. 95, 20781, 1990.

[58] Walsh, D., F.T. Haddock, and H.F. Schulte: Cosmic radio intensities at 1.225 and 2.0 Mc measured up to an altitude of 1700 km, in: Space Res., 4, edited by P. Muller, pp. 935–959, North Holland Publishing Company, Amsterdam, 1964.

[59] Weatherwax, A.T., J. LaBelle, M.L. Trimpi, and R. Brittain: Ground-based observations of radio emissions near $2f_{ce}$ and $3f_{ce}$ in the auroral zone, Geophys. Res. Lett. 20, 1447, 1993.

[60] Weatherwax, A.T., J. LaBelle, and M.L. Trimpi: A new type of auroral radio emission observed at medium frequencies ($\sim$1350–3700 kHz) using ground-based receivers, Geophys. Res. Lett. 21, 2753, 1994.

[61] Weatherwax, A. T., J. LaBelle, M. L. Trimpi, R. A. Treumann, J. Minow, and C. Deehr: Statistical and case studies of radio emissions observed near $2f_{ce}$ and $3f_{ce}$ in the auroral zone, J. Geophys. Res. 100, 7745, 1995.

[62] Weatherwax, A.T., P.H. Yoon, and J. LaBelle: Interpreting observations of MF/HF radio emissions: Unstable wave modes and possibilities to passively diagnose ionospheric densities, J. Geophys. Res. 107, 1213, doi:10.1029/2001JA000315, 2002.

[63] Willes, A.J. and I.H. Cairns: Banded frequency structure from linear mode conversion in inhomogeneous plasmas, Phys. Plasmas 10, 4072, 2003.

[64] Yoon, Peter H., A.T. Weatherwax, T.J. Rosenberg, and J. LaBelle, Lower ionospheric cyclotron maser theory: A possible source of $2f_{ce}$ and $3f_{ce}$ auroral radio emission, J. Geophys. Res. 101, 27,015, 1996.

[65] Yoon, P.H., A.T. Weatherwax, and T.J. Rosenberg: On the generation of auroral radio emissions at harmonics of the lower ionospheric electron cyclotron frequency: X, O, and Z mode maser calculations, J. Geophys. Res. 103, 4071, 1998a.

[66] Yoon, P.H., A.T. Weatherwax, T.J. Rosenberg, J. LaBelle, and S.G. Shepherd: Propagation of medium frequency (1–4 MHz)auroral radio waves to the ground via the Z-mode radio window, J. Geophys. Res. 103, 29267, 1998b.

[67] Yoon, P.H., A.T. Weatherwax, and J. LaBelle: Discrete electrostatic eigenmodes associated with ionospheric density structure: Generation of auroral roar fine frequency structure, J. Geophys. Res. 105, 27580, 2000.

[68] Yoon, P.H. and J. LaBelle, Discrete Langmuir waves in density structure, J. Geophys. Res. 110, A11308, doi:10.1029/2005JA011186, 2005.

9

Theoretical Studies of Plasma Wave Coupling: A New Approach

D.-H. Lee[1], K. Kim[2], E.-H. Kim[1], and K.-S. Kim[1]

[1] Department of Astronomy and Space Science, Kyung Hee University, Yongin, Kyunggi 449-701, Korea,
dhlee@khu.ac.kr
[2] Department of Molecular Science and Technology, Ajou University, Suwon, Kyunggi 443-749, Korea

Abstract. New numerical and analytical methods are applied to wave coupling in inhomogeneous plasma. It is found that the X-mode feeds energy into the upper hybrid resonance at plasma inhomogeneities oriented perpendicular to the ambient magnetic field. The results are consistent with previous studies using other methods. When a finite pressure is introduced, the upper hybrid waves are no longer stationary and start propagating. They can form cavity modes and emit a small fraction of O (or X) waves. Limitations of the present study are the neglect of collisions and plasma pressure effects which might limit the growth of the upper hybrid waves; furthermore, this study concentrates on the case for which the density gradient is perpendicular to the magnetic field, a condition that is valid near the equator.

Key words: Mode coupling, propagation in inhomogeneous plasma, resonant absorption of X mode waves, upper hybrid resonance, cavity modes

9.1 Introduction

Plasma waves become often coupled owing to inhomogeneity in space. A certain mode is reflected at the cutoff region, and changes the polarization at the crossover region. At resonances, one mode can be converted into the other resonant mode where wave energy is irreversibly transferred into the resonances.

Since mode conversion is often associated with singular solutions, the subject of plasma wave coupling has difficulties in both analytical and numerical aspects. For instance, analytical solutions can provide only asymptotic approximations near the resonances or approximate global solutions by adopting the WKB method [4, 5, 16]. When wave equations are strongly coupled, it becomes difficult to treat the coupled equations in an exact manner.

Even numerical studies based on eigenmode analysis also meet similar difficulties in such coupled systems. Strictly speaking, no pure eigenmodes

D.-H. Lee et al.: *Theoretical Studies on Plasma Wave Coupling: A New Approach*, Lect. Notes Phys. **687**, 235–249 (2006)
www.springerlink.com

exist owing to the singular behavior in wave coupling, which should limit the application of the eigenmode analysis. To avoid singularities, damping is often introduced and expressed using a complex frequency $\omega = \omega_r + i\gamma$. For finite γ, it is expected that all wave modes will decay via this damping decrement. It should be noted that γ represents an average value resulting from a Fourier transform over infinite time. However, we have interests in finite time histories in reality and the damping rates among different modes can be different for a finite time period.

Thus, it is still an important issue to further investigate the process of plasma wave coupling. In this study, we introduce new numerical and analytical techniques, which compensate for such shortcomings mentioned above. One technique is a time-dependent numerical model, which allows multi-fluid components in an arbitrarily inhomogeneous 3-D plasma. This time-dependent model can be very useful in the coupled wave problem [Lee and Lysak 12] since the mode conversion is often associated with singular solutions, which provide only asymptotic approximations. The other technique is an analytical tool, which is called the invariant imbedding method (IIM) [3]. Using the invariant imbedding equations, we are able to calculate the reflection and transmission coefficients, and the wave amplitudes for the propagation of arbitrary number of coupled waves in arbitrarily-inhomogeneous stratified media [10].

In this study, we will show the application of these numerical and analytical techniques to one of the plasma wave coupling problems: the coupling of ordinary (O), extraordinary (X) waves, and upper hybrid resonances (UHR). Linear mode conversion of O and X waves into upper hybrid (UH) waves is investigated numerically by adopting a time-dependent numerical model, and analytically by adopting IIM, respectively.

9.2 Numerical Model

We study the wave coupling at the UHR region by using a 3-D multi-fluid numerical model [9, 11]. Our approach differs from previous studies in the sense that plasma waves are studied with time histories of electric and magnetic field components. In a cold plasma, the linearized electron waves can be obtained by Maxwell equations, Ohm's law and the equation of motion. For simplicity, we consider only the motion of electron fluids and assume the cold plasma approximation in this work.

$$\nabla \times \mathbf{E} = -\frac{\partial \mathbf{B}}{\partial t} \tag{9.1}$$

$$\nabla \times \mathbf{B} = \mu_o \mathbf{J} + \frac{1}{c^2} \frac{\partial \mathbf{E}}{\partial t} \tag{9.2}$$

$$\mathbf{J} = -e n_0(x) \mathbf{v_e} \tag{9.3}$$

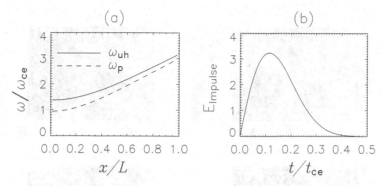

Fig. 9.1. (a) The upper hybrid and plasma frequency profiles assumed in the model. The *solid line* represents the local upper hybrid frequency, $\omega_{uh}(x)$. The *dashed line* represents the local plasma frequency, $\omega_p(x)$. (b) The applied impulse at $x = 0$, which is assumed on the E_z (or E_y) component

$$m_e \frac{\partial \mathbf{v_e}}{\partial t} = -e(\mathbf{E} + \mathbf{v_e} \times \mathbf{B_0}) \qquad (9.4)$$

where $\mathbf{E}, \mathbf{B}, \mathbf{J}, \mathbf{v}$ and n represent electric and magnetic fields, current density, velocity and number density, respectively.

In order to solve (9.1)–(9.4) as an initial-valued problem, the finite difference method is used in time and space. In our box model, a uniform magnetic field $\mathbf{B_0}$ is assumed to be parallel to the z-axis and the density gradient is introduced along the x-axis. Details of this numerical model are referred to in Kim and Lee [9] and Kim et al. [11]. The size of box used in this calculation is $10^2 \times 10^2 \times 10^2 \, \text{km}^3$. The profile of plasma frequency (ω_p) and UH frequency (ω_{uh}) assumed in our model is given in Fig. 9.1a. Frequencies are normalized to the electron cyclotron frequency, $\omega_{ce}(= 6 \times 10^3 \, \text{rad/sec})$. Length is normalized to the radial distance, $L = 100$ km.

The impulsive input is assumed on E_z and E_y for O wave and X wave initial inputs, respectively, at $x = 0$. Fundamental harmonic wave numbers are assumed in both y and z directions. The EM impulse used in the simulations is given by Fig. 9.1b. Time histories of electric and magnetic field components at a line of grid points along the x direction are recorded. The boundaries are assumed to become perfect reflectors after the impulsive stimulus ends. Thus the total wave energy inside the box model remains constant in time, which enables us to easily examine the energy transfer among different wave modes. We start the simulation with the impulsive input on the E_z (or E_y) component to represent the O wave (or X wave) impulse.

9.3 Numerical Results

The power spectra of electric and magnetic field components are obtained through the FFT at each grid point. Figure 9.2 shows the spectra, where the

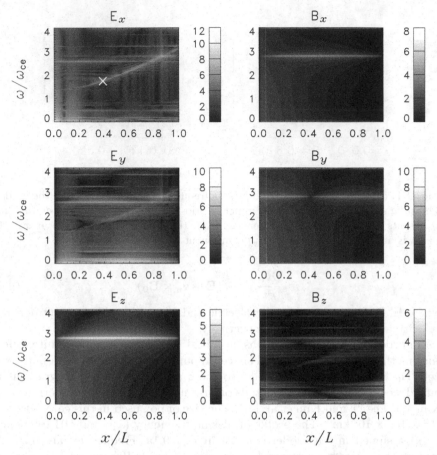

Fig. 9.2. The power spectra of perturbation components **E** and **B** when the impulse of O waves is assumed

amplitudes are represented by the degree of brightness after they are scaled logarithmically. In Fig. 9.2, in the spectrum of E_x, one continuous band appears, which corresponds to the local electrostatic (ES) upper hybrid waves in Fig. 9.1a. The E_z component can effectively represent the O wave since the wave vector is almost perpendicular to $\mathbf{B_0}$ by assuming relatively small k_y and k_z. The electromagnetic (EM) waves globally appear in all E and B components since they freely propagate inside the box when an impulsive input excites EM waves. The EM modes show a few cavity modes in Fig. 9.2 owing to the perfect reflecting boundary conditions. This feature is well confirmed in the spectra of B_x, B_y and B_z which are purely EM wave components.

To examine the coupling between ES and EM modes at the UHR, we select one point (x_1) at a certain UH frequency marked by X in Fig. 9.2. Figure 9.3 shows time histories of the electric and magnetic fields at this resonant point which are obtained by applying the inverse Fourier transform at the given

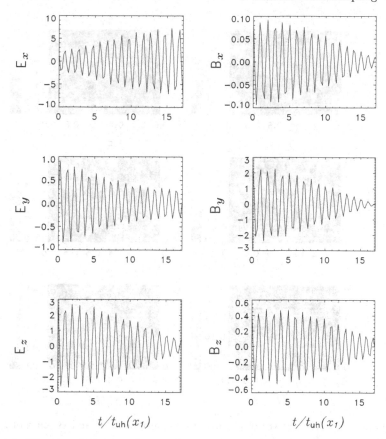

Fig. 9.3. The time histories of perturbation components **E** and **B** (case of O wave incidence). The electric fields are normalized to an arbitrary value E_A and the magnetic fields are normalized to $B_A = E_A/c$, respectively

frequency. Figure 9.3 indicates that the ES wave grows in time, while the other EM wave damped. Therefore it is found that the EM waves excited by the O wave impulse are mode-converted into UH waves.

We now consider an EM impulse of X wave by assuming the impulse above on the E_y component. Figure 9.4 shows the power spectra of E and B. In the spectrum of E_x, one continuous band appears again, which corresponds to the local UH frequency. The spectral feature in Fig. 9.4 is similar to that of Fig. 9.2 except that relatively large power is found in both E_x and E_y compared to E_z, while Fig. 9.2 shows relatively large power in E_z. This difference arises because the E_z input would produce relatively large O wave power, while the E_y would produce more X wave power inside the box.

Figure 9.5 shows the time histories at the resonant point (x_2) marked by X in Fig. 9.4. Our results show the growth of the ES wave E_x and the decay

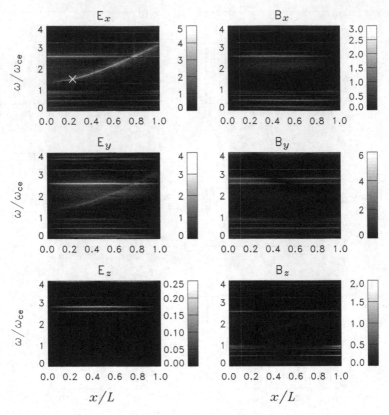

Fig. 9.4. The power spectra of perturbation components **E** and **B** (case of X wave incidence)

of the other EM wave (B_x, B_y, B_z, E_y and E_z) at this UH frequency, which are pretty similar to the results of the O wave impulse.

9.4 Invariant Embedding Method

In recent work of Kim et al. [10], a new invariant embedding theory was presented for studying the propagation of coupled waves in inhomogeneous stratified media. We consider N coupled waves propagating in a stratified medium, where physical parameters depend on only one coordinate. We take this coordinate as the z axis and assume the medium of thickness L lies in $0 \leq z \leq L$. We also assume that all waves propagate in the xz plane. The x component of the wave vector, q, is then a constant. In a variety of problems, the wave equation of N coupled waves has the form

$$\frac{d^2\psi}{dz^2} - \frac{d\mathcal{E}}{dz}\mathcal{E}^{-1}(z)\frac{d\psi}{dz} + \left[\mathcal{E}(z)K^2\mathcal{M}(z) - q^2 I\right]\psi = 0 , \qquad (9.5)$$

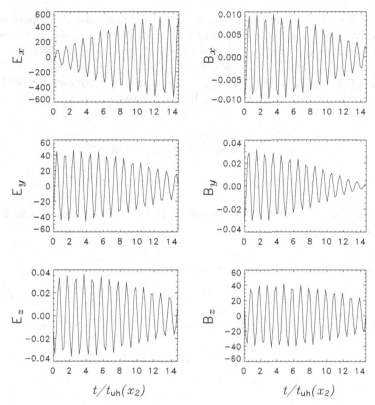

Fig. 9.5. The time histories of perturbation components **E** and **B** (case of X wave incidence)

where $\psi = (\psi_1, \cdots, \psi_N)^T$ is a vector wave function and $\mathcal{E}$ and $\mathcal{M}$ are $N \times N$ matrix functions. We assume that the waves are incident from the vacuum region where $z > L$ and transmitted to another vacuum region where $z < 0$. I is a unit matrix and K is a diagonal matrix such that $K_{ij} = k_i \delta_{ij}$, where k_i is the magnitude of the vacuum wave vector for the i-th wave. By assigning $\mathcal{E}(z)$ and $\mathcal{M}(z)$ suitably, (9.5) is able to describe various kinds of waves in a number of stratified media. A wide variety of mode conversion phenomena observed in space and laboratory plasmas can also be studied using this equation [4, 7, 13, 16].

We generalize (9.5) slightly, by replacing the vector wave function ψ by an $N \times N$ matrix wave function Ψ, the j-th column vector $(\Psi_{1j}, \cdots, \Psi_{Nj})^T$ of which represents the wave function when the incident wave consists only of the j-th wave. We are interested in the reflection and transmission coefficient matrices $r = r(L)$ and $t = t(L)$. Let us introduce a matrix

$$g(z, z') = \begin{cases} \mathcal{T} \exp\left[i \int_{z'}^{z} dz'' \, \mathcal{E}(z'')P\right], & z > z' \\ \tilde{\mathcal{T}} \exp\left[-i \int_{z'}^{z} dz'' \, \mathcal{E}(z'')P\right], & z < z' \end{cases} \tag{9.6}$$

where T and $\tilde{T}$ are the time-ordering and anti-time-ordering operators, respectively. When applied to a product of matrices which are functions of z, T $(\tilde{T})$ arranges the matrices in the order of increasing (decreasing) z. For instance, $T\left[\mathcal{E}(z_2)\mathcal{E}(z_1)\right] = \mathcal{E}(z_1)\mathcal{E}(z_2)$, if $z_2 > z_1$. The matrix P is a diagonal matrix satisfying $P_{ij} = p_i\delta_{ij}$ and p_i is the negative z component of the vacuum wave vector for the i-th wave. It is easy to prove that $g(z, z')$ satisfies the equations

$$\frac{\partial}{\partial z}g(z, z') = i\,\mathrm{sgn}(z - z')\,\mathcal{E}(z)Pg(z, z'),$$

$$\frac{\partial}{\partial z'}g(z, z') = -i\,\mathrm{sgn}(z - z')\,g(z, z')\mathcal{E}(z')P\,. \tag{9.7}$$

Using (9.6) and (9.7), the wave equation (9.5) is transformed to an integral equation

$$\Psi(z, L) = g(z, L) - \frac{i}{2}\int_0^L dz'\,g(z, z')$$
$$\times\left[\mathcal{E}(z')P - P\mathcal{M}(z') - q^2P^{-1}\mathcal{M}(z') + q^2P^{-1}\mathcal{E}^{-1}(z')\right]\Psi(z', L)\,. \tag{9.8}$$

We take a derivative of this equation with respect to L and obtain

$$\frac{\partial\Psi(z, L)}{\partial L} = i\Psi(z, L)\alpha(L)\,, \tag{9.9}$$

where

$$\alpha(L) = \mathcal{E}(L)P - \frac{1}{2}\Psi(L, L)$$
$$\times\left[\mathcal{E}(L)P - P\mathcal{M}(L) - q^2P^{-1}\mathcal{M}(L) + q^2P^{-1}\mathcal{E}^{-1}(L)\right]\,. \tag{9.10}$$

Taking now the derivative of $\Psi(L, L)$ with respect to L, we obtain

$$\frac{d\Psi(L, L)}{dL} = i\mathcal{E}(L)P\left[r(L) - I\right] + i\Psi(L, L)\alpha(L)\,. \tag{9.11}$$

Since $\Psi(L, L) = I + r(L)$, we find the invariant embedding equation satisfied by $r(L)$:

$$\frac{dr}{dL} = i\left[r(L)\mathcal{E}(L)P + \mathcal{E}(L)Pr(L)\right] - \frac{i}{2}[r(L) + I]$$
$$\times\left[\mathcal{E}(L)P - P\mathcal{M}(L) - q^2P^{-1}\mathcal{M}(L) + q^2P^{-1}\mathcal{E}^{-1}(L)\right][r(L) + I]\,. \tag{9.12}$$

Similarly by setting $z = 0$ in (9.9), we find the invariant embedding equation for $t(L)$:

$$\frac{dt}{dL} = it(L)\mathcal{E}(L)P - \frac{i}{2}t(L)$$
$$\times\left[\mathcal{E}(L)P - P\mathcal{M}(L) - q^2P^{-1}\mathcal{M}(L) + q^2P^{-1}\mathcal{E}^{-1}(L)\right][r(L) + I]\,. \tag{9.13}$$

These equations are supplemented with the initial conditions, $r(0) = 0$ and $t(0) = I$. We solve the coupled differential equations (9.12) and (9.13) numerically using the initial conditions and obtain the reflection and transmission coefficient matrices r and t as functions of L. The invariant embedding method can also be used in calculating the field amplitude $\Psi(z)$ inside the medium. Rewriting (9.9), we get

$$\frac{\partial \Psi(z,l)}{\partial l} = i\Psi(z,l)\mathcal{E}(l)P - \frac{i}{2}\Psi(z,l)$$
$$\times \left[\mathcal{E}(l)P - P\mathcal{M}(l) - q^2 P^{-1}\mathcal{M}(l) + q^2 P^{-1}\mathcal{E}^{-1}(l) \right] \left[r(l) + I \right]. \quad (9.14)$$

For a given z $(0 < z < L)$, the field amplitude $\Psi(z,L)$ is obtained by integrating this equation from $l = z$ to $l = L$ using the initial condition $\Psi(z,z) = I + r(z)$.

9.5 Application of IEM to the Mode-Conversion of O and X Waves

Equations (9.12), (9.13) and (9.14) are the starting point in our exact analysis of a variety of wave coupling and mode conversion phenomena. In the rest of this paper, we demonstrate the utility of our invariant embedding equations by applying them to the high frequency wave propagation and mode conversion in cold, magnetized plasmas.

We assume that the plasma density varies only in the z direction and the external magnetic field $\mathbf{B_0}$ $(= (B_0 \sin\theta, 0, B_0 \cos\theta))$ is directed parallel to the xz plane and makes an angle θ with the z axis. The cold plasma dielectric tensor, ϵ, for high frequency waves in the present geometry is written as

$$\epsilon = \begin{pmatrix} \epsilon_1 + (\epsilon_3 - \epsilon_1)\sin^2\theta & i\epsilon_2 \cos\theta & (\epsilon_3 - \epsilon_1)\sin\theta\cos\theta \\ -i\epsilon_2 \cos\theta & \epsilon_1 & i\epsilon_2 \sin\theta \\ (\epsilon_3 - \epsilon_1)\sin\theta\cos\theta & -i\epsilon_2\sin\theta & \epsilon_1 + (\epsilon_3 - \epsilon_1)\cos^2\theta \end{pmatrix}, \quad (9.15)$$

where

$$\epsilon_1 = 1 - \frac{\omega_p^2 (\omega + i\nu)}{\omega \left[(\omega + i\nu)^2 - \omega_c^2 \right]},$$

$$\epsilon_2 = \frac{\omega_p^2 \omega_c}{\omega \left[(\omega + i\nu)^2 - \omega_c^2 \right]},$$

$$\epsilon_3 = 1 - \frac{\omega_p^2}{\omega (\omega + i\nu)}. \quad (9.16)$$

The constant ν is the phenomenological collision frequency. The spatial inhomogeneity of plasmas enters through the z dependence of the electron number density n.

For monochromatic waves of frequency ω, the wave equation satisfied by the electric field in cold magnetized plasmas has the form

$$\nabla^2 \mathbf{E} + \frac{\omega^2}{c^2} \epsilon \cdot \mathbf{E} = 0 . \tag{9.17}$$

In this paper, we restrict our interest to the case where plane waves propagate parallel to the z axis. In this situation, we can eliminate E_z from (9.17) and obtain two coupled wave equations satisfied by $E_x = E_x(z)$ and $E_y = E_y(z)$, which turn out to have precisely the same form as (9.5) with $q = 0$ and

$$\psi = \begin{pmatrix} E_x \\ E_y \end{pmatrix}, \quad K = \frac{\omega}{c} I, \quad \mathcal{E} = I,$$

$$\mathcal{M} = \frac{1}{\epsilon_1 + (\epsilon_3 - \epsilon_1) \cos^2 \theta}$$

$$\times \begin{pmatrix} \epsilon_1 \epsilon_3 & i\epsilon_2 \epsilon_3 \cos \theta \\ -i\epsilon_2 \epsilon_3 \cos \theta & \epsilon_1 \epsilon_3 - \left[(\epsilon_3 - \epsilon_1) \epsilon_1 + \epsilon_2^2 \right] \sin^2 \theta \end{pmatrix} . \tag{9.18}$$

We use (9.12), (9.13) and (9.18) in calculating the reflection and transmission coefficients. In our notation, $r_{11}(r_{21})$ is the reflection coefficient when the incident wave is E_x (that is, linearly polarized in the x direction) and the reflected wave is $E_x(E_y)$. Similarly, $r_{22}(r_{12})$ is the reflection coefficient when the incident wave is E_y (that is, linearly polarized in the y direction) and the reflected wave is $E_y(E_x)$. Similar definitions are applied to the transmission coefficients. The reflectances and transmittances are defined by $R_{ij} = |r_{ij}|^2$ and $T_{ij} = |t_{ij}|^2$. When the dielectric permittivities of the incident region and the transmitted region are the same, we can calculate the wave absorption by $A_j \equiv 1 - R_{1j} - R_{2j} - T_{1j} - T_{2j}$ $(j = 1, 2)$. If a mode conversion occurs, this quantity is nonzero even in the limit where the damping constant ν goes to zero. We will call A_j the mode conversion coefficient.

For a specific calculation, we assume that the electron density profile is given by

$$n(z) = 10^4 \left[9 \left(1 - \frac{z}{L} \right)^2 + 1 \right] \text{ m}^{-3} \tag{9.19}$$

where the plasma thickness L is equal to 10^5 m. The electromagnetic wave is incident from the vacuum region where $z > L$ and transmitted to the vacuum region where $z < 0$. We also assume that the electron cyclotron frequency is equal to $\omega_c = 6 \times 10^3$ rad/s.

In Fig. 9.6, we plot the reflectances R_{11}, R_{12} $(= R_{21})$, R_{22}, the transmittances T_{11}, T_{12}, T_{21}, T_{22} and the mode conversion coefficients A_1, A_2 as functions of the normalized frequency $\omega L/c$, when a wave is incident perpendicularly on the stratified plasma. The external magnetic field is directed at $\theta = 45°$ from the z axis. The resonance associated with the extraordinary wave component occurs for $\omega_{e1} \le \omega \le \omega_{e2}$ where ω_{e1} and ω_{e2} are the minimum and maximum values of

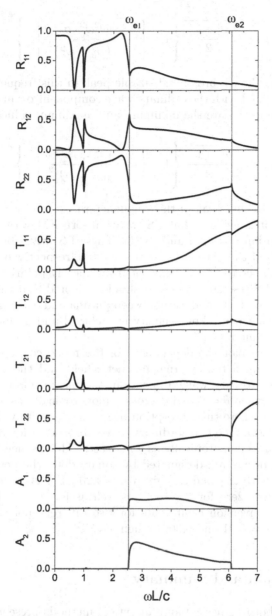

Fig. 9.6. Reflectances, transmittances and the mode conversion coefficients when the external magnetic field is directed at 45° from the z axis and the wave is incident parallel to the z axis. The mode conversion coefficients are nonzero for $\omega_{e1} \leq \omega \leq \omega_{e2}$, where ω_{e1} and ω_{e2} are the minimum and maximum values of ω_e defined by (9.20) and for $\omega_{o1} \leq \omega \leq \omega_{o2}$, where ω_{o1} and ω_{o2} are the minimum and maximum values of ω_o defined by (9.21)

$$\omega_e \equiv \sqrt{\frac{\omega_p^2 + \omega_c^2}{2}} \left\{ 1 + \left[1 - \frac{4\omega_p^2 \omega_c^2 \cos^2 \theta}{\left(\omega_p^2 + \omega_c^2\right)^2} \right]^{\frac{1}{2}} \right\}^{\frac{1}{2}}. \tag{9.20}$$

Both A_1 and A_2 show a broad and sizable peak in this frequency range. The resonance associated with the ordinary wave component occurs for $\omega_{o1} \leq \omega \leq \omega_{o2}$ where ω_{o1} and ω_{o2} are the minimum and maximum values of

$$\omega_o \equiv \sqrt{\frac{\omega_p^2 + \omega_c^2}{2}} \left\{ 1 - \left[1 - \frac{4\omega_p^2 \omega_c^2 \cos^2 \theta}{\left(\omega_p^2 + \omega_c^2\right)^2} \right]^{\frac{1}{2}} \right\}^{\frac{1}{2}}. \tag{9.21}$$

A_1 and A_2 show a tiny peak in this frequency range.

In Fig. 9.6, it is evident that ES waves absorb EM wave energy via the resonant absorption at (9.20) and (9.21). These ES waves become the UHR when $\theta \approx \pi/2$ and the plasma oscillations $\theta \approx 0$, respectively. Thus A_1 (A_2) denotes the mode conversion coefficient from linearly polarized EM waves of E_x (E_y) to the ES resonant modes. It should be noted that these absorptions are not affected by the boundaries between plasma and vacuum since both A_1 and A_2 disappear when inhomogeneity is excluded and uniform plasmas are assumed in the same region.

In Fig. 9.7, we plot the dependence of the mode conversion coefficients on the angle between the external magnetic field and the z axis. Both for A_1 and A_2, the ordinary wave absorption, the magnitude of which is quite small, is largest when $\theta = 0$ and decreases monotonically as θ increases. The frequency ranges where this absorption occurs agree quite well with ω_{o1} and ω_{o2} given by (9.21). The extraordinary wave absorption for A_2 is zero when $\theta = 0$ and increases monotonically as θ increases. The frequency ranges where this absorption occurs are designated by square dots, the positions of which agree precisely with ω_{e1} and ω_{e2} given by (9.20). For A_1, the extraordinary wave absorption is zero for $\theta = 90°$, as well as for $\theta = 0$. In Fig. 9.8, we plot the mode conversion coefficients for two different normalized damping coefficients $\nu L/c = 0.01$ and 0.0001 when $\theta = 45°$.

9.6 Discussion and Summary

Our results of both numerical and analytical methods presented above show that both O and X waves are likely to give energy into the UHR when the inhomogeneity lies perpendicular to the ambient magnetic field. It is suggested that the EM electron waves are mode-converted to the ES electron waves via the resonant absorption at the UH frequency. Our simulation results are found to be also consistent with previous studies such as investigation of the Budden problem by White and Chen [19], Grebogi et al. [6], Lin and Lin [14], Antani et al. [1, 2], and Ueda et al. [18].

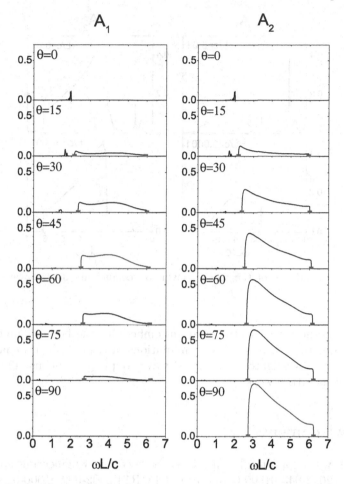

Fig. 9.7. Dependence of the mode conversion coefficients on the angle between the external magnetic field and the z axis. Square dots represent the positions of ω_{e1} and ω_{e2}

However, there are several limitations of our calculations. We neglect the effects of collisions and plasma pressure which might limit the growth of UHR amplitude [White and Chen, 19]. In addition, when the finite pressure is introduced, the ES upper hybrid waves are able to propagate and no longer stationary. Under certain circumstances such as the density striations these ES waves can form the cavity modes and emit a small fraction of O (or X) waves into space [8, 15, 17, 20].

In this study the density gradient is assumed to be perpendicular to the background magnetic field. This condition is valid probably only at the equatorial region and it should be extended to the case of arbitrary angles with respect to the Earth's magnetic field in the high-latitude region. Recent

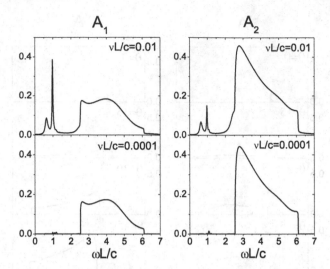

Fig. 9.8. Dependence of the mode conversion coefficients on the wave damping parameter ν

analytical methods such as the invariant embedding method of Kim et al. [10] as well as our time-dependent 3-D simulations in Kim et al. [11] enable us to quantitatively investigate such general wave coupling problems. This subject will be left as future work.

Acknowledgments

This work was supported by the Korea Science and Engineering Foundation grant R14-2002-043-01000-0 and in part by R14-2002-062-01000-0.

References

[1] Antani, S.N., N.N. Rao, and D.J. Kaup: Geophys. Res. Lett. 18, 2285 (1991).
[2] Antani, S.N., D.J. Kaup, and N.N. Rao: J. Geophys. Res. 101, 27,035 (1996).
[3] Bellman, R. and G.M. Wing: *An Introduction to Invariant Imbedding* (Wiley, New York, 1976).
[4] Budden, K.G.: *The Propagation of Radio Waves* (Cambridge, Cambridge, 1985).
[5] Ginzburg, V.L.: *Propagation of electromagnetic waves in plasmas* (Pergamon Press, New York, 1964).
[6] Grebogi, C., C.S. Liu, and V.K. Tripathi: Phys. Rev. Lett. 39, 338 (1977).
[7] Hinkel-Lipsker, D.E., B.D. Fried, and G.J. Morales: Phys. Fluids B 4, 559 (1992).
[8] Hughes, J.M. and J. LaBelle: Geophys. Res. Lett, 28, 123 (2001).

[9] Kim, E.-H. and D.-H. Lee: Geophys. Res. Lett. 30, 2240 (2003).

[10] Kim, K., D.-H. Lee, and H. Lim: Europhys. Lett. 69, to appear (2005).

[11] Kim, K.-S., D.-H. Lee, E.-H. Kim, and K. Kim: Geophys. Res. Lett., submitted (2005)

[12] Lee, D.-H. and R.L. Lysak: J. Geophys. Res. 94, 17,097 (1989).

[13] Lee, D.-H., M.K. Hudson, K. Kim, R.L. Lysak, and Y. Song: J. Geophys. Res. 107, 1307 (2002).

[14] Lin, A.T. and C.C. Lin: Phys. Fluids 27, 2208 (1984).

[15] Shepherd, S.G., J. LaBelle, and M.L. Trimpi: Geophys. Res. Lett. 24, 3161 (1997).

[16] Swanson, D.G.: *Theory of Mode Conversion and Tunneling in Inhomogeneous Plasmas* (Wiley, New York, 1998).

[17] Weatherwax, A.T., J. LaBelle, M.L. Trimpi, R.A. Treumann, J. Minow, and C. Deehr: J. Geophys. Res. 100, 7745 (1995).

[18] Ueda, H.O., Y. Omura, and H. Matsumoto: Ann. Geophysicae 16, 1251 (1998).

[19] White, R.B. and F.F. Chen: Plasma Phys. 16, 565 (1974).

[20] Yoon, P.H., A.T., Weatherwax, and T.J. Rosenberg: J. Geophys. Res. 103, 4071 (1998).

Plasma Waves Near Reconnection Sites

A. Vaivads[1], Yu. Khotyaintsev[1], M. André[1], and R.A. Treumann[2,3]

[1] Swedish Institute of Space Physics, Uppsala, Sweden
andris@irfu.se, yuri@irfu.se, mats.andre@irfu.se
[2] Geophysics Section, The University of Munich, Munich, Germany
tre@mpe.mpg.de
[3] Department of Physics and Astronomy, Dartmouth College, Hanover, NH, USA

Abstract. Reconnection sites are known to be regions of strong wave activity covering a broad range of frequencies from below the ion gyrofrequency to above the electron plasma frequency. Here we explore the observations near the reconnection sites of high frequency waves, frequencies well above the ion gyrofrequency. We concentrate on in situ satellite observations, particularly on recent observations by the Cluster spacecraft and, where possible, compare the observations with numerical simulations, laboratory experiments and theoretical predictions. Several wave modes are found near the reconnection sites: lower hybrid drift waves, whistlers, electron cyclotron waves, Langmuir/upper hybrid waves, and solitary wave structures. We discuss the role of these waves in the reconnection onset and supporting the reconnection, in anomalous resistivity and diffusion, as well as a possibility for using these waves as a tool for remote sensing of reconnection sites.

Key words: Reconnection, Hall region, separatrix physics, lower hybrid waves, whistlers, high-frequency waves, electron holes, anomalous resistance, structure of reconnection region, wave signatures

10.1 Background

Collisionless magnetic reconnection in space plasma has the two important properties of converting the available free magnetic energy into kinetic energy of charged particles in large regions of space and causing significant mass and energy transfer across the boundaries that separate the interacting plasmas. The regions where the energy conversion takes place, e.g. the auroral zone, ionosphere, shocks, etc., emit waves and generate turbulence over a wide frequency range. The occurrence of reconnection is not exceptional. To the contrary, reconnection is abundant in collisionless plasmas taking place everywhere where sufficiently thin current sheets are generated. Understanding the role of waves and turbulence in the energy conversion, energy trans-

A. Vaivads et al.: *Plasma Waves Near Reconnection Sites*, Lect. Notes Phys. **687**, 251–269 (2006)
www.springerlink.com

port, and structure formation of the reconnection sites is thus an important and challenging task.

Astrophysical environments allow studies of reconnection regions only from remote by observing the emitted electromagnetic radiation. For instance, the electron beams which cause solar and interplanetary type III radio bursts at the local plasma frequency and its harmonics are believed to originate from regions near reconnection sites in the solar corona [see, e.g., Cane et al., 9]. Remote studies of the emission provide solely average information about the reconnection sites with the averaging proceeding over large spatial volumes. They therefore suffer from severe limitations on the spatial resolution. As a consequence the information obtained about local conditions and micro processes in the reconnection region is very limited and in most cases no information can be extracted at all. The only places where reconnection sites can be studied in detail are in the laboratory and the Earth's magnetosphere or other accessible environments in our solar system that have been visited by spacecraft, such as the solar wind, some of the other magnetized planets, comets, and the outer heliosphere. Spacecraft observations give a much more detailed picture of the plasma dynamics than any of the laboratory experiments. This is due mainly to the possibility of resolving the particle distribution functions and electromagnetic fields at small scales, in some cases down to the smallest electron scales. In the laboratory, in addition, it is practically impossible to reproduce the collisionless and dilute conditions at the temperatures prevailing in space and astrophysical plasmas. Observations in space are thus uniquely suited for the understanding of reconnection. However, since reconnection involves many processes at different spatial and temporal scales, numerical simulations serve as a superior tool for understanding the environment and physical processes in the vicinity of reconnection sites.

The two main regions in the Earth's magnetosphere where the processes of reconnection have so far been studied are the subsolar magnetopause and the magnetotail current sheet. Under normal conditions in those regions the plasma is overdense, with $f_{pe} \gg f_{ce}$. Reconnection in the magnetotail proceeds in a relatively symmetric way, in the sense that the plasmas to both sides of the tail current sheet have similar properties. Quite an opposite situation is encountered at the magnetopause. Here reconnection is manifestly asymmetric. Another important difference between magnetopause and tail reconnection is that the typical spatial scales, e.g. the ion inertial length and the ion gyro radius, are usually a factor of ten smaller at the magnetopause than in the tail. This is important for any in situ studies where the instrumental resolution enters as a limiting factor. Altogether, a significant number of studies deal with high frequency waves at the magnetopause and in the magnetotail, but only in a few cases have attempts been made to find a direct relation between the observed waves and the reconnection process, even though in some cases the existence of such a relationship has been put forth. In the present paper we summarize in situ observations of high frequency waves under conditions when reconnection signatures are well defined as well as observations where one merely speculates about a relation of the observed waves to possible reconnection going on at a distance from the location of the observations.

Figure 10.1 shows a sketch of the reconnection site. Two oppositely directed magnetic fields in the *inflow* regions merge in the diffusion region forming an *X-line*. The magnetic field lines connected to the X-line are the *separatrices*. We call the regions close to the separatrices *separatrix regions*. Plasma containing reconnected magnetic fields is flowing away from the X-line in the *outflow* regions, where it escapes in the form of jets. This sketch draws a rather simplified two-dimensional picture of the reconnection process. In reality, the reconnection site has a considerably more complicated structure, consisting of multiple X-lines and exhibiting a complex and three-dimensional configuration. However, in many cases the simple 2D picture may serve as a lucid and sufficient approximation to a reconnection site. A counterexample is the complex structure arising from spontaneous antiparallel reconnection where patchy X-lines form within a narrow current sheet. On the other hand, when a small guide field is added, well ordered X-lines develop, and the 2D description can become sufficiently accurate to serve as an approximate description of the reconnection process [see, e.g., Scholer et al., 33].

Another general property of a single reconnection site is its pronounced inhomogeneity rendering almost all homogeneous plasma models of scales of the order of the spatial extension of the reconnection site invalid. Density and temperature gradients as well as non-Maxwellian particle distribution functions in the vicinity of a reconnection site result in the generation of various plasma wave modes. All of them contribute to the processes of particle acceleration and energy redistribution, generation of transport coefficients, and the wanted diffusion of magnetic field and plasma near the X-line which is necessary in order to maintain the merging of the oppositely directed magnetic fields. Space observations allow for the direct in situ observation of the wave-generation and wave-particle interactions in these merging processes at the reconnection sites.

10.1.1 Observations of Different Wave Modes

Some of the high frequency wave modes which are usually observed in the vicinity of a reconnection site are sketched in Fig. 10.1. The locations where these modes are observed are rather speculative because there is very limited knowledge so far on the relative locations of the different modes. Emissions with the strongest electric fields are ordinarily detected near or below the lower hybrid frequency f_{LH}, indicating the presence of lower hybrid drift waves (LHD). These emissions have both strong electric and magnetic field components. They seem to assume their highest amplitudes just near the steep density gradients like the ones found in the separatrix region.

In general the separatrix region seems to be the location of very strong wave emissions of several different wave modes in a wide frequency range.

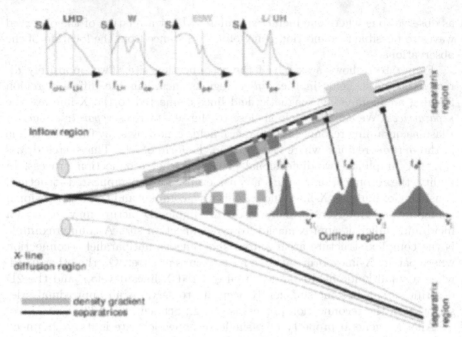

Fig. 10.1. Sketch of the reconnection site. Separatrices with density gradients are marked. A the top different kinds of wave spectra are sketched that are commonly observed near reconnection sites. Some common places to observe those waves are marked in different gray shadowing. Typical electron distribution functions in the vicinity of the separatrix are indicated as well

Strong electric fields are found around the electron plasma frequency f_{pe}. These emissions are usually believed to be Langmuir (L) waves or, if oblique, upper hybrid (UH) waves. Often, electrostatic solitary waves (ESW) can also be associated with reconnection. Whistler emissions (W) are identified from narrow spectral peaks in the frequency range $f_{LH} < f < f_{ce}$ between the lower hybrid frequency and the local electron gyro-frequency f_{ce}.

Despite the applicability of such waves to astrophysical environments, there have been very few observations of radio emissions from reconnection sites at frequencies around and above the electron plasma frequency even though one expects their presence in the separatrix region where they should be generated by the fast electron beams escaping from the X line. This is probably due to the weakness of the electromagnetic signals in comparison with the electrostatic waves.

In the rest of this article we summarize the in situ observations of the above mentioned wave modes and discuss their possible generation mechanisms and relations to reconnection. Before doing so we also summarize some of the results obtained from numerical simulations dealing with high frequency waves or being relevant to the discussion in this paper.

10.2 Numerical Simulations

The generation of high frequency waves involves the dynamics of electron and electron kinetic effects which in most cases are very important. Therefore, full-particle and Vlasov simulations are best qualified for this kind of study. Such simulations are also computationally expensive. So far relatively few wave modes have been addressed. An additional difficulty is that in some cases, for example in the simulation of lower hybrid drift waves, it is absolutely essential to simulate the phenomena in all three spatial dimensions.

The available numerical simulations indicate that the separatrix regions are the most dynamically active regions [see, e.g., Cattell et al., 10]. Intense electron beams are generated in the reconnection process along the separatrices by parallel electric fields which are distributed along the separatrices as shown by Pritchett [30] and Hoshino et al. [17]. In addition, electron conics and shell distributions can form due to diverging magnetic flux tubes close to the reconnection site. High-frequency wave modes that have been studied in detail by numerical simulations are the following:

- *Lower hybrid drift waves* (LHD). It has been suggested that these waves play a crucial role in the narrowing of the current sheets with the subsequent onset of reconnection [see, e.g., 11, 33, 35]. LHD waves tend to be electrostatic ($\delta E/\delta B > V_A$). Their largest amplitudes are found at the edges of the current sheet. They interact efficiently with both electrons and ions and can cause a significant anomalous resistivity and corresponding anomalous diffusion. It has been realized that these waves possess a significant magnetic component in the center of the current sheet that also contributes to anomalous resistivity [Silin et al., 35].

- *Solitary waves* (SW) and *double layers* (DL). Solitary wave and double layer generation due to electron beams have been studied in great detail [e.g., Omura et al., 29]. Electron beams form mainly close to the separatrices as shown by Hoshino et al. [17] and Pritchett and Coroniti [31] and particularly under guide field conditions. In this case strong double layers can be generated at the reconnection site [Drake et al., 13]. Solitary structures and lower hybrid waves may couple as well, as has been found in the same simulations.

High frequency waves that have been studied in numerical simulations much less frequently. The main types investigated are:

- *Whistlers.* Numerical simulations show that the Hall term in the Generalized Ohm's law is important for the onset of fast reconnection [Birn et al., 8]. The Hall term also introduces whistlers into the system, and it has been speculated that the region close to the reconnection site is some kind of a standing whistler. Observations indicate that electromagnetic modes in the whistler frequency range are observed also at distances far from the reconnection site, but this type of emission has received very little attention in numerical studies.

- *Waves at the plasma frequency.* While observations indicate the presence of strong Langmuir/upper hybrid waves in the separatrix regions, these waves have not been studied in any of the reconnection simulations.
- *Electron cyclotron waves.* Similarly, observations indicate the presence of electron cyclotron waves and speculate about their relation to reconnection, but in the numerical simulations they have not been studied yet.
- *Radio emissions.* Free-space modes are usually generated at the local plasma frequency or above. They can freely propagate out into space. They are of primary importance in the astrophysical application of reconnection. There exists a large theoretical effort in studying these waves but the amount of numerical simulations is very limited and has mainly been done for astrophysical plasma conditions.

10.3 Lower Hybrid Drift Waves

10.3.1 Observations

Large wave-electric fields at the magnetopause and in the magnetotail are usually observed at frequencies near the local lower-hybrid frequency f_{LH} [3, 10]. The strongest peak-to-peak amplitudes are of the order $\delta\tilde{E} \geq 1$, where $\delta\tilde{E}$ is the normalized root mean square electric field. At the magnetopause the waves are strongest on the magnetospheric side where the inflow Alfvénic speed is high. Often, these waves are located in those regions (narrow sheets with spatial scale less than an ion gyroradius) where the DC electric field reaches high values, about half of the peak-to-peak amplitude of the waves.

At this time there is no statistical analysis available of the occurrence rates of these waves, but it seems that the highest-amplitude lower-hybrid waves are located in the regions of steepest density gradients. Numerical simulations and analytical calculations suggest that the observed waves are lower-hybrid drift (LHD) waves even though the modified two-stream instability could generate waves with similar properties. An example of LHD wave observations is shown in Fig. 10.2.

Particular characteristics of LHD waves are their short wavelengths, $k\rho_e \sim 1$, a broadband spectrum extending from frequencies well below to well above f_{LH}, perpendicular wave numbers $k_\perp \gg k_\parallel$, a phase velocity of the order of the ion thermal velocity, and a coherence length of the order of one wavelength. The observations also suggest that the wave potential can be close to the electron thermal energy [Bale et al., 6].

Based on spectral and interferometric results, some satellite observations support the LHD-interpretation [Vaivads et al., 38]. However, more detailed studies are required in order to confirm the lower-hybrid drift nature of the observed waves. Observations also show that these waves have a significant magnetic component with $c \gg \delta E/\delta B \gg v_A$ and can have a preferential direction of propagation along the ambient magnetic field [Vaivads et al., 38].

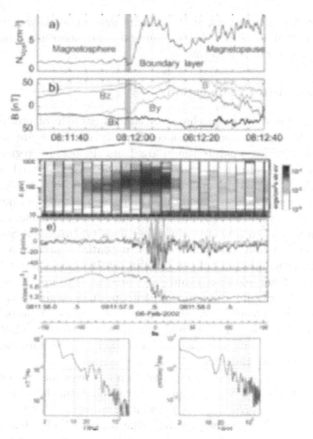

Fig. 10.2. Example of LHD wave observations near the separatrices far from the reconnection X-line. The figure is adopted from Vaivads et al. [38] and André et al. [4]. (**a**), (**b**): Large-scale density (obtained from the satellite potential) and magnetic field observations. The rest of the panels are zoomed-in on the time interval marked bright when the strongest electric fields were observed. This time is identified as a possible separatrix region [André et al., 4]. (**c**): Electron spectrogram showing electron beams along B. (**d**): Electric field normal and tangential to the magnetopause. (**e**): Plasma density. (**f**): Magnetic field spectra, and (**g**): Electric field spectra (Reprinted with permission of American Geophysical Union)

10.3.2 Generation Mechanisms

In a simplified picture, the driving force for the LHD waves is a density gradient with relative flow between electrons and ions due to their different diamagnetic drifts. In the case of the modified-two stream instability (MTSI) it is the cross-field drift of the electrons with respect to the ions that presents the driving force. Strong DC electric fields on scales smaller than ion gyroradius are almost always seen in association with LHD waves. From the energetic point of view it has been shown that the observed waves can also be generated by the electron beams present at the density gradients [Vaivads et al., 37]. It is well-known that in places like the auroral zone electron beams do indeed generate intense lower-hybrid waves.

10.3.3 Relation to Reconnection

Reconnection is one of the mechanisms that is capable of producing strong and narrow density gradients in space. Density gradients are formed along the separatrices which separate the inflow from the outflow regions. The separatrices contain local density dips which have been observed in numerical simulations [see, e.g., 34] . Numerical simulations and observations also suggest that the separatrices can maintain their steep density gradient structure over distances very far away from the reconnection site (tens of λ_i). It has not yet been explained properly why these density dips exist. Such dips can, however, evolve when a static field-aligned potential is applied along the separatrix which evacuates part of the plasma locally. The separatrices are also regions where strong electron beams are present [Hoshino et al., 17]. Such beams have a longitudinal pressure anisotropy and therefore are hardly capable of excluding plasma from the separatrix regions. A probable cause is a magnetic-field-aligned electric field which accelerates the electrons into a beam thereby evaporating the plasma. Nevertheless, it remains unclear which mechanism maintains the required pressure balance.

The role of these beams in the generation of waves has not been fully explored. It is not clear, moreover, which other mechanisms besides reconnection could produce such narrow (few λ_i) density gradients. There are speculative ideas, like "peeling" or "snow-plowing" due to, e.g., FTEs, but no clear understanding exists as yet of how such a process would work. Thus, while we expect strong LHD waves near the reconnection site along the separatrices it is not yet clear whether all or most of the intense LHD-wave observations are related to ongoing reconnection.

Lower hybrid waves can affect reconnection in several important ways:

- Through anomalous resistivity: Usually, in simulations and observations the most intense LHD waves generate anomalous collision frequencies for electrons of the order of $\nu_{an}^{e^-} \sim 2\pi f_{LH}$ [Silin et al., 35].
- Through electron acceleration: While the phase velocity of lower hybrid waves in the direction perpendicular to the magnetic field $\mathbf{B}$ is comparable to the ion thermal speed, the phase velocity along the magnetic field becomes comparable to the electron thermal velocity thus enabling the LHD waves to resonate with thermal electrons and efficiently accelerate these electrons. For the same reason electron beams can be efficient in generating lower hybrid waves by the inverse resonance process.
- Through current sheet bifurcation, thinning and reconnection onset: Numerical simulations have shown that LHD waves apparently play a crucial role in the reconnection onset within thin current sheets [11, 33].

They evolve due to the steep plasma gradients at the current sheet boundary and in fact tend to broaden the current sheet. When propagating into the sheet they contribute to anomalous resistivity, heating and the diffusivity necessary for reconnection. This issue is controversial and has not been settled.

10.4 Solitary Waves and Langmuir/Upper Hybrid Waves

10.4.1 Observations

Other quasi-stationary structures containing large electric fields that have been observed close to the reconnection sites are electrostatic solitary waves (ESW), broad-band electrostatic noise (BEN) and Langmuir/upper hybrid waves [e.g., Deng et al., 12]. They all tend to appear approximately in the same region and have similar amplitudes, which are usually several or many times smaller than the amplitudes of the LHD waves. Part of BEN observations are due to ESW passing over the spacecraft thus giving rise to a broadband spectrum. Observations show that the strongest emissions are observed along the separatrices [10, 14, 38]. The emissions change their character rapidly but it seems that narrow-band emissions at the Langmuir frequency and broadband emissions do not appear simultaneously [Khotyaintsev et al., 22]. So far only a rough comparison has been possible between the electron distribution functions and the wave characteristics [Deng et al., 12]. In the magnetotail these waves are usually located at the boundary between lobe and plasmasheet. It has been suggested that they are related to the reconnection process [Kojima et al., 24].

In order to distinguish Langmuir waves from upper hybrid waves one must study the polarization of the waves. It has been found that near the electron plasma frequency the wave electric fields are often polarized at large angles with respect to the ambient magnetic field indicating that the observed waves are upper hybrid waves [12, 14, 21, 23] rather than Langmuir waves. Examples of observations are shown in Fig. 10.3.

10.4.2 Generation Mechanisms

It is known from numerical simulations, [e.g. Omura et al., 29], that Langmuir modes are usually driven by the weak-beam instability. ESW can be the result of saturation of the nonlinear bump-on-tail instability or the two-stream instability. Upper hybrid waves can also be generated by beams, however they can also be generated by loss cone and shell distributions. Such distributions form preferentially in diverging magnetic fields either close to the reconnection site or when magnetic flux tubes approach the Earth.

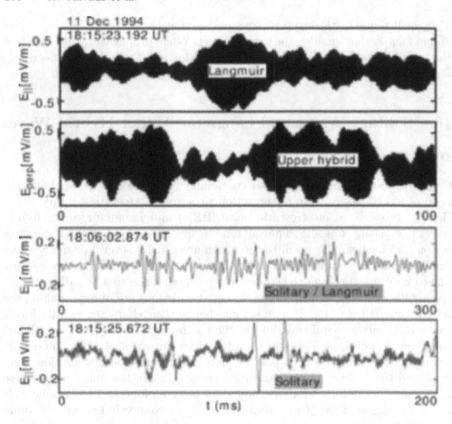

Fig. 10.3. Example of wave observations near the separatrices of a reconnection X-line. (**a, b**) Monochromatic waves which can be interpreted as Langmuir waves when $E_\parallel \gg E_\perp$ and as upper hybrid waves when $E_\perp \gg E_\parallel$. (**c**) Example of a mixture between electrostatic solitary waves and Langmuir waves. (**d**) Electrostatic solitary waves [figure adapted from Deng et al., 12]

10.4.3 Relation to Reconnection

One of the major questions of reconnection is how parallel electric fields are distributed near the reconnection site. These fields are required to create changes in the magnetic field-line topology that is associated with reconnection. ESW with a net potential drop can be one source for parallel electric fields. At the same time ESW and L/UH waves interact efficiently with electrons and generate high energy tails on the electron distribution functions. ESW also contribute to the anomalous resistivity in field-aligned currents.

10.5 Whistlers

10.5.1 Observations

There are many observations related to whistler emission in regions that are directly related to reconnection, such as the plasma sheet boundary layer [see, e.g., Gurnett et al., 16] and the magnetopause [LaBelle and Treumann, 26] and more recently [Stenberg et al., 36]. Whistler wave modes can be identified on the basis of the observed frequency of the waves (in between the electron and ion gyro-frequencies), the presence of a strong magnetic component that can reach ~0.1 nT [Gurnett et al., 16], or the direct measurement of wave polarization [Zhang et al., 39]. Narrow spectral peaks in a wide region of frequencies between the lower hybrid and electron cyclotron frequencies are typical for these waves. However, broadband spectra extending over whistler frequencies are also often observed [LaBelle and Treumann, 26]. Such waves can consist of whistlers or they can be associated with the magnetic component of lower hybrid drift waves. Figure 10.4 gives an example of wave measurements in the high-altitude magnetopause/cusp region showing indications that the reconnection process proceeds in a high-beta plasma [Khotyaintsev et al., 22].

10.5.2 Generation Mechanisms

In addition to the temperature anisotropy or loss-cone instabilities by which it is conventionally known, whistlers can be excited as a consequence of the perpendicular anisotropy of the electron distribution function, or ion beams [Akimoto et al., 1]. Electron beams [Zhang et al., 39] have also been suggested as possible generators for whistlers under conditions when the loss cone is absent and the plasma is isotropic. This becomes possible since electrons or ions of sufficiently high speed can undergo resonance with whistlers which not only Landau damps but, under certain circumstances, also excites whistlers.

In this spirit it has been shown that whistler waves in the magnetotail are most probably generated by electron beams through the cyclotron resonance and not through the temperature anisotropy or other non-beam instability mechanisms [Zhang et al., 39]. The estimated resonance energy of the electron beam is about 10 keV suggesting reconnection as the most probable source of the beam. The instability mechanism producing broad-band magnetic turbulence is less clear. Laboratory experiments are consistent with the modified two-stream instability producing the emissions, since the waves propagate in the same direction of the electron flow [Ji et al., 20]. Part of these emissions can be associated with the magnetic component of lower hybrid drift waves.

10.5.3 Relation to Reconnection

The most direct evidence that whistlers can play a crucial role in the reconnection process comes from laboratory experiments which show that the reconnection rate correlates with the amplitude of obliquely propagating broad-band

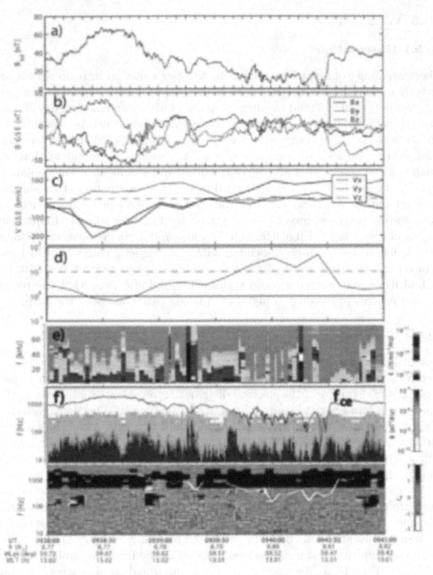

Fig. 10.4. Example of high frequency wave observations in the high-altitude magnetopause/cusp [adopted from Khotyaintsev et al., 22]. (**a**, **b**): The magnetic field magnitude and components, (**c**): convection velocity $\mathbf{E} \times \mathbf{B}$, (**d**): plasma beta, (**e**): spectra of the electric field in the 2–80 kHz range, (**f**): spectra of the magnetic field in the 8–4000 Hz range, and electron-cyclotron frequency, (**g**): polarization of wave magnetic field with respect to the ambient magnetic field. Whistlers are identified from right-hand polarization

whistler waves inside the reconnecting current sheet [cf., Ji et al., 20]. Also, numerical simulations suggest that the magnetic fluctuations in the whistler band can cause a significant anomalous resistivity [Silin et al., 35]. Such observations are yet to be confirmed by the observations of reconnection in space. So far, space observations indicate that whistlers are associated with the reconnection processes [Stenberg et al., 36] and most probably are related to electron beams generated during reconnection [Zhang et al., 39], however much more detailed studies are required to confirm that the whistlers are not related secondarily but play an essential role in the process of reconnection itself. An important aspect of whistlers is their ability to propagate over large distances away from the reconnection site without appreciable damping. This property makes whistlers a perfect tool for use in remote sensing of reconnection sites. In addition, whistlers by this particular property may transport information from the reconnection site to other places in the plasma.

10.6 Electron Cyclotron Waves

10.6.1 Observations

Electron cyclotron waves are commonly observed in the inner regions of the magnetosphere, i.e. the polar cusp, auroral zone, and plasmasphere. The observations from the outer magnetosphere are not as abundant. Electron cyclotron waves have been observed in association with flux transfer events [2, 25] and the emerging of energetic plasma in the magnetotail [Gurnett et al., 16]. Both electrostatic and electromagnetic cyclotron waves have been observed at the magnetopause (Anderson et al. 1982). Observations in the cusp and close to the magnetopause indicate that electron cyclotron waves tend to be generated on open field lines [Menietti et al., 28]. In addition, observations show that there can be a close correspondence between observations of electron cyclotron waves and solitary waves [Menietti et al., 28].

10.6.2 Generation Mechanisms

The simplest way to excite electron cyclotron waves is again by transverse temperature anisotropies or loss cones which excite diffuse electron cyclotron waves between the harmonic bands. This has been recognized early. Purely transverse electron cyclotron waves are Bernstein modes which normally are damped and represent merely resonances. However, again, when other kinds of distribution functions are present, the resonance can be inverted and waves can be excited instead of being damped. This is, in particular, the case when electron beams pass the plasma and has been suggested as a possible generation mechanism based on particle observations and close association of electron cyclotron waves with solitary waves [Menietti et al., 28]. Other possible sources of instability as for instance loss-cones, temperature anisotropies, and horse-shoe distributions exist as well. The different excitation mechanisms have been discussed in LaBelle and Treumann [27]. In case of electron beams, it is crucial that in addition to the beam a cold electron population is present in order for the waves to become unstable. Unfortunately, however, direct measurements of such a component in relation to reconnection are in most cases missing [Menietti et al., 28].

10.6.3 Relation to Reconnection

It has been suggested that electron beams generating electron cyclotron waves originate in the reconnection region. The presence of these waves in association with flux transfer events as well as mainly on open field lines indicates that the reconnection process is important in creating unstable electron distribution functions. However, observations of electron cyclotron waves close to the reconnection site are so far missing. It has been suggested that the steep magnetic field gradients encountered near the reconnection site should preclude the generation of electron cyclotron waves or inhibit them from developing significant amplitudes.

10.7 Free Space Radiation

10.7.1 Observations

Electromagnetic radio emissions above the electron plasma frequency can propagate freely throughout the plasma and thus be detected at large distance from the source. This makes the observation of radiation a perfect tool for remote diagnostics of reconnection whenever reconnection sites emit such radiation. Observation of radiation emitted from reconnection is thus of high importance in particular in the astrophysical application, e.g. in the interpretation of the radiation emitted from the solar corona [see, eg., the compilation given in Aschwanden, 5]. In fact, most of the solar radio emission at meter wavelengths is believed to be generated in some coherent emission process that is somehow related to ongoing reconnection in the solar corona [Bastian et al., 7]. These emissions can be classified into different classes of which the most important for our purposes are those which are emitted close to the local plasma frequency f_{pe} and become free space modes as they propagate away from the generation region. Coherent radio emissions from other astronomical objects, such as stellar flares and brown dwarfs, are believed to be generated in similar ways [Güdel, 15]. The impossibility of performing in situ measurements in the coronal reconnection regions raises the importance of observations at the accessible magnetospheric reconnection sites. Unfortunately, so far there are no in situ studies of electromagnetic wave generation close to reconnection sites in the magnetosphere. Until now, most attention has been paid to the electromagnetic emission generation near the bow-shock where these waves are strongest. However, since reconnection sites are sources of fast particles which are injected into the environment one expects that they are also sources of radio wave emission.

10.7.2 Generation Mechanisms

The generation mechanisms of possible free space modes in reconnection are not entirely clear. Several possibilities have been suggested. For example, transverse free space modes (T) can be generated by mode conversion of Langmuir $(L$-waves) at steep density gradients, by mode coupling of Langmuir waves with ion sound (S) or other Langmuir (L') waves according to the relations $L + L' \rightarrow T, L + S' \rightarrow T$ or through direct electron gyro-resonance emission. Of these mechanisms, the latter is the least probable in reconnection as it depends on two conditions. First, the plasma has to be relativistic or at least weakly relativistic. Second, and even more crucial, the plasma has to be underdense with $f_{pe} < f_{ce}$. Close to the reconnection site the involved plasma is, however, overdense as stated in the introduction. In this case cyclotron damping of the free space modes inhibits their excitation. Thus, it seems improbable that reconnection sites would radiate by the gyro-resonance mechanism. In situ measurements close to the reconnection sites are required to distinguish between the remaining possible generation mechanisms. Finally, the large number of energetic electrons generated in reconnection might be another source of nonthermal synchrotron radiation under conditions when the electron energies reached are high.

10.7.3 Relation to Reconnection

The localization of electromagnetic radiation generation with respect to the reconnection site has not been investigated. Numerical simulations have studied how electron beams generated in the reconnection process can generate Langmuir waves which, in their turn, can mode convert to the emission of electromagnetic radiation [see, e.g. Sakai et al., 32]. This is the most probable radiation mechanism since electron beams are involved. In addition this mechanism is non-thermal. Depending on the available number of electron beam electrons the emission coming from one single reconnection site might still be below the detection threshold in the magnetosphere or at the magnetopause. For astrophysical applications like the sun and stars the ejected electron beams in type III radiation generating plasma waves are dense and intense enough to provide observable intensities. In remote astrophysical objects, however, synchrotron emission is more important as a nonthermal emission mechanism [Jaroschek et al., 18, 19]. Though it is very weak, the large numbers of particles injected from the reconnection site into a large volume and distributed there increase the emission measure in proportion to the involved volume, making such radiation a good candidate for observation even though it will not provide information about the microscopic scale of the involved astrophysical reconnection sites.

10.8 Summary and Outlook

We have summarized the in situ observations of high frequency waves, at frequencies near the lower hybrid frequency and up to the plasma frequency, generated near the reconnection sites in the Earth magnetosphere. There are many observational studies dealing with the most intense waves, such as lower hybrid drift waves, solitary waves, and Langmuir waves. Some of the observations suggest that these waves are most intense along the separatrices emanating from the reconnection sites. However detailed studies of the wave locations are still missing. Electron beams generated in the reconnection process seem to be a major free energy source that can generate different waves, but also density gradients or different kinds of distribution functions (e.g., loss-cone or horse-shoe) are important.

Electromagnetic waves such as whistlers and radio emissions are important for remote diagnostic possibilities of the reconnection sites. In the case of radio emissions there are direct astrophysical applications. However, in situ studies of both these modes in association with reconnection are very limited.

We identify several topics of high importance for further study in the near future:

- *Wave location.* How are different wave modes located with respect to the inner structure of the current sheet and the separatrices, and how does this location depend on the reconnection parameters, such as density gradients, velocity shear, plasma beta, temporal evolution? Could some of the wave modes be used to determine the distance to the reconnection X-line? Possible candidates are the intense solitary waves and Langmuir/upper hybrid waves.
- *Wave-particle interaction.* Which wave modes are most important for particle acceleration, heating, and formation of energetic tails on the electron distribution functions? Are electron beams generated in the wave particle interaction or are they generated in prompt electron acceleration in reconnection-generated electric fields?
- *Anomalous resistivity.* There exist first estimates of the anomalous resistivity and anomalous diffusion for lower hybrid drift turbulence in connection with reconnection. These results should be confirmed for different reconnection conditions. Moreover, the electromagnetic part of the anomalous resistivity needs to be studied more closely both theoretically and experimentally.
- *Radio emissions.* Where and by which mechanism is free space radiation generated near the reconnection sites? How could it be used to remotely sense the reconnection site properties such as the stationarity of reconnection, extension of the reconnection line, etc.?

References

[1] Akimoto, K., S.P. Gary, and N. Omidi: Electron/ion whistler instabilities and magnetic noise bursts, J. Geophys. Res. 92, 11209, 1987.

[2] Anderson, R.R., T.E. Eastman, C.C. Harvey, M.M. Hoppe, B.T. Tsurutani, and J. Etcheto: Plasma waves near the magnetopause, J. Geophys. Res. 87, 2087, 1982.

[3] André, M., R. Behlke, J.-E. Wahlund, A. Vaivads, A.-I. Eriksson, A. Tjulin, T.D. Carozzi, C. Cully, G. Gustafsson, D. Sundkvist, Y. Khotyaintsev, N. Cornilleau-Wehrlin, L. Rezeau, M. Maksimovic, E. Lucek, A. Balogh, M. Dunlop, P.-A. Lindqvist, F. Mozer, A. Pedersen, and A. Fazakerley: Multi-spacecraft observations of broadband waves near the lower hybrid frequency at the earthward edge of the magnetopause, Ann. Geophysicæ 19, 1471, 2001.

[4] André, M., A. Vaivads, S.C. Buchert, A.N. Fazakerley, and A. Lahiff: Thin electron-scale layers at the magnetopause, Geophys. Res. Lett. 31, 3803, 2004.

[5] Aschwanden, M.J.: Physics of the Solar Corona: An Introduction, Springer, 2004.

[6] Bale, S.D., F.S. Mozer, and T. Phan: Observation of lower hybrid drift instability in the diffusion region at a reconnecting magnetopause, Geophys. Res. Lett. 29, 33, 2002.

[7] Bastian, T.S., A.O. Benz, and D.E. Gary: Radio Emission from Solar Flares, Ann. Rev. Astron. Astrophys. 36, 131, 1998.

[8] Birn, J., J.F. Drake, M.A. Shay, B.N. Rogers, R.E. Denton, M. Hesse, M. Kuznetsova, Z.W. Ma, A. Bhattacharjee, A. Otto, and P.L. Pritchett: Geospace Environmental Modeling (GEM) magnetic reconnection challenge, J. Geophys. Res. 106, 3715, 2001.

[9] Cane, H.V., W.C. Erickson, and N.P. Prestage: Solar flares, type III radio bursts, coronal mass ejections, and energetic particles, J. Geophys. Res. 107, 14, 2002.

[10] Cattell, C., J. Dombeck, J. Wygant, J.F. Drake, M. Swisdak, M.L. Goldstein, W. Keith, A. Fazakerley, M. André, E. Lucek, and A. Balogh: Cluster observations of electron holes in association with magnetotail reconnection and comparison to simulations, J. Geophys. Res. 110, 1211, 2005.

[11] Daughton, W., G. Lapenta, and P. Ricci: Nonlinear Evolution of the Lower-Hybrid Drift Instability in a Current Sheet, Phys. Rev. Lett. 93, 105004, 2004.

[12] Deng, X.H., H. Matsumoto, H. Kojima, T. Mukai, R.R. Anderson, W. Baumjohann, and R. Nakamura: Geotail encounter with reconnection diffusion region in the Earth's magnetotail: Evidence of multiple X lines collisionless reconnection?, J. Geophys. Res. 109, 5206, 2004.

[13] Drake, J.F., M. Swisdak, C. Cattell, M.A. Shay, B.N. Rogers, and A. Zeiler: Formation of electron holes and particle energization during magnetic r, Science 299, 873, 2003.

[14] Farrell, W.M., M.D. Desch, M.L. Kaiser, and K. Goetz: The dominance of electron plasma waves near a reconnection X-line region, Geophys. Res. Lett. 29, 8, 2002.

[15] Güdel, M.: Stellar Radio Astronomy: Probing Stellar Atmospheres from Protostars to Giants, Ann. Rev. Astron. Astrophys. 40, 217, 2002.

[16] Gurnett, D.A., L.A. Frank, and R.P. Lepping: Plasma waves at the distant magnetotail, J. Geophys. Res. 81, 6059, 1976.

[17] Hoshino, M., T. Mukai, T. Terasawa, and I. Shinohara: Suprathermal electron acceleration in magnetic reconnection, J. Geophys. Res. 106, 25979, 2001.

[18] Jaroschek, C.H., R.A. Treumann, H. Lesch, and M. Scholer: Fast reconnection in relativistic pair plasmas: Analysis of particle acceleration in self-consistent full-particle simulations, Phys. Plasmas 11, 1151, 2004a.

[19] Jaroschek, C.H., H. Lesch, and R.A. Treumann, Relativistic kinetic reconnection as the possible source mechanism for high variability and flat spectra in extragalactic radio sources, Astrophys. J. 605, L9, 2004b.

[20] Ji, H., S. Terry, M. Yamada, R. Kulsrud, A. Kuritsyn, and Y. Ren: Electromagnetic Fluctuations during Fast Reconnection in a Laboratory Plasma, Phys. Rev. Lett. 92, 115001, 2004.

[21] Kellogg, P.J. and S.D. Bale: Nearly monochromatic waves in the distant tail of the Earth, J. Geophys. Res. 109, 4223, 2004.

[22] Khotyaintsev, Y., A. Vaivads, Y. Ogawa, B. Popielawska, M. André, S. Buchert, P. Décréau, B. Lavraud, and H.Rème: Cluster observations of high-frequency waves in the exterior cusp, Ann. Geophysicæ 22, 2403, 2004.

[23] Kojima, H., H. Furuya, H. Usui, and H. Matsumoto: Modulated electron plasma waves observed in the tail lobe: Geotail waveform observations, Geophys. Res. Lett. 24, 3049, 1997.

[24] Kojima, H., K. Ohtsuka, H. Matsumoto, Y. Omura, R.R. Anderson, Y. Saito, T. Mukai, S. Kokubun, and T. Yamamoto: Plasma waves in slow-mode shocks observed by Geotail Spacecraft, Adv. Space Res. 24, 51, 1999.

[25] LaBelle, J., R.A. Treumann, G. Haerendel, O.H. Bauer, and G. Paschmann: AMPTE IKRM obsevations of waves assosicated with flux transfer events in the magnetosphere J. Geophys. Res., 92, 5827, 1987.

[26] LaBelle, J. and R.A. Treumann: Plasma waves at the dayside magnetopause, Space Sci. Rev., 47, 175, 1988.

[27] LaBelle, J. and R.A. Treumann: Auroral radio emissions, 1. Hisses, roars, and bursts, Space Sci. Rev. 101, 295, 2002.

[28] Menietti, J.D., J.S. Pickett, G.B. Hospodarsky, J.D. Scudder, and D.A. Gurnett: Polar observations of plasma waves in and near the dayside magnetopause/magnetosheath, Planet. Space Sci. 52, 1321, 2004.

[29] Omura, Y., H. Matsumoto, T. Miyake, and H. Kojima: Electron beam instabilities as generation mechanism of electrostatic solitary waves in the magnetotail, J. Geophys. Res. 101, 2685, 1996.

[30] Pritchett, P.L.: Collisionless magnetic reconnection in a three-dimensional open system, J. Geophys. Res., 106, 25961, 2001.

[31] Pritchett, P.L. and F.V. Coroniti: Three-dimensional collisionless magnetic reconnection in the presence of a guide field, J. Geophys. Res. 109, 1220, 2004.

[32] Sakai, J.I., T. Kitamoto, and S. Saito: Simulation of Solar Type III Radio Bursts from a Magnetic Reconnection Region, Astrophys. J. Lett. 622, L157, 2005.

[33] Scholer, M., I. Sidorenko, C.H. Jaroschek, R.A. Treumann, and A. Zeiler: Onset of collisionless magnetic reconnection in thin current sheets: Three-dimensional particle simulations, Phys. Plasmas 10, 3521, 2003.

[34] Shay, M.A., J.F. Drake, B.N. Rogers, and R.E. Denton: Alfvénic collisionless magnetic reconnection and the Hall term, J. Geophys. Res. 106, 3759, 2001.

[35] Silin, I., J. Büchner, and A. Vaivads: Anomalous resistivity due to nonlinear lower-hybrid drift waves, Phys. Plasmas 12, submitted, 2005.

[36] Stenberg, G., T. Oscarsson, M. André, M. Backrud, Y. Khotyaintsev, A. Vaivads, F. Sahraoui, N. Cornilleau-Wehrlin, A. Fazakerley, R. Lundin, and P. Décréau: Electron-scale structures indicating patchy reconnection at the magnetopause? J. Geophys. Res. 110, submitted, 2005.

[37] Vaivads, A., M. André, S.C. Buchert, J.-E. Wahlund, A.N. Fazakerley, and N. Cornilleau-Wehrlin: Cluster observations of lower hybrid turbulence within thin layers at the magnetopause, Geophys. Res. Lett. 31, 3804, 2004.

[38] Vaivads, A., Y. Khotyaintsev, M. André, A. Retinò, S.C. Buchert, B.N. Rogers, P. Décréau, G. Paschmann, and T.D. Phan: Structure of the Magnetic Reconnection Diffusion Region from Four-Spacecraft Observations, Phys. Rev. Lett. 93, 105001, 2004.

[39] Zhang, Y., H. Matsumoto, and H. Kojima: Whistler mode waves in the magnetotail, J. Geophys. Res. 104, 28633, 1999.

Part III

High-Frequency Analysis Techniques and Wave Instrumentation

11

Tests of Time Evolutions in Deterministic Models, by Random Sampling of Space Plasma Phenomena

H.L. Pécseli[1,3] and J. Trulsen[2,3]

[1] University of Oslo, Institute of Physics, Box 1048 Blindern, 0316 Oslo, Norway
  hans.pecseli@fys.uio.no
[2] University of Oslo, Institute of Theoretical Astrophysics, Box 1029 Blindern, 0315 Oslo, Norway
  jan.trulsen@astro.uio.no
[3] Centre for Advanced Study, Drammensveien 78, 0271 Oslo, Norway

Abstract. We discuss general ideas, which can be used for estimating models for coherent time-evolutions by random sampling of data. They turn out to be particularly useful for interpreting data from instrumented spacecraft. These "new methods" are applied to examples of localized bursts of lower-hybrid waves and correlated density depletions observed on the FREJA satellite. In particular, lower-hybrid wave collapse is investigated. The statistical arguments are based on three distinct elements. Two are purely geometric, where the chord length distribution is determined for given cavity scales, together with the probability of encountering those scales. The third part of the argument is based on the actual time variation of scales predicted by the collapse model. The cavities are assumed to be uniformly distributed along the spacecraft trajectory, and it is assumed that they are encountered with equal probability at any time during the dynamical evolution. Cylindrical and ellipsoidal cavity models are discussed. It turns out that the collapsing cavity model can safely be ruled out on the basis of disagreement of data with the predicted cavity lengths and evolution time scales. Application to Langmuir wave collapse is suggested in order to check its reality and relevance.

Key words: Wave packet dynamics, lower-hybrid wave collapse, Freja observations, nonlinear dynamics, random data sampling

11.1 Introduction

When studying data from rocket or satellite observations, one might often encounter situations where the results invite an interpretation in terms of analytical models for the space-time evolution of some physical phenomena, nonlinear plasma waves for instance. Very often the spacecraft velocity is

H.L. Pécseli and J. Trulsen: *Tests of Time Evolutions in Deterministic Models, by Random Sampling of Space Plasma Phenomena*, Lect. Notes Phys. **687**, 273–297 (2006)
www.springerlink.com

so large that the observations have to be interpreted as "snap-shots" of the phenomena, and a direct comparison between analytical results for the time evolution and the observations is not feasible. The problem seems almost a dead-lock, by the implication that time evolutions can under *no* circumstances be observed under such conditions. This situation easily stimulates speculations leading to unsubstantiated claims, in the sense that the interpretation of data has been extended beyond what can be justified by the observations alone.

In the present communication we will describe a general method which allows tests of models for deterministic time evolutions, provided the database covers sufficiently many observations. The basic idea is that one can predict the probability density of observable quantities analytically, based on the deterministic model together with some plausible assumptions, and that these distributions can then be tested experimentally [1, 2]. The results cannot *prove* a theory to be correct (logically, this might not be possible anyhow), but can certainly be useful in *disproving* certain models. We give an outline of the statistical arguments, and illustrate the ideas by an analysis of data for lower-hybrid waves as observed by the Freja satellite. For this case an interpretation in terms of lower-hybrid wave collapse might be tempting and has indeed been suggested. Our analysis indicates that this interpretation is in error.

Later on we will present more detailed examples, but as an introduction it might here suffice to consider a simple (over-)idealized case, where we assume that spherical voids are forming in a plasma. Such a void is assumed to be embedded in the plasma with density n_0, and have a density depletion of Δn. Any void forms spontaneously, and its radius has a deterministic time variation as, say,

$$\mathcal{R}(t) = r_0 \frac{t - t_0}{\mathcal{T}} \left(1 - \frac{t - t_0}{\mathcal{T}}\right), \qquad (11.1)$$

for $t_0 < t < t_0 + \mathcal{T}$ and $\mathcal{R} = 0$ otherwise. There is a well defined initial time, t_0, for the formation of individual voids, but for the ensemble of voids, these times are random, and mutually independent. The time evolution of the voids is thus completely deterministic, but the formation time (and consequently the collapse times as well) are statistically distributed. When such an ensemble of voids is sampled by a spacecraft, the sampling process is itself associated with statistically distributed quantities. At a certain time, t_*, the width of the void is $2\mathcal{R}(t_*)$, but if it is determined by analyzing the data obtained by a satellite moving along a straight line orbit, we find a chord-length ℓ, which is in general shorter. We can argue that a chord length distribution can be determined analytically, with the assumption that we encounter a cavity at any position in its cross section with equal probability: any other assumption will imply that the formation of the void is correlated with the presence of the satellite! It is then a simple matter to obtain the probability density for the observed chord-lengths, ℓ.

We first consider the impact parameter b as measured from the center of the void. We note that with the given assumptions, the probability for having an impact parameter in the interval $\{b, b + db\}$, is obtained as the ratio of the two areas $2\pi b\, db$ and $\pi \mathcal{R}^2(t_*)$, assuming that the void does not change appreciably during the passage of the detector. We have $P(b)db$ being the probability of finding an impact parameter in the interval $\{b, b+db\}$, and also the relation $b = \sqrt{\mathcal{R}^2(t_*) - \ell^2/4}$. Using $P(b)db = P(\ell)d\ell$ we find

$$P(\ell|\mathcal{R}) = \frac{\ell}{2\mathcal{R}^2(t_*)}, \tag{11.2}$$

for $0 < \ell \le 2\mathcal{R}(t_*)$ and $P(\ell|\mathcal{R}) = 0$ otherwise. Since the maximum value of $\mathcal{R}$ is $\mathcal{R}_m = r_0/4$, we have the maximum value of ℓ being $\ell_m = r_0/2$. Implicit in the argument is, as said, the assumption that the void develops slowly, i.e. it does not change appreciably during the transit time of the satellite. The result (11.2) is conditional, as indicated, in the sense that $\mathcal{R}$ is assumed given at the relevant time.

The information concerning the density in the void is trivial in the present model, since we assumed $\Delta n = \text{const.}$

In addition to ℓ, also the time of interception, t_*, is statistically distributed with respect to the formation time t_0. Again, we can safely argue that the satellite intercepts the time-evolving void at a time randomly distributed in the interval $\{0, T\}$, i.e. $P(t_*)dt_* = dt_*/T$ for $0 < t_* < T$. For the simple example in (11.1), the time variation is monotonic and symmetric in the two intervals $\{0, T/2\}$ and $\{T/2, T\}$, so we need only be concerned with one of them, say the first one. The variation of $\mathcal{R}$ is restricted to the interval $\{0, r_0/4\}$.

We want to determine the statistical distribution of radii in the voids, $\mathcal{R}$, which is a consequence of the random distribution of detection times in the relation (11.1). We have $P(\mathcal{R})d\mathcal{R} = dt_*/T$. With (11.1) we find $d\mathcal{R}/dt_* = r_0/T - 2r_0 t_*/T^2$, giving

$$P(\mathcal{R}) = \frac{1}{r_0} \frac{2}{\sqrt{1 - 4\mathcal{R}/r_0}}. \tag{11.3}$$

For $\mathcal{R} > r_0/4$, we have $P(\mathcal{R}) = 0$.

Each realization of the plasma has "blobs" randomly distributed in all different stages of their time-evolution, and we encounter them with a probability depending on their cross section,

$$\sigma = \pi \mathcal{R}^2(t_*), \tag{11.4}$$

which gives

$$P(\ell) = \frac{1}{C} \int_{\ell/2}^{r_0/4} \sigma(\mathcal{R}) P(\ell|\mathcal{R}) P(\mathcal{R}) d\mathcal{R}, \tag{11.5}$$

where $\mathcal{C}$ is a normalization constant, introduced because the probability density obtained by introducing σ is not automatically normalized. The expression (11.5) is readily solved for this case to give

$$P(\ell) = \frac{15}{r_0} \frac{\ell}{r_0} \sqrt{1 - 2\frac{\ell}{r_0}} \; , \tag{11.6}$$

and $P(\ell) = 0$ for $\ell > r_0/2$. The result (11.6) is the probability density which can be tested experimentally. An estimate for the probability density (11.5) could be obtained experimentally, and we might then hope to find support (if not proof) for the proposed model, including the time variation it implies.

The model described here is oversimplified in many respects, in particular also by assuming all structures to develop identically. As a minimum requirement to make the model at least *slightly* convincing, we have to relax the assumption of identical r_0's, and allow the reference magnitude, r_0, of the "blobs" to be randomly distributed, in general over the interval $\{0; \infty\}$. We can assume that this distribution is also random, and assign a probability density $P(r_0)$. The result for the probability density (11.5) or (11.6) of ℓ is then conditional, $P(\ell|r_0)$ and we find

$$P(\ell) = \int_{2\ell}^{\infty} P(\ell|r_0) P(r_0) dr_0 \; , \tag{11.7}$$

for a given $P(r_0)$. Unfortunately we may not know $P(r_0)$ a priori. The problem can, however, be "inverted", and we might see whether it is possible to propose a physically realistic $P(r_0)$, which gives agreement with the observed probability densities. If not, we have a contradiction, and have to reject the model, and may be a little wiser, at least in this respect. If reasonable agreement *is* found, we might proceed with the model, by making it more detailed and explore its limitations.

Finally, before entering into discussions of more realistic cases and their associated models, we might draw attention to a "practical" problem. Even if the phenomena we consider are persistent, we might be in need of data. Thus, we would like to make a data-basis consisting of statistically independent observations, and might be tempted to retain only one observation from each satellite pass, for instance. This seems a safe and sound idea, but it leaves many observations redundant, and the estimate of the probability densities may become uncertain. In addition, the plasma parameters are likely to change from one pass to the next, and having to model statistically distributed plasma as well may become troublesome. It is preferable to use as many data as possible from consistent plasma parameters. If the observations are abundant in some orbits (fortunately, they *are*, sometimes), we might like to know to what extent, if at all, these observations form a set of *independent* data? This question is not easily answered, but we can test at least *one* hypothesis against the observations. By randomly distributed events in space, we mean that the position of one is statistically independent of the position of all the others. The probability of finding a structure in a short interval dx measured along the satellite orbit is taken to be μdx, with μ being the density of structures along this line (the dimension of μ is $length^{-1}$). This assumption, we know, leads to a Poisson distribution of the number, N, of structures in an interval L, i.e. $P(N, L) = (\mu L)^N \exp(-\mu L)/N!$. This distribution can be tested, usually by simply checking $(\langle N^2 \rangle - \langle N \rangle^2)/\langle N \rangle = 1$, as valid for the Poisson distribution. More extensive tests can also be carried out [3].

In the following Sect. 11.2 we extend the model discussions, by considering the consequences of having more general density variations than the simple "inverted top hat" density depletion model used for illustrations in this Introduction. Then, in Sect. 11.3, we consider one particular problem, namely collapse of lower hybrid waves, and apply some of the ideas developed here.

11.2 Model Discussions

Assume that we have an analytical model for the time evolution of a physical phenomenon, predicting a quantity $\Psi(\mathbf{r}, t)$, where Ψ might denote the space-time varying plasma density, or the potential or something else. As an example, we might have Ψ representing the space time evolution of a sound wave pulse, phase space vortex, a shock, or similar. Often, we can assume that the phenomenon is associated with one or more characteristic lengths. We can take the width at half-amplitude, for instance. Let such a length scale be $\mathcal{L}_p(t)$, which is in general a function of time. We predict $\mathcal{L}_p(t)$, but measure $\mathcal{L}_p(t = t_0)$, where t_0 is the time we encounter the structure. We do not know what time this is in relation to the beginning of the time evolution of the phenomenon, and this is basically one of the roots of our problem. Another is that we predict a characteristic length as the width of a wave pulse, but the trajectory of the spacecraft need not cross the structure at its maximum diameter. If we later cross a similar object in precisely the same stage of its development we will measure a different width, simply because we are likely to cross the object along a different trajectory in different trials. The problem thus has *two* basic statistically distributed quantities, one associated with the *temporal*, and another with the *spatial* variable. There is a further statistical distribution associated with the fact that different events are not identical, but originate from statistically distributed initial conditions.

11.2.1 Spatial Sampling with One Probe Available

Let us first discuss the distribution in a characteristic quantity as a length scale, due to the random distribution of sampling trajectories. In general, we may not have any a priori knowledge about the actual shape of the structures, i.e. $n = n(x, y)$ in this case. A model can be proposed, however, and subsequently be tested against the data. This will be feasible in particular when the detecting spacecraft is equipped with two or more spatially separated probes.

As a sufficiently general generic form, useful as an illustration, we consider a density pulse in a simple two dimensional model

$$n(x, y) \equiv n_0 - \widetilde{n}$$
$$= n_0 - \Delta n \, e^{-\frac{1}{2}(x^2 + y^2)^m / R^{2m}} . \tag{11.8}$$

This model contains an inverted top-hat model as the limiting case $m \to \infty$, and a rotationally symmetric Gaussian model corresponding to $m = 1$. We might let Δn have either sign, corresponding to density "humps" or depletions. We can use the same exponent on both x and y if a symmetry condition is satisfied, i.e. in the absence of a preferred direction.

The chord length, ℓ, corresponding to the $1/e$-width of the density depletion, as obtained along the satellite orbit, is easily obtained from (11.8). For $m > 1$ an increase in the impact parameter implies a *decrease* in measured chord length. For $m < 1$ the opposite impact parameter variation is obtained; a somewhat counter-intuitive result. The depth of cavities as detected along the sector determined by the satellite trajectory can also be obtained from (11.8), giving an expression varying with impact parameter y. A straight forward parametric representation of the normalized depth – chord length relation can be obtained as

$$\{\widetilde{n}/\Delta n, \ell/R\} = \left\{ e^{-\frac{1}{2}\xi^{2m}}, 2\sqrt{(2 + \xi^{2m})^{1/m} - \xi^2} \right\}, \tag{11.9}$$

with $\xi \equiv y/R$ being the normalized impact parameter. In particular, for $m \to \infty$, we have $\xi^{2m} \to 0$, so that $\ell/R = 2\sqrt{1 - \xi^2}$.

The expression (11.9) gives the peak cavity depth and chord length for a given normalized impact parameter y for cavities defined by (11.8), assuming Δn, R and m to be known. When examining an actual record, the impact parameter is in general varying from cavity to cavity, with y being statistically distributed. The only acceptable assumption concerning its probability density is a uniform distribution, $P(|y|) = 1/L_M$ for $0 < |y| < L_M$ and zero otherwise. Here L_M is an assumed maximum impact parameter from the satellite to a cavity. Eventually we let $L_M \to \infty$. With this assumption we can in principle obtain the probability densities $P(\ell)$ and $P(\widetilde{n})$. For instance, using $P(\widetilde{n})d\widetilde{n} = P(|y|)d|y|$ we readily find

$$P\left(\frac{\widetilde{n}}{\Delta n}\right) = \frac{R}{mL_M} \frac{\Delta n}{\widetilde{n}} \left| 2\ln\left(\frac{\widetilde{n}}{\Delta n}\right) \right|^{-1 + 1/(2m)}, \tag{11.10}$$

for $e^{-\frac{1}{2}(L_M/R)^{2m}} < \widetilde{n}/\Delta n \leq 1$. The probability density (11.10) is normalized for all finite L_M, but the expression may for some values of m give trivial results in the limit of $L_M \to \infty$.

The corresponding results for $P(\ell)$ are not so easily obtained in a closed form, but can be determined for selected values of m.

1) For $m = 1/2$, we find, using $P(\ell)d\ell = P(y)dy$

$$P(\ell) = \frac{\ell}{8RL_M} , \qquad (11.11)$$

for $4 < \ell/R < 4\sqrt{1 + L_M/R}$, see the second half of the expression (11.9).

2) For $m = 1$ we have

$$P(\ell) = \delta\left(\ell - 2R\sqrt{2}\right) . \qquad (11.12)$$

3) For $m = 2$ we have $(y/R)^2 = [32 - (\ell/R)^4]/[8(\ell/R)^2]$ giving

$$P(\ell) = \frac{1}{L_M 2\sqrt{2}} \left(\frac{R}{\ell}\right)^2 \frac{32 + (\ell/R)^4}{\sqrt{32 - (\ell/R)^4}} , \qquad (11.13)$$

with $2\sqrt{(2 + (L_M/R)^4)^{1/2} - (L_M/R)^2} < \ell/R < (32)^{1/4}$. Models with other values of m can be analyzed similarly. It is evident that the results depend sensitively on the model, here the exponent m, and a statistical analysis for determining this and other parameter values is worthwhile.

All these foregoing results are *conditional* in the sense that they assume all cavities to be identical, i.e. given Δn and R all cavities are of the same constant form (11.8). Given a priori knowledge, or at least a qualified guess, of the probability density of the characteristic parameters, we can then use Bayes' theorem to obtain unconditioned distributions.

4) For the case where $m \to \infty$ the analysis becomes more lengthy. We consider two cases, a cylindrical and an ellipsoidal model.

Cylindrical Model

As an illustration assume first that the cavity has a cylindrical shape with a circular cross section of radius $L_\perp(t)$, while $L_\parallel$ is for the moment considered infinite. The observations are then essentially restricted to a plane. As a simple case, we here assume that we have an "inverted top-hat" density depletion, i.e. $n = n_0$ outside the cavities, and $n = n_0 - \Delta n$ inside. The geometrical cross-section for the cavity is thus $\sigma = 2L_\perp$ with dimension "*length*" in this planar approximation. The probability for actually encountering a cavity in an interval of length dx along the spacecraft trajectory is proportional to $\mu(L_\perp) 2L_\perp \, dL_\perp \, dx$, where, again, $\mu(L_\perp)dL_\perp$ is the density of cavities with perpendicular radii within $L_\perp, L_\perp + dL_\perp$. We have $\int_0^\infty \mu(L_\perp)dL_\perp = \mu_c$, where μ_c is the cavity density irrespective of diameter. For the present model, the dimensions of μ_c are $length^{-2}$, i.e. the dimensions of $\mu(L_\perp)$ are $length^{-3}$. Assuming the cavities to appear and disappear at random, the dynamics can be assumed to be time-stationary in a statistical sense, with the density of cavities $\mu(L_\perp)$ being constant, even if $L_\perp$ for the individual cavity varies with time. In a given realization there are many different scale sizes present at the same time. The probability for encountering one particular value for $L_\perp$ depends on the relative density of cavities with that particular diameter. Assuming that the spacecraft *has* encountered a cavity it is evident that the probability of its radius being $L_\perp$ is proportional to the density of cavities with that particular radius as well as to the corresponding geometrical cross-section. We can, more generally, derive the probability density for the radii of observed cavities $\mathcal{P}(L_\perp)$, as illustrated later on in (11.25).

Let the angle between the spacecraft trajectory and the magnetic field be θ. A straight-line trajectory intercepts cavities of radius $L_\perp$ along chords at a constant y-value in the ellipse

$$\frac{x^2}{L_\perp^2/\sin^2\theta} + \frac{y^2}{L_\perp^2} = 1,$$

all $|y|$-values in the interval $\{0, L_\perp\}$ being equally probable. With the foregoing assumption, the distribution of chord lengths, ℓ, is readily obtained as

$$P(\ell|L_\perp) = \frac{1}{2L_\perp}\frac{\ell\sin^2\theta}{\sqrt{4L_\perp^2 - \ell^2\sin^2\theta}} \tag{11.14}$$

for $0 < \ell < 2L_\perp/\sin\theta$ for a given fixed $L_\perp$. The angle θ is considered as a constant. It was also here assumed that the satellite speed is so large that the cavity does not change appreciably during its passage. Evidently (11.14) predicts that there is a large probability of finding $\ell \sim 2L_\perp/\sin\theta$. For distributed cavity radius $L_\perp$ we have

$$P(\ell) = \int_0^\infty P(\ell|L_\perp)\mathcal{P}(L_\perp)dL_\perp \tag{11.15}$$

$$= \ell\sin^2\theta \int_{\ell\sin\theta/2}^{L_\perp\mathrm{max}} \frac{1}{2L_\perp}\frac{\mathcal{P}(L_\perp)}{\sqrt{4L_\perp^2 - \ell^2\sin^2\theta}}dL_\perp$$

with the actual probability density $\mathcal{P}(L_\perp)$ for the cavity radii to be inserted.

Ellipsoidal Model

The foregoing result for $P(\ell|L_\perp)$, the probability density of observed chord lengths, was derived for a cylindrical form of the cavity. For a spherical cavity we have the simple expression (11.2) already derived in the introduction. More generally a rotationally symmetric ellipsoidal model (i.e. a "cigar"), with half axes $L_\perp$ perpendicular to $\mathbf{B}$ and $L_\parallel$ parallel to $\mathbf{B}$, can be used for the density cavities. We can generalize (11.8) to

$$n = n_0 - \Delta n\, e^{-\frac{1}{2}[(x^2+y^2+(z/L_\parallel)^2 L_\perp^2)^m/L_\perp^{2m}]}, \tag{11.16}$$

Again, the case $m = 1$ is relatively simple, so we consider the simple "inverted-top-hat" model, with $m \to \infty$.

Because of the rotational symmetry of the problem with respect to the magnetic field, (11.16) is presumably an adequate model. The centers of the ellipsoids are assumed to be randomly and uniformly distributed in three dimensions. The cross-section of a given cavity for a spacecraft moving at an angle θ to the major axis is

$$\sigma(L_\parallel, L_\perp) = \pi L_\perp \sqrt{L_\perp^2 \cos^2 \theta + L_\parallel^2 \sin^2 \theta}.$$

For $\theta = 0$, i.e. for satellite propagations along the major axis of the ellipsoid, which will generally be the magnetic field direction, we find $\sigma = \pi L_\perp^2$, as expected. The probability density $\mathcal{P}(L_\perp, L_\parallel)$ of encountering a certain cavity specified by $(L_\parallel, L_\perp)$, is proportional to $\sigma(L_\parallel, L_\perp)$ and to the appropriate density of cavities, $\mu(L_\parallel, L_\perp)$, as discussed before. We have $\mu(L_\parallel, L_\perp)dL_\parallel dL_\perp = \mu_c P(L_\perp, L_\parallel)dL_\parallel dL_\perp$, with $\int_0^\infty \int_0^\infty \mu(L_\parallel, L_\perp)dL_\parallel dL_\perp = \mu_c$ being the density of cavity centers, irrespective of cavity widths, as before. The probability density $\mathcal{P}(L_\perp, L_\parallel)$ can not off-hand be equated to $\mathcal{P}(L_\perp)\mathcal{P}(L_\parallel)$ since the time evolutions of $L_\parallel$ and $L_\perp$ are in general not statistically independent. As an illustrative model, which will be useful later on with (11.24), we assume $L_\parallel = \beta L_\perp^2$, where the constant β is determined by initial conditions. Assuming the probability density $P(L_\perp)$ to be known (an example will be analytically determined in (11.25), for a specific model) as well as the cross section obtained previously, we obtain

$$\mathcal{P}(L_\perp, L_\parallel) = \delta(L_\parallel - \beta L_\perp^2)\, P(L_\perp)\sigma(L_\parallel, L_\perp)$$
$$\times \left[\int_0^{L_\perp \mathrm{max}} \int_0^{L_\parallel \mathrm{max}} \delta(L_\parallel - \beta L_\perp^2) P(L_\perp)\sigma(L_\parallel, L_\perp) \right]^{-1},$$

where $L_\parallel \mathrm{max} = \beta L_\perp^2 \mathrm{max}$. The normalizing factor in the angular parentheses can in many cases be calculated analytically [1], but might become a rather lengthy expression.

We now consider the conditional probability density for observed chord lengths for a *given* cavity specified by $(L_\parallel, L_\perp)$. After some calculations we obtain

$$P(\ell \mid L_\perp, L_\parallel) = \ell \frac{\mathcal{L}^2}{2L_\parallel^2 L_\perp^2}, \tag{11.17}$$

for $0 < \ell < 2L_\parallel L_\perp/\mathcal{L}$ with $\mathcal{L}^2 = L_\perp^2 \cos^2 \theta + L_\parallel^2 \sin^2 \theta$. For propagation across the major axis of the ellipsoid, i.e. $\theta = 90°$, we have $P(\ell \mid L_\perp, L_\parallel) = \ell/(2L_\perp^2)$ as a particular case of (11.17), independent of $L_\parallel$. This result is identical to the one obtained for a sphere with radius $L_\perp$, and can be understood by simple geometrical arguments. For distributed values of $L_\perp$ and $L_\parallel$ we find $P(\ell)$ as in (11.14)

$$P(\ell) = \frac{\ell}{2} \iint \left(\frac{\cos^2 \theta}{L_\parallel^2} + \frac{\sin^2 \theta}{L_\perp^2} \right) \mathcal{P}(L_\perp, L_\parallel)dL_\perp dL_\parallel, \tag{11.18}$$

with the integration restricted to that part of the $L_\perp,L_\parallel$-plane where $2L_\parallel^2 L_\perp^2 > \ell^2(L_\perp^2 \cos^2\theta + L_\parallel^2 \sin^2\theta)$ and $0 < L_\perp < L_{\perp\text{max}}$, $0 < L_\parallel < L_{\parallel\text{max}}$. The parenthesis in (11.16) originates from $P(\ell \mid L_\perp, L_\parallel)$. The maximum values $L_{\parallel\text{max}}$ and $L_{\perp\text{max}}$, are also here assumed to be given. In the actual situation they will differ from case to case, and the corresponding probability density must be included in the analysis at a later stage.

11.2.2 Spatial Sampling with Two Probes Available

The models can be tested to a greater accuracy in cases where the spacecraft is equipped with two or more probes for measuring fluctuations in the plasma density. We can then have *two* cross sections for the same structure, and be able to determine, or at least estimate, parameters in (11.8). Assume that the two probes are separated by a distance Δ_s in the direction perpendicular to the velocity vector of the satellite. Referring to the model (11.8) we let again ξ be the normalized impact parameter for probe 1, so the normalized chord length detected by this probe is $\ell_1/R = 2\sqrt{(2+\xi^{2m})^{1/m} - \xi^2}$. The normalized chord length detected by probe 2 crossing the same cavity will be one of the two values $\ell_{2\pm}/R = 2\sqrt{(2 + (\xi \pm \Delta)^{2m})^{1/m} - (\xi \pm \Delta)^2}$ with equal probability, since the satellite can impact the structure on either side with equal probability. We introduced the normalized probe separation $\Delta = \Delta_s/R$. If it is possible to eliminate ξ from the expressions for ℓ_1 and $\ell_{2\pm}$, we can obtain a distribution for the two measurable chord-lengths.

If, in particular, we have $m = 1$, we find $\ell_1 = \ell_{2\pm}$ irrespective of ξ, emphasizing the special properties of structures with Gaussian shape.

For $m = 1/2$ we have $\xi = (\ell_1/4R)^2 - 1$ to give

$$
\ell_{2\pm} = 2\sqrt{(2 + (\ell_1/4R)^2 - 1 \pm \Delta)^2 - ((\ell_1/4R)^2 - 1 \pm \Delta)^2}
$$
$$
= 4\sqrt{1 + (\ell_1/4R)^2 - 1 \pm \Delta}\,. \tag{11.19}
$$

For the limiting case $m \to \infty$ we have

$$
\ell_{2\pm} = 2\sqrt{1 - \left(\sqrt{1 - \ell_1^2/(4R^2)} \pm \Delta\right)^2}\,, \tag{11.20}
$$

although this latter model is complicated by the possibility that one probe crosses the structure, while the other one does not, i.e. $\xi < 1$ but $\xi + \Delta > 1$, for instance. If we take an expression like (11.19) or (11.20) to be a hypothesis, it can be directly tested against available data.

11.2.3 Temporal Sampling

Independent observations happen at different times of the evolution of the structures. If we, for instance, were to analyze whistler wave-packets as they are often excited in the ionosphere by lightning strokes, we might detect the individual whistler wave-packets at different times after the lightning, this time difference being statistically distributed. Since the arrival of the space-craft is safely assumed to be independent of the generation mechanism, we can assume the time of observation to be uniformly distributed in a (large) time interval $\{0, T\}$, where we might eventually let $T \to \infty$. We have $P(t) = 1/T$, which gives an immense simplification of the analysis, as already illustrated in the Introduction. This assumption is consistent with our other basic assumptions, namely that spatial structures are traversed at positions which are uniformly distributed in the plane perpendicular to the spacecraft orbit.

11.3 Nonlinear Lower-Hybrid Wave Models

A number of spacecraft observations demonstrated the existence of small density depletions in the Earth's ionosphere [4, 5, 6], see Fig. 11.1. The structures were associated with a localized enhanced wave activity, where the waves were identified as electrostatic lower-hybrid waves [7]. The observations had a close resemblance to what could be expected from an intermediate stage of wave-collapse, and the observations evidently received considerable attention. However, it is evident that a simple visual inspection is not sufficient for identifying a collapse phenomenon, and a more detailed analysis had to be performed [1, 3, 8]. As the first part of such an analysis it is necessary to find a model for the (deterministic) time evolution of observable quantities.

A set of nonlinear model equations for the lower-hybrid wave dynamics has been proposed [9, 10, 11]. They are based on a set of nonlinear equations for the wave potential ϕ and the slowly varying density perturbation $\overline{n}$

$$-2i \frac{1}{\omega_{LH}} \frac{\partial}{\partial t} \nabla^2 \phi - \Lambda^2 \nabla^4 \phi + \frac{M}{m} \frac{\omega_{LH}^2}{c^2} \left(\frac{\omega_{pe}}{\Omega_{ce}} \right)^2 \phi$$

$$+ \frac{M}{m} \frac{\partial^2}{\partial z^2} \phi = -i \frac{M}{m} \frac{\omega_{LH}}{\Omega_{ce}} \left(\nabla \phi \times \nabla \frac{\overline{n}}{n_0} \right) \cdot \hat{\mathbf{z}} \qquad (11.21)$$

with

$$\Lambda^2 = \left(3 \frac{T_i}{\omega_{LH}^2 M} + 2 \frac{T_e}{\Omega_{ce}^2 m} \frac{\omega_{pe}^2}{\omega_{pe}^2 + \Omega_{ce}^2} \right),$$

where $\omega_{LH}^2 = \Omega_{ci}^2 + \omega_{pi}^2/(1 + \omega_{pe}^2/\Omega_{ce}^2)$. The time evolution of $\overline{n}$ is, in one limiting case of the model, given by

$$\left(\frac{\partial^2}{\partial t^2} - C_s^2 \nabla^2 \right) \frac{\overline{n}}{n_0} = i \frac{\varepsilon_0 \omega_{pe}^2}{4 n_0 M \Omega_{ce} \omega_{LH}} \nabla^2 (\nabla \phi^* \times \nabla \phi) \cdot \hat{\mathbf{z}} . \qquad (11.22)$$

A driving term in (11.21) is not included as the waves are frequently observed in regions where such a mechanism cannot be identified. The magnetic field is

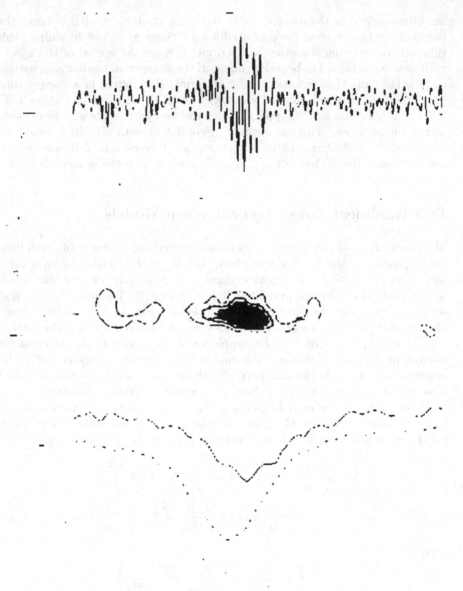

Fig. 11.1. Example of lower hybrid wave cavity detected by instruments on the FREJA satellite. The lower hybrid wave electric field is shown in the upper frame, giving the medium frequency band of the detecting probe circuits. The signal has a bandwidth of 0–16 kHz and a sampling rate of $32 \cdot 10^3$ samples/s. The bottom frame shows the low frequency relative plasma density variations in percent as obtained from two Langmuir probes with an 11.0 m separation. The middle frame shows a wavelet transform of the upper frame [1]. The spin-period of the satellite is approximately 6 s, so its spin phase can be considered constant during a given data sequence

ignored in (11.22) for the slow plasma variation because it is anticipated that the appropriate time scale is shorter than the ion gyro-periods, approximately 2.5 ms for H^+ and 35–40 ms for O^+ for the present conditions. The assumption has to be verified a posteriori, but in case the assumption is violated, the equations are readily modified. For time-stationary conditions this equation is reduced to

$$\frac{\overline{n}}{n_0} = i\frac{\varepsilon_0\omega_{pe}^2}{4n_0 T_e \Omega_{ce}\omega_{LH}}(\nabla\phi \times \nabla\phi^*)\cdot\hat{\mathbf{z}}\,. \tag{11.23}$$

Contrary to the case of Langmuir oscillations, the lower-hybrid waves can be localized within density wells, $\overline{n} < 0$, as well as density "humps", $\overline{n} > 0$ [11]. This apparent symmetry is broken by small terms left out in the analysis. The properties of the model equations (11.21)–(11.22) have been studied in great detail [11]. Most important is the observation of collapsing solutions. For sufficiently large wave intensities the nonlinearities form wave-filled plasma cavities that collapse into singularities within a finite time with time variations of their perpendicular and parallel diameters given by

$$L_\perp \sim (t_c - t)^{1/2} \quad \text{and} \quad L_\parallel \sim (t_c - t)\,, \tag{11.24}$$

where t_c is the collapse time and t_0 an arbitrarily chosen initial time for the process, $t_0 < t < t_c$. At the same time the electric field amplitude at the center of the collapsing cavity increases without bounds. The result in (11.24) is based on a sub-sonic model for the collapse. The wave cavity is strongly **B**-field aligned with $L_\parallel > L_\perp\sqrt{M/m}$. There is no natural propagation velocity associated with the cavity. We introduce the initial maximum length scales as $L_\perp = L_{\perp max}$ and $L_\parallel = L_{\parallel max}$ at $t = t_0$.

Ultimately, for very large field amplitudes and small cavity sizes, the model equations (11.21)–(11.22) become unphysical also because they lack a proper dissipative mechanism. With linear damping being small or negligible, we expect transit time damping to be the dominating mechanism [12, 13]. Here, particles from the surrounding plasma pass through the collapsing cavity and interact with the intensified electric fields there. A particle passing through this region in less than one period of the oscillating electric field will on average gain energy, giving rise to a corresponding damping of the electric field. For parameters $\Omega_e > \omega_{pe}$ discussed here, the collapse will be damped by electrons with small pitch angles passing through the collapse region in the direction parallel to the magnetic field. Significant damping and eventual arrest of the collapse process are expected when the parallel length of the density depletion becomes comparable to the distance travelled by a thermal electron during one period of the lower-hybrid wave. For the present parameters we find a corresponding minimum length scale of $L_{\parallel min} \sim 500$ m. Electrons gain energy (on average) at a *single* pass when propagating along the magnetic field. Fast, light, ions moving with a large Larmor radius from the surrounding plasma into the cavity can also contribute to the damping by being accelerated in that fraction of their gyroperiod they spend inside the cavity. This contribution to the wave damping is somewhat more complicated to describe analytically, and it is effective only for perpendicular cavity widths smaller than or comparable to the ion Larmor radius, $L_\perp \leq \rho_L$. Due to their gyration, ions interact many times with an intensified field in a long field aligned cavity. Since $\omega_{LH} > \Omega_{ci}$ we can assume that the wave phase is randomly varying at each ion encounter with the cavity. This multiple interaction gives a multiplication factor of the order of $L_\parallel/(v_\parallel \tau_i)$, where $v_\parallel$ is the **B**-parallel ion velocity and $\tau_i = 2\pi/\Omega_{ci}$. Thus, even for cases where ions (on average) obtain only a small energy increment at a single pass through a cavity, the net result can be significant. This mechanism clearly favors transverse ion energization where $v_\parallel$ is small.

In agreement with the analytical results based on (11.21) and (11.22), we would expect lower-hybrid wave fields to build up by some mechanism, e.g. a linear instability. When a sufficiently large amplitude has been reached, a modulational instability sets in, breaking the wavetrain up into wave cavities which collapse and ultimately dissipate the wave energy at small scales. In the case where the waves are generated at short wavelengths the model outlined here may have to be modified by inclusion of parametric decay processes.

Two basically different scenarios can be envisaged; one where the lower-hybrid waves are continuously maintained, essentially rendering the cavity formation a statistically time stationary process. Alternatively, we can imagine a situation where lower-hybrid waves are excited in a large region of space as a burst, which eventually breaks up into cavitons and finally dissipates. It is not possible to discriminate between these two scenarios on the basis of the available data. However, large amplitude lower-hybrid waves are frequently observed at medium frequencies, and it is likely that they constitute a background wave component for extended periods of time.

11.4 Probability Densities for Observables

In this section we discuss the observations with the a priori assumption that their explanation is to be sought in a collapse of lower-hybrid waves, without reference to the wave generation mechanism. As it is not possible to study the full space-time evolution of individual events in detail, we attempt to derive probability densities for observable quantities within such a model. These can then be compared with those obtained from the data. In the early investigations of Freja observations, when data were sparse, a simple "inverted-top-hat" density depletion was used to model the density cavities [1]. Later, more detailed, studies [8] demonstrated that a more simpler Gaussian depletion model was the most accurate.

In particular, we discuss here the probability density for the cavity width determined from the data, as this is one of the quantities which is most easily and accurately obtained. This width coincides with the width of the localized wave packets, so these need not be analyzed separately in this context. The measurements of the electric field intensities are uncertain and give underestimates, and a statistical analysis of the lower-hybrid wave amplitudes is not really meaningful.

At early stages a cavity is presumably large and irregular and will hardly be distinguished from the background of low frequency density fluctuations. Assume that after reaching a significant amplitude, a cavity can be effectively recognized in a time interval $\{t_0, t_c\}$, and that the time variation here can be described by (11.24) with t_c being the collapse time. Evidently, we can assume that the spacecraft intercepts a cavity at times uniformly distributed in $\{t_0, t_c\}$, i.e. the probability density for the time of interception is $P(t) = 1/(t_c - t_0)$ for $t_0 < t < t_c$. The distribution of length scales is obtained from $P(L)dL = P(t)dt$. By use of (11.24) the results are

$$P(L_\perp) = \frac{2\,L_\perp}{L_{\perp\mathrm{max}}^2} \qquad \text{and} \qquad P(L_\|) = \frac{1}{L_{\|\mathrm{max}}}, \qquad (11.25)$$

for $0 < L_\perp < L_{\perp\mathrm{max}}$ and $0 < L_\| < L_{\|\mathrm{max}}$, respectively. For a uniform random temporal distribution of collapse times t_c, the density of cavities having a transverse scale in the interval $\{L_\perp, L_\perp + dL_\perp\}$ is given by $\mu(L_\perp)dL_\perp = \mu_c P(L_\perp)dL_\perp$, where μ_c is the spatial density of collapse centers. It was here implicitly assumed that all cavities start out with essentially the same value for L_{max}. Later we relax this condition.

As the satellite moves (essentially in the direction perpendicular to **B**) we expect predominantly *large* scales to be detected, observations of the actual collapse being statistically improbable. Although the collapse itself is thus unlikely to be detected (its "cross-section" is too small), the entire time evolution preceding it will in principle be reflected in the probability distribution of length scales. The cavity spends a comparatively large time in a state with large diameters, where it also has the largest cross-section for observation.

For the simple cylindrical Gaussian model, with $m = 1$ in (11.8), we have

$$P(\ell|L_{\perp\mathrm{max}}) = \int_0^{L_{\perp\mathrm{max}}} \delta(\ell - 2L_\perp\sqrt{2})P(L_\perp)dL_\perp = \frac{\ell}{2L_{\perp\mathrm{max}}^2\sqrt{2}} \qquad (11.26)$$

using (11.25). For $\ell > 2L_{\perp\mathrm{max}}\sqrt{2}$, we have $P(\ell|L_{\perp\mathrm{max}}) = 0$. If we have a statistical distribution of $L_{\perp\mathrm{max}}$ for the ensemble of cavities, we evidently have

$$P(\ell) = \frac{\ell}{\mathcal{C}}\int_{\ell/2}^\infty \sigma(L_{\perp\mathrm{max}})\frac{P(L_{\perp\mathrm{max}})}{L_{\perp\mathrm{max}}^2}\,dL_{\perp\mathrm{max}}.$$

The integral, and the normalizing quantity $\mathcal{C}$, can readily be solved for realistic choices of $P(L_{\perp\mathrm{max}})$, using that $\sigma(L_{\perp\mathrm{max}}) \sim 2\,L_{\perp\mathrm{max}}$ for the present cylindrical model.

Within an "inverted top hat" cylindrical density depletion model, $m \to \infty$ in (11.8), and the time evolution for $L_\perp$ given by (11.24) and (11.25) as appropriate for a lower hybrid collapse, the probability density of observed chord lengths becomes

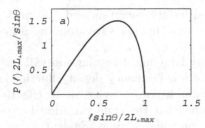
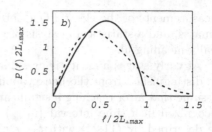

Fig. 11.2. Probability densities for chord lengths, ℓ, in the case of randomly distributed time-stationary cylindrical cavities in (**a**), and ellipsoidal ones in (**b**). *Solid lines* show cases where all cavities are identical, the *dashed line* illustrates the effect of distributed diameters, according to a model distribution. The results are obtained by explicitly using the properties of an "inverted top-hat" density depletion, $m \to \infty$ in (11.8)

$$P(\ell|L_{\perp\mathrm{max}}) = \frac{3\,\ell \sin^2 \theta}{4\,L_{\perp\mathrm{max}}^2} \sqrt{1 - \left(\frac{\ell \sin \theta}{2L_{\perp\mathrm{max}}}\right)^2} \qquad (11.27)$$

for $\ell \sin \theta < 2L_{\perp\mathrm{max}}$ and $P(\ell|L_{\perp\mathrm{max}}) = 0$ otherwise. This result is shown in Fig. 11.2a. In case there are reasons to expect that cavities have a significant distribution in scale sizes, $L_{\perp\mathrm{max}}$, the averaging over the appropriate probability density is easily included in (11.27).

Still with an "inverted top hat" density depletion model and the time evolution for $L_\perp$ given by (11.24), we consider a three dimensional case with randomly distributed ellipsoids. For simplicity we consider the case $\theta \sim 90°$, i.e. the satellite propagates essentially perpendicular to **B**. Inserting the previously obtained $\mathcal{P}(L_\perp, L_\parallel)$ we obtain

$$P(\ell) = \frac{\ell}{L_{\perp\mathrm{max}}^2} \left(1 - \left(\frac{\ell}{2L_{\perp\mathrm{max}}}\right)^2\right). \qquad (11.28)$$

This result is shown in Fig. 11.2b with a full line. This result, as well as that in Fig. 11.2a, is applicable for the case where the spread in the values of $L_{\perp\mathrm{max}}$ is small. The dashed line on the figure indicates a corresponding result for a *large* spread, where for the sake of argument, we assumed a probability density of the form

$$P(L_{\perp\mathrm{max}}) = \frac{64}{L_0 3\sqrt{2\pi}} \left(\frac{L_{\perp\mathrm{max}}}{L_0}\right)^4 \exp(-2(L_{\perp\mathrm{max}}/L_0)^2) ,$$

where L_0 is a typical scale length for the conditions. (For the form of $P(L_{\perp\mathrm{max}})$ used here, we actually find $\langle L_{\perp\mathrm{max}}\rangle = 8/(3\sqrt{2\pi})L_0 \approx 1.06L_0$.) We have no a priori physical arguments for choosing the probability density given here, but find it sufficiently general to accommodate realistic problems. Now, (11.27) as well as (11.28) assigns a finite probability for measuring very small chord lengths, corresponding to cases when the satellite trajectory crosses almost at the boundary of the cavity. In *reality* such cases are unlikely to be properly recognized in the background noise level and these contributions will be under represented in the experimental estimate for the probability density. It is interesting that (11.27) as well as (11.28) gives a flat maximum for $P(\ell)$ for ℓ in the range $\sim (1 - 2)L_{\perp max}$, see Fig. 11.2. Also, we find it comforting that the cylindrical model and the ellipsoidal one in Figs. 11.2 a) and b) are similar, this means that the result is relatively robust and does not depend on fine details in the models.

In the foregoing discussion, the length scale $L_{\perp max}$ entered simply as a parameter. It can be related to other basic parameters for the wave-field. Thus, the typical length-scale for lower-hybrid cavities formed by the modulational instability can be estimated by balancing the nonlinear term with the dispersive term [11], using (11.23) in (11.21). The estimate for the resulting maximum, **B**-transverse length scale is given by

$$
L_{\perp max} \sim \Lambda \left(\frac{4n_0 T_e}{\varepsilon_0 \, |\nabla \phi|_0^2} \frac{m}{M} \frac{\Omega_{ce}^2}{\omega_{pe}^2} \right)^{1/2} , \tag{11.29}
$$

where $|\nabla \phi|_0^2$ is the square of the electric field in the center of the cavity at the initial stage.

In case the high-frequency field is turbulent with a broad wavenumber range, we expect that the scale-length obtained here have to be shorter than the correlation length of the fluctuations. The **B**-parallel scale length becomes

$$
L_{\parallel max} \sim \frac{L_{\perp max}^2}{\Lambda} \sqrt{\frac{M}{m}} . \tag{11.30}
$$

Also a characteristic time-scale for the collapse process can be estimated by balancing the time derivative with the nonlinear term in (11.21) and using (11.23), giving

$$
t_{LH} \sim \omega_{LH}^{-1} \frac{m}{M} \frac{\Omega_{ce}^2}{\omega_{pe}^2} \frac{4n_0 T_e}{\varepsilon_0 \, |\nabla \phi|_0^2} . \tag{11.31}
$$

The time scale obtained in this way may, of course, be a somewhat crude estimate. A more accurate expression can be obtained from the growth rate γ_{LH} for the modulational instability which can be obtained with some algebra. It was found [11] that $t_{LH} \sim \gamma_{LH}^{-1}$, at least for order of magnitude estimates. A characteristic contraction rate for the cavity evolution based on the analytical estimates is given by the ratio $L_{\perp max}/t_{LH}$. This contraction rate is below the ion-sound speed consistent with sub-sonic collapse.

Assuming that we have determined a characteristic shape for the structures, for instance by estimating an optimum value for m in (11.8), we can attempt to obtain a distribution of the peak absolute value (with positive or negative sign) of the density detected by a satellite pass of a structure. For instance for $m = 1$ we find distributions illustrated in Fig. 11.3 by using (11.10).

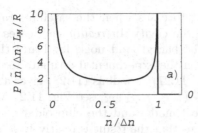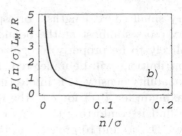

Fig. 11.3. The conditional probability density for normalized density depletions, $\tilde{n}/\Delta n$ is shown in (**a**), obtained for the preferred model with $m = 1$ in (11.10), with a given Δn. Assuming a Rayleigh distribution for the peak densities Δn we obtain the result shown in (**b**)

Assuming that we deal with density depletions we have in Fig. 11.3a the distribution of detected density minima, assuming that all cavities are identical, i.e. Δn is the same for all. The two singularities in $P(\tilde{n}/\Delta n)$ are easily understood by simple geometrical arguments. In Fig. 11.3b we illustrate the effect of Rayleigh distributed values for Δn.

11.5 Observations

As already mentioned, our emphasis will be on the statistical properties of the observed density depletions associated with lower hybrid wave cavitation, as detected by the Freja satellite. First we want to investigate to what extent we might assume these cavities to be randomly distributed along the spacecraft orbit. The experimentally obtained distribution of the number of cavities in intervals of given length are shown in Fig. 11.4, together with the probability density for the relative distance between cavities. For the present case we estimate the density of cavities to be approximately $\mu = 2.5 \times 10^{-3}$ m^{-1}. The distributions are very reproducible, also when data are combined from orbits obtained with more than a year time separation. Very large distances are under-represented in the data because of the finite sample duration (usually 0.75 s). The dots show an exponential fit, $\exp(-x/\zeta)$, where ζ corresponds to approximately 200 m. The exponential fit indicates that the probability of finding a cavity in a small interval dx is proportional to dx itself, with a constant of proportionality given by the density, μ, of cavities along the spacecraft trajectory. In particular, we should like to point out that a model with spatially uniformly distributed, statistically independent density depletions results in a Poisson distribution, $P(N) = (\mu L)^N \exp(-\mu L)/N!$, for the number of cavities, N, in an interval, L, along the spacecraft trajectory, as discussed in Sect. 11.1.

In order to give an estimate for the *shape* of the cavities, we note that very often we have *two* probes for density measurements active on Freja. It

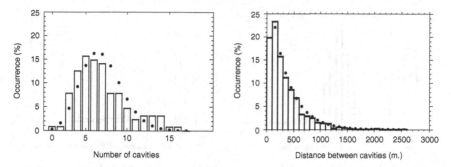

Fig. 11.4. *Left*: Experimental estimate for the probability density for the number of cavities in a segment of duration 0.375 s (corresponding to a distance of 2330 m along the spacecraft trajectory). *Right*: distribution of distance between cavities. A slight data gap appears for cavity separations smaller than a typical cavity width. The small filled circles indicate the appropriate results for a Poisson distribution in both figures

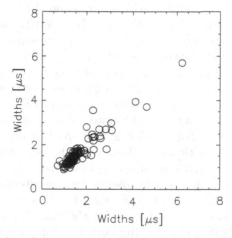

Fig. 11.5. Scatter plot of the two corresponding chord lengths (ℓ_1, ℓ_2) as detected by the two Langmuir probes crossing a cavity at two separated positions. The figure contains observations of 130 cavities. All points are essentially located on the center line, indicating that the chord-lengths detected by the two probes are independent of the probe separation, giving an indication for the validity of the Gaussian model with $m = 1$ in (11.8).

is therefore feasible to follow the ideas outlined in Sect. 11.2.2. In Fig. 11.5 we thus show the two dimensional distribution of chord-lengths, as obtained by simultaneous observations of the same cavity by both probes, irrespective of the spin-phase of the satellite, i.e. independent of the component of the probe separation in the direction of the satellite's velocity vector. We note basically that all observation points are located on the line $\ell_1 = \ell_2$, indicating

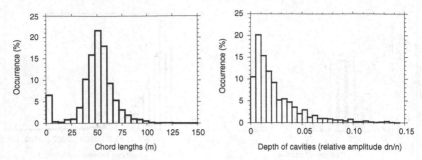

Fig. 11.6. *Left*: Distribution of chord lengths (not to be confused with cavity widths if $m \neq 1$ in (11.8)) obtained from the data. *Right*: Distribution of cavity depths as detected along the spacecraft orbit

an excellent fit to the model assuming $m = 1$ in (11.10). We can thus argue that the depletions have a Gaussian shape to a good approximation [8].

The distribution of cavity widths obtained on the basis of data from one selected orbit is shown in Fig. 11.6. Here chord lengths are identified as the separation between the two points of maximum curvature at the baseline of the signal. The distribution shown in Fig. 11.6 is very robust; it is well reproduced for data obtained with a time separation of more than a year.

We can also obtain the distribution of the maximum depth of the density depletion along the probe trajectories, with results shown in Fig. 11.6. We note a convincing similarity with the results shown in Fig. 11.3b as obtained for the Gaussian model-shape, which is supported by Fig. 11.5, provided that we include a wide statistical distribution of the peak values of the density depletions over the ensemble of cavity realizations.

In Fig. 11.7, we show a scatter-plot of the distribution of density depletions and chord lengths, as detected along the spacecraft orbit. As already demonstrated, we have solid evidence for the Gaussian model, with $m = 1$ in (11.8) to be representative for the density depletions associated with the cavities [8]. Then the abscissa on Fig. 11.7 in effect gives the cavity width, just as Fig. 11.6.

11.6 Discussions

By comparing the observed probability densities for chord lengths with those derived on the basis of a simple collapse process we find that the shape of the theoretical probability density are not readily made to agree with the observed distributions. The disagreement is most conspicuous in Fig. 11.7, where we readily note that the deepest cavities are associated with the intermediate scales, and not with the most narrow ones, as one would expect by a collapse model. This observation can be supported even more strongly by data from other passes [8]. We note also nontrivial disagreements when

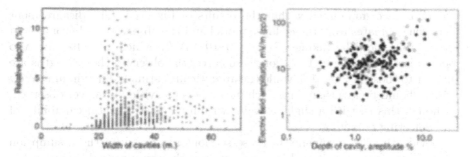

Fig. 11.7. *Left*: Scatter plot for distribution of density depletions and chord lengths (which for the $m = 1$ model equals cavity widths), as detected along the spacecraft orbit. The vertical "stripes" in the figure are due to the temporal sampling of the data. *Right*: scatter plot of corresponding values for peak observed relative density variations, $\tilde{n}/n_0$ in percent, and the available electric field components obtained from the peak electric field detected at medium frequencies by the antennas within the cavities. The electric fields are measured as the half peak-to-peak value $(pp/2)$, measured at maximum. For a given electric field component, we usually have two measurements for the density depletion, one from each probe. We plot both points, shown with small open and filled circles, respectively

characteristic times, t_{LH}, and scale sizes, $L_{\perp max}$, from theory are compared with observations [8]. The seeming disagreement with an interpretation based on collapsing lower-hybrid waves can be substantiated also by use of the analytical expressions (11.21) and (11.22).

11.6.1 Length Scales

The observational results speak in favor of a situation where the lower-hybrid waves are of a somewhat "bursty" nature, and it is the distance between the bursts which would determine the separation, while the amplitude and spatial extent of the bursts set the value for $L_{\perp max}$ in the individual cavitations. This interpretation is physically quite plausible and is in itself not contradictory to the collapse model. Concerning individual cavities, the data are not able to provide a direct estimate of $L_{\parallel}$, so are here only concerned with the characteristic value for the perpendicular length scale. From Figs. 11.6 and 11.7 we estimate $\langle L_{\perp} \rangle \approx 60$ m, while $L_{\perp max} \sim 150-200$ m. A typical relative density depletion is approximately 25%.

It should be brought in mind that the analytical estimate (11.29) can be used at any stage of the collapse [10], i.e. the instantaneous value of the cavity width can be estimated as $L_{\perp est} \sim \Lambda \left[4n_0 T_e m \Omega_{ce}^2 / (\varepsilon_0 |\nabla\phi|^2 M \omega_{pe}^2) \right]^{1/2}$ as an order of magnitude, with $\nabla\phi$ now being the actual value of the electric field in the center of the cavity. Taking the value of 40 mV/m from Fig. 11.1, we find $L_{\perp est} \sim 2$ m. The electric field is a *lower* limit and $L_{\perp est}$ derived from it is then an *upper* limit for the cavity diameter. This value cannot by any means be accommodated within the results of Fig. 11.1. The disagreement between estimates from the collapse model and the observations become even more pronounced by taking $|\nabla\phi| \sim 70\text{--}100\,\mathrm{mV/m}$ which are, after all, also being observed, see Fig. 11.7. For these cases, the observed chord lengths are similar to those in Fig. 11.6, while the analytically estimated maximum cavity width should be less than 1 m. Such narrow deep cavities are never observed, although this is, with a slight margin, within the instrumental capability of the detecting circuits.

Evidently, the arguments in these sections are based on the assumption that the cavity parameters $L_\perp$ and $L_\parallel$ do not change appreciably during the time it takes the spacecraft to traverse a cavity. The slow time variation in the density signals from the two probes confirm that this requirement is fully satisfied. It is, however, not so for an arbitrary collapse scenario predicted from the model equations (11.21) and (11.22).

It could be argued that the transit time damping arrests the contraction of cavities by damping out the lower-hybrid waves at a certain small scale. This argument would, however, imply that predominantly the smallest cavities should be void of wave activity, in disagreement with observations.

11.6.2 Time Scales

From the observations (Fig. 11.2) a typical size of the observed cavities perpendicular to the magnetic field is $\sim$60 m. The satellite (with velocity $\sim$ 6 km/s in the **B**-perpendicular direction) is traversing this distance in approximately 10 ms. Cavities are frequently traversed without indications of any deformation during the time of passage, as judged from the two density probe signals, see for instance Fig. 11.1. In the case of a significant cavity contraction during the passage of the satellite, the time record of the density depletions and corresponding wave-envelopes should appear skew, or significantly non-symmetric. Significantly skew density variations are observed only very rarely, and when they occur their skewness can have both signs.

On the basis of the experimental data, we thus argue that the characteristic time for the cavity evolution must be significantly larger than 10 ms, a time-scale of 100 ms is probably an underestimate. This means that at least the H^+-component is magnetized on relevant time scales. The characteristic time scale, t_{LH}, for a wave-collapse process estimated from the analytical result (11.31) is, on the other hand, comparable to or smaller than 10 ms and therefore too short to be in agreement with the experimental results.

The probability densities derived on the basis of a collapse model are giving unfair representation of the smallest scales, i.e. no collapse model is expected to remain valid when the length scales approaching scales where transit time damping becomes important. This argument is not significant for our interpretation because such small scales will escape observation due to their small cross section. The smallest scales thus have negligible weight in our results. The essential element of the collapsing time-variation of the cavity widths, as given by (11.24), is that the time variation of *large* scales is slow, while *small* scales change rapidly. Significant modifications of the obtained statistical distributions will only be obtained if these conditions are reversed, thus making the *small* scales being those most likely to be observed. This type of time evolution can, however, not be accommodated within a description based on wave collapse.

11.7 Conclusions

In this paper we discussed general ideas which can be used for estimating models for coherent time-evolutions by random sampling of data. The ideas may turn out to be particularly useful for interpreting data from instrumented spacecrafts, and were here illustrated by discussions of localized bursts of lower-hybrid waves and correlated density depletions observed on the FREJA satellite [5, 6]. Particular attention was given to an explanation in terms of wave-collapse. The statistical arguments are based on three distinct elements. Two are purely geometric, where the chord length distribution is determined for given cavity scales, together with the probability of encountering those scales. The third part of the argument is based on the actual time variation of scales predicted by the collapse model. The statistical assumption is basically that the cavities are uniformly distributed along the spacecraft trajectory and that they are encountered with equal probability at any time during the dynamic evolution. We believe that the cylindrical and ellipsoidal models discussed are sufficiently general to accommodate actual forms of a collapsing cavity. The time variation given by (11.24) is only an approximation at early times of the evolution of large cavities. This uncertainty cannot be of importance as the large scales seem not to be significantly represented in the data, in spite of their large cross-section. Since the measured electric fields are somewhat uncertain [1, 3, 8], we have not discussed the statistics of this quantity in any greater detail. We concluded that the interpretation in terms of wave collapse in its simplest form can be ruled out on the basis of a pronounced disagreement between the length and time scales predicted by the collapse-model and those observed in the data.

The experimentally observed distribution of chord lengths in the cavities was explained best by assuming a large number of **B**-elongated cavities, uniformly distributed in space, with values of 40–80 m for the diameters in the direction perpendicular to the ambient magnetic field [14]. We proposed [1] a mechanism where the cavities start out with a **B**-perpendicular scale size very close to the one they end up with. It is assumed that the waves give up energy to ions as well as electrons. This seems to be the best candidate for explaining this characteristic length scale, which is then a consequence of the thermal expansion of the plasma, with electrons streaming along **B** while ions expand across the magnetic field lines for transverse structures smaller than or comparable to twice the ion Larmor diameter of ions. It was demonstrated

[1] that based on this model, a probability density for chord-lengths can be derived which agrees well with observations.

With minor modifications, the statistical analysis presented in the present work can be generalized also for studies of the possible evidence of *Langmuir* wave collapse in rocket or satellite data. In particular, the ellipsoid approximation discussed here contains also the "pancake" model which is relevant for the Langmuir problem in weak magnetic fields [15].

It is self evident that the analysis summarized in the present communication refers to *coherent* phenomena. In case the structures of interest are embedded into a turbulent background, a filtering of the data might be advantageous. Several such methods have been discussed in the literature, conditional sampling for instance [16], but also matched filters might be useful.

Acknowledgments

The present work was carried out as a part of the project "Turbulence in Fluids and Plasmas," conducted at the Centre for Advanced Study (CAS) in Oslo in 2004/05.

References

[1] Pécseli, H. L. et al.: J. Geophys. Res. 101, 5299, 1996.

[2] Kofoed-Hansen, O., H. L. Pécseli, and J. Trulsen: Phys. Scr. 40, 280, 1989.

[3] Kjus, S. H. et al.: J. Geophys. Res. 103, 26633, 1998.

[4] Vago, J. L. et al.: J. Geophys. Res. 97, 16935, 1992.

[5] Dovner, P. O., A. I. Eriksson, R. Boström, and B. Holback: Geophys. Res. Lett. 21, 1827, 1994.

[6] Eriksson, A. I. et al.: Geophys. Res. Lett. 21, 1843, 1994.

[7] Schuck, P. W., J. W. Bonnell, and P. M. J. Kintner: IEEE Trans. Plasma Sci. 31, 1, 2003.

[8] Høymork, S. H. et al.: J. Geophys. Res. 105, 18519, 2000.

[9] Musher, S. L. and B. I. Sturman: Pis'ma Zh. Eksp. Teor. Fiz. 22, 537, 1975, english translation in: JETP lett. 22, 265 1975.

[10] Sotnikov, V. I., V. D. Shapiro, and V. I. Shevchenko:, Fiz. Plazmy 4, 450, 1978.

[11] Shapiro, V. D. et al.: Phys. Fluids B5, 3148, 1993.

[12] Robinson, P. A.: Phys. Fluids B3, 545, 1991.

[13] Skjæraasen, O. et al.: Phys. Plasmas 6, 1072, 1999.

[14] McBride, J. B., E. Ott, J. P. Boris, and J. H. Orens: Phys. Fluids 15, 2367, 1972.

[15] Krasnosel'skikh, V. V. and V. I. Sotnikov: Fiz. Plazmy 3, 872, 1977, see also Sov. J. Plasma Phys. 3, 491, 1977.

[16] Johnsen, H., H. L. Pécseli, and J. Trulsen: Phys. Fluids 30, 2239, 1987.

12

Propagation Analysis of Electromagnetic Waves:
Application to Auroral Kilometric Radiation

O. Santolík[1] and M. Parrot[2]

[1] Faculty of Mathematics and Physics, Charles University, Prague, Czech
    Republic; also at IAP/CAS, Prague, Czech Republic.
    ondrej.santolik@mff.cuni.cz
[2] LPCE/CNRS, Orléans, France.
    mparrot@cnrs-orleans.fr

Abstract. We give a brief tutorial description of techniques for determination of
wave modes and propagation directions in geospace, based on multi-component measurements of the magnetic and electric field fluctuations. One class of analysis methods is based on the assumption of the presence of a single plane wave and can be
used to determine the direction of the wave vector. If the wave field is more complex, containing waves which simultaneously propagate in different directions and/or
wave modes, the concept of the wave distribution function becomes important. It is
based on estimation of a continuous distribution of wave energy with respect to the
wave-vector direction. This concept can furthermore be generalized to the distribution of energy in different wave modes. As an example of analysis of satellite data,
our emphasis is on application of these techniques to the high-frequency waves, for
example to Auroral Kilometric Radiation (AKR). We analyze multicomponent data
of the MEMO instrument obtained using multiple magnetic and electric antennas
onboard the Interball 2 spacecraft. Results of different analysis techniques are compared. The intense structured AKR emission is found to propagate predominantly
in the R-X mode with wave energy distributed in relatively wide peaks at oblique
angles with respect to the terrestrial magnetic field. As expected, the AKR sources
correspond to multiple active regions on the auroral oval.

Key words: Electromagnetic waves in plasmas, direction finding, wave distribution function, auroral kilometric radiation

12.1 Introduction

Waves in space plasmas can often simultaneously propagate in different modes
(with different wavelengths) at a given frequency. To trace these waves back to
their original source regions and to estimate their source mechanisms, recognition of their modes and propagation directions in the anisotropic plasma

O. Santolík and M. Parrot: *Propagation Analysis of Electromagnetic Waves: Application to
Auroral Kilometric Radiation*, Lect. Notes Phys. **687**, 297–312 (2006)
www.springerlink.com   © Springer-Verlag Berlin Heidelberg 2006

medium is often crucial. Experimental analysis of propagation modes and directions of waves in plasmas is easier and more reliable if we measure several components of the fluctuating magnetic and electric fields at the same time. Such measurements using multiple antennas on spacecraft have been first proposed by Grard [12] and Shawhan [66]. The complete set of three Cartesian components of the vector of magnetic field fluctuations and three Cartesian components of the electric field fluctuations (and, in some cases, also the density fluctuations) would be the basis of an ideal data set. However, most often some of those measurements are missing for various, predominantly technical, reasons. For example, it is technically difficult to place a long electric antenna in the direction of the spin axis of a spin-stabilized spacecraft.

Several analysis methods applicable to the multi-component measurements have been first developed for the ground-based geophysical data [e.g. 2, 35, 36, 54, 55]. These methods and other newly developed techniques [e.g. 8, 25, 58, 64, 71, 72] have been later used for analysis of data of spacecraft missions carrying instruments for multi-component measurements of the wave magnetic and electric fields in various frequency ranges, such as GEOS, Aureol 3, Freja, Polar, Interball 2, Cluster, Double Star, and DEMETER. Although these missions were designed to investigate different regions of the geospace, similar analysis methods have been used for their wave measurements [e.g. 29, 30, 31, 32, 33, 45, 46, 47, 48, 56, 57, 58, 59, 60, 61, 62, 73]. Analysis methods which are described in this chapter rely on this heritage. We will concentrate on their application to the high-frequency waves, namely to Auroral Kilometric Radiation (AKR).

AKR is a strong emission of radio waves at frequencies from a few tens of kHz up to 600–800 kHz. It was first observed more than 30 years ago by Benediktov et al. [5] and Dunckel et al. [9] but still is a subject of active research. The first proper interpretation was done by Gurnett [15], who demonstrated the electromagnetic nature of the waves and found the origin of the radiation in the terrestrial auroral zone at altitudes of a few Earth radii (R_E). A widely accepted generation mechanism is based on the relativistic cyclotron maser interaction with a "horse-shoe" or "shell" electron distribution function [34, 51, 53, 74] connected to active auroral regions [15, 20, 39]. The AKR emissions consist of many narrow-band components with varying center frequencies [see, e.g., Gurnett et al., 16]. The bandwidth of these components is typically 1 kHz but it could be as low as 5–10 Hz [see, e.g., Baumback and Calvert, 3]. This fine structure could be explained by a wave amplification in a resonator represented by small field-aligned structures in the auroral region [6]. Measurements inside the AKR source have shown that the radiation originates in field-aligned density depletions with transverse dimensions of the order of 100 km, filled by hot and tenuous plasmas [7, 10, 11, 19]. Theory considering relativistic effects on the wave dispersion relation [Pritchett, 51, 52] and small scale gradients [see, e.g., Le Quéau and Louarn, 26] appears to be necessary to explain all the details of the wave generation and subsequent propagation from the source cavity.

Since AKR propagates at frequencies higher than all the characteristic frequencies of the plasma medium (above the plasma frequency and the electron cyclotron frequency) it can leave the source region in one of the two basic propagation modes: R-X (right-handed/extraordinary) and L-O (left-handed/ordinary) [67]. The two modes can be recognized using measurements of phase shifts between the different components of the electric and/or magnetic field fluctuations. Kaiser et al. [22] analyzed direct measurements of the polarization sense made by multiple electric antennae of Voyager 1 and 2. Supposing that the satellites are always in the northern magnetic hemisphere they found R-X mode in more than 80% of cases.

Analysis of wave propagation directions was used in many AKR studies to localize the source region or to verify the generation mechanism. Gurnett [15], using Imp 6 and Imp 8 data, first estimated the direction of AKR propagation analyzing the spin pattern of a single electric antenna. At a distance of about 30 R_E, he found the wave vector within 6° from the Earthward direction. Kurth et al. [24] confirmed his results by a detailed study based on large Hawkeye-1 and Imp-6 data sets. To find the propagation direction they extended the spin pattern method using the least-squares fit of the modulation envelope of the electric antenna. However, they had to work with 1-hour averages to achieve sufficient angular resolution. Similar results were also obtained by Alexander and Kaiser [1] who analyzed the RAE-2 recordings of lunar occultations of the Earth at radio frequencies to locate the emission region in three dimensions. Gurnett et al. [16] proposed to use a phase interferometry method using two-point measurements of AKR waveforms onboard ISEE-1 and ISEE-2 satellites, and Baumback et al. [4] found by this method that the AKR source has a diameter between 10 and 20 km. James [21] applied the spin-pattern method to the data of ISIS-1 sounder receiver. He measured wave normals declined by 90°–140° from the terrestrial magnetic field, and in a ray-tracing study he inferred originally downward propagation and a subsequent reflection. Calvert [6] developed a direction finding method based on DE-1 onboard correlation measurements. He used signals from two orthogonal electric dipoles, and with a model of their phase difference with respect to the spin phase he obtained the wave-normal direction.

This method was also used by Mellott et al. [37] who studied different propagation of R-X and L-O mode emissions. Several studies using the Viking data have again used a simple spin-pattern method. For instance, de Feraudy et al. [7] found that the spin-pattern phase abruptly changes by about 80° at the boundary of the AKR source region. Morioka et al. [38] used five-component measurement of the Akebono satellite to calculate the Poynting flux vector in the satellite frame. With onboard narrow-band receivers they studied phase relations between the field components in a band of 100 Hz inside an AKR emission. Gurnett et al. [13] and Mutel et al. [39, 40] implemented the very long baseline interferometer (VLBI) technique to determine locations of individual AKR bursts. They used a triangulation method based on differential delays from cross-correlated wide-band electric field waveforms recorded by the WBD instruments on board the four Cluster spacecraft and confirmed that the AKR bursts are generally located above the auroral zone with a strong preference for the evening sector.

Multi-component measurements of the MEMO (Mesures Multicomposantes des Ondes) instrument onboard the Interball 2 spacecraft have been analyzed by Lefeuvre et al. [32] and Parrot et al. [46], showing that the wave normals of the right-hand quasi-circularly polarized (R-X mode) waves have wave vectors inclined by approximately 30° from the direction of the terrestrial magnetic field. Santolík et al. [58] showed for another AKR case measured by the same instrument at an altitude of 3 R_E that the analysis of wave propagation indicates a source region located at an altitude of 1.2 R_E above the northern auroral zone. Schreiber et al. [65] performed, based on similar Interball-2 measurements of the MEMO and POLRAD [17] instruments, a three-dimensional ray tracing study using the analysis of wave-normal directions at the spacecraft position. Rays traced back toward the sources imply the existence of two different large active regions, as seen at the same time by the ultra-violet camera on board the Polar spacecraft.

Analysis presented in this chapter is also based on the data set of the MEMO instrument. The chapter is organized as follows. A short tutorial description of the analysis methods is given in Sect. 12.2. Subsection 12.2.1 introduces methods based on the approximation of the single wave-normal direction. Subsection 12.2.2 then briefly describes methods of estimation of the wave distribution function. We show that this method can be naturally extended to estimate also a distribution of energy between different wave modes. Section 12.3 follows showing results of analysis of AKR measured by multiple antennas onboard the Interball 2 spacecraft.

12.2 Analysis Methods

The multi-component measurements of the wave magnetic and electric fields allow us to determine, for example, the average Poynting flux. Supposing the presence of a single plane wave, the direction of the wave vector can be determined. For a more complex wave field we can estimate a continuous distribution of wave energy with respect to the wave-vector direction (wave distribution function). These results are useful for the localization of sources of observed emissions.

12.2.1 Plane Wave Methods

Supposing the presence of a single plane wave at a frequency f with a wave vector k, the magnetic field B as a function of time t and position x can be written

$$B(t, x) = B_0 + \Re\left\{ \mathcal{B}(f, k)\exp\left[i(2\pi ft - k \cdot x)\right] \right\}, \qquad (12.1)$$

where B_0 is the ambient stationary magnetic field (terrestrial magnetic field in the magnetosphere),$\Re$ means the real part, and B is the "magnetic vector complex spectral amplitude" for a given frequency f and wave vector k. Under the same circumstances, similar expression can be used for the fluctuating electric field E using the "electric vector complex spectral amplitude" $\mathcal{E}$, and similarly also for the "current density complex spectral amplitude" $\mathcal{J}$ and "charge density complex spectral amplitude" ϱ. Using the SI units with $1/\sqrt{\varepsilon_0\mu_0} = c$ (speed of light), the Maxwell's equations can be written for the complex spectral amplitudes

$$k \times B = \left(\mathrm{i}\,\mu_0\,\mathcal{J} - 2\pi f\,\mathcal{E}/c^2 \right), \tag{12.2}$$

$$k \times \mathcal{E} = 2\pi f B, \tag{12.3}$$

$$k \cdot \mathcal{E} = \varrho/\varepsilon_0, \tag{12.4}$$

$$k \cdot B = 0. \tag{12.5}$$

Equation (12.3) (Faraday's law) implies that B is always perpendicular to both wave vector k and $\mathcal{E}$,

$$k \cdot B = 0 \tag{12.6}$$

$$\mathcal{E} \cdot B = 0 \tag{12.7}$$

Note that the first of these conditions (12.6) is equivalent also to the fourth Maxwell's equation (12.5). If we now write it in Cartesian coordinates, where

$$B = (B_1, B_2, B_3), \tag{12.8}$$

and multiply (12.6) successively by three Cartesian components of the complex conjugate B^* we obtain a set of six mutually dependent real equations

$$\mathrm{A} \cdot k = \begin{pmatrix} \Re S_{11} & \Re S_{12} & \Re S_{13} \\ \Re S_{12} & \Re S_{22} & \Re S_{23} \\ \Re S_{13} & \Re S_{23} & \Re S_{33} \\ 0 & -\Im S_{12} & -\Im S_{13} \\ \Im S_{12} & 0 & -\Im S_{23} \\ \Im S_{13} & \Im S_{23} & 0 \end{pmatrix} \cdot \begin{pmatrix} k_1 \\ k_2 \\ k_3 \end{pmatrix} = 0, \tag{12.9}$$

where $\Re$ means the real part, $\Im$ means the imaginary part, and the components of the Hermitian magnetic spectral matrix S_{ij} are obtained from the three Cartesian components of the magnetic vector complex amplitude using the relation

$$S_{ij} = B_i\,B_j^*, \qquad i,j = 1\ldots3. \tag{12.10}$$

Note that the homogeneous set of equations (12.9) can be multiplied by any real coefficient. Consequently, this set cannot be used to determine the modulus of the unknown vector k. It only can determine the direction of this vector. Note also that the set of equations (12.9) naturally contains only two independent real equations corresponding to the single complex equation

(12.6). The importance of this expansion, however, becomes evident when it is used with experimental data.

Using the experimentally measured multi-component signals of the wave magnetic field

$$\hat{\boldsymbol{B}} = (\hat{B}_1, \hat{B}_2, \hat{B}_3) \tag{12.11}$$

we can use spectral analysis methods (for example, fast Fourier transform or wavelet analysis) to estimate, at a given frequency, the components of the "magnetic vector complex spectral amplitude" $\hat{\boldsymbol{B}}$ and, subsequently, the magnetic spectral matrix,

$$\hat{S}_{ij} = \langle \hat{B}_i \hat{B}_j^* \rangle, \qquad i, j = 1 \ldots 3, \tag{12.12}$$

where $\langle \rangle$ means average value.

The homogeneous set of six equations (12.9) can be then rewritten,

$$\hat{\mathsf{A}} \cdot \hat{\boldsymbol{\kappa}} = 0, \tag{12.13}$$

where the matrix $\hat{\mathsf{A}}$ is composed of the superposed imaginary and real parts of the experimental spectral matrix $\hat{S}_{ij}$ (instead of the idealized spectral matrix S_{ij}), and $\hat{\boldsymbol{\kappa}}$ is an unknown unit vector defining the estimate of the wave vector direction,

$$\hat{\boldsymbol{\kappa}} = \hat{\boldsymbol{k}}/|\hat{\boldsymbol{k}}| . \tag{12.14}$$

Using Cartesian coordinates connected to the principal axis of symmetry of the plasma medium where the waves propagate (the direction of the ambient stationary magnetic field $\boldsymbol{B}_0$), the wave vector direction can be defined by two angles θ and ϕ, where θ is the deviation from the $\boldsymbol{B}_0$ direction and ϕ is an azimuth centered, for instance, to the plane of the local magnetic meridian (see Fig. 12.1). The wave vector direction then reads

$$\hat{\boldsymbol{\kappa}} = (\sin\theta\cos\phi, \sin\theta\sin\phi, \cos\theta) . \tag{12.15}$$

The unknown unit vector $\hat{\boldsymbol{\kappa}}$ thus reduces to two real unknowns θ and ϕ. As a consequence, the system (12.13) is over-determined, containing six

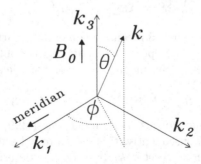

Fig. 12.1. Cartesian coordinate system for the determination of the wave vector direction $\hat{\boldsymbol{\kappa}} = \boldsymbol{k}/|\boldsymbol{k}|$

equations for two unknowns. Generally, these 6 equations are independent. This is different compared to the case of the ideal set of equations (12.9). The reason is that the matrix $\hat{A}$ is composed of experimental data which can contain natural and/or experimental noise and do not necessarily exactly correspond to an ideal plane wave.

Since we only have two unknowns, a subset of any two independent equations picked up from the set (12.9) is sufficient to obtain a unique solution for θ and ϕ. This is the basis of several analysis methods. The method of Means [36] is based on imaginary parts of three cross-spectra and the procedure is equivalent to solving any two of the last three equations in (12.13). The method of Samson and Olson [55] (their equation 11) is equivalent to finding a unique solution from another subset of equations selected in (12.13). The method of McPherron et al. [35] uses the first three equations and finds a unique solution using the eigenanalysis of the real part of the spectral matrix.

However, the choice of the subset of equations is rather arbitrary and for a different subset we do not generally obtain the same result from given experimental data. Other methods thus attempt to estimate an "average" solution of the entire set (12.9) using different techniques. Samson [54], again using the eigenanalysis, presented methods of decomposition of the entire complex spectral matrix. Santolík et al. [64] used a singular value decomposition (SVD) technique to estimate a solution of the entire set of equations (12.13) in the "least-squares" sense, decomposing the matrix $\hat{A}$

$$\hat{A} = U \cdot W \cdot V^T , \tag{12.16}$$

where U is a matrix 6×3 with orthonormal columns, W is a diagonal matrix 3×3 of three non-negative singular values, and V^T is a matrix 3×3 with orthonormal rows. Note that the SVD algorithm can often be found in numerical libraries [e.g., 50]. The "least-squares estimate" for $\hat{\kappa}$ is then directly found as the row of V^T corresponding to the minimum singular value at the diagonal of W.

The results of the plane-wave analysis often allow a straightforward interpretation of results. This was found useful, for example, in the analysis of sub-auroral ELF hiss emissions from the measurements of the Aureol 3 and Freja spacecraft [33, 62]. These different methods often provide us also with estimates of the validity of the initial assumption of the presence of a single plane wave. Different definitions of such an estimator ("degree of polarization", "polarization percentage" or "planarity") have been introduced, based on different descriptions of the coherence of the magnetic components and their confinement to a single polarization plane [49, 54, 64]. Similar techniques can also allow us to estimate the sense of the magnetic polarization with respect to the ambient stationary magnetic field B_0. This has been used, for example, to analyze electromagnetic emissions in the auroral region by Lefeuvre et al. [30, 31], Santolík et al. [59], and Santolík and Gurnett [56], using the data of the Aureol 3, Interball 2, and Polar spacecraft, respectively.

The above mentioned SVD technique can also be used with both the measured magnetic and electric components. In that case, an "average" solution to an over-determined set of 36 equations derived from equation (12.3) is estimated. This allows us to determine also the sign of $\hat{\kappa}$, i.e., to distinguish between the two antiparallel wave vector directions [for more details, see 64]. This technique also allows us to estimate the validity of the plane-wave assumption, but this time it is defined as a measure of closeness of the observed wave fields to equation (12.3). This determination of the "electromagnetic planarity" was, for example, used in [57] to estimate the dimension of the source of chorus emissions from the data of the four Cluster spacecraft.

12.2.2 Wave Distribution Function

The concept of wave distribution function is necessary when the wave field is more complex, for example when waves from multiple distant sources are simultaneously detected. The wave distribution function (WDF) is defined as a continuous distribution of wave energy with respect to the wave-vector direction; see the review by Storey [69]. It was first introduced by Storey and Lefeuvre [70]. The theoretical relation of the WDF to the experimentally measurable spectral matrix has been called the WDF direct problem. Supposing a continuous distribution of elementary plane waves at a frequency f having no mutual coherence and a narrow bandwidth Δf, the relationship between the spectral matrix $S_{ij}(f)$ and the WDF $G_m(f, \theta, \phi)$ is given by

$$S_{ij}(f) = \sum_m \oint a_{mij}(f, \theta, \phi) \, G_m(f, \theta, \phi) \, d^2\kappa \, , \qquad (12.17)$$

where m represents the different simultaneously present wave modes. The integration is carried out over the full solid angle of wave-normal directions κ, and for a given wave mode m the integration kernels a_{mij} are calculated from

$$a_{mij}(f, \theta, \phi) = \Delta f \, \frac{\xi_{mi}(f, \theta, \phi) \, \xi_{mj}^*(f, \theta, \phi)}{u_m(f, \theta, \phi)} \, , \qquad (12.18)$$

where ξ_{mi}, ξ_{mj} are complex spectral amplitudes of the ith and jth elementary signals of the wave electric or magnetic fields, and u_m is the energy density. All these quantities correspond to an elementary plane wave propagating in a mode m with a normal direction defined by θ and ϕ. f represents the Doppler-shifted frequency in the spacecraft frame. The complex spectral amplitudes of the wave electric or magnetic fields can be calculated by considering the physical properties of the medium. This calculation requires the knowledge of the theoretical solutions to the wave dispersion relation. Characteristics of the particular wave experiment should also be taken into account.

The theory of the WDF direct problem for the cold-plasma approximation has been developed by Storey and Lefeuvre [71, 72], and revisited by Storey [68]. This basic theory has been used by Lefeuvre [27], Lefeuvre and

Delannoy [28] and Delannoy and Lefeuvre [8] to develop practical methods for estimation of the WDF from the spacecraft measurements (the WDF inverse problem). Using a slightly different definition of the WDF and abandoning the explicit dependence of the WDF on the wave frequency, Oscarsson and Rönnmark [42, 43] and Oscarsson [41] introduced the hot plasma theory into the WDF reconstruction techniques. With the wave-vector dependent WDF they also introduced the Doppler effect in a natural way. Santolík and Parrot [60] used the hot plasma theory for the frequency-dependent WDF and further investigated the influence of the Doppler effect in this more complex situation. Santolík and Parrot [63] compared different techniques for resolution of the WDF inverse problem, mainly based on the minimization of the least-squares type merit function, in the context of the plane wave estimates.

The WDF techniques have been used in numerous studies with both ground based and spacecraft data. For instance, Lefeuvre and Helliwell [29], Parrot and Lefeuvre [45], Hayakawa et al. [18], and Storey et al. [73] used the multi-component measurements of the GEOS spacecraft to characterize the WDF of the ELF chorus and hiss emissions on the equatorial region. Based on the data of the Aureol 3, Akebono, and Freja spacecraft, Lefeuvre et al. [33], Kasahara et al. [23] and Santolík and Parrot [61, 63], respectively, estimated the WDF of the down-coming ELF hiss in the sub-auroral and auroral regions. Oscarsson et al. [44] compared the different reconstruction schemes for the data of the Freja spacecraft. The up-going funnel-shaped auroral hiss has been investigated by Santolík and Gurnett [56] using the WDF analysis of measurements of the Polar spacecraft. Parrot et al. [46] estimated the WDF for auroral kilometric radiation observed on board the Interball spacecraft, compensating at the same time for the a priori unknown experimentally induced phase between the electric and magnetic signals. Simultaneous WDF estimation of the Z-mode and the whistler mode in the auroral region has been done by Santolík et al. [59], based on the data of the Interball 2 spacecraft.

12.3 Analysis of Auroral Kilometric Radiation

The MEMO instrument on board the Interball 2 spacecraft [32] had two basic modes of measurement: burst mode and survey mode. During the survey mode, low-resolution overview spectrograms were recorded. In its burst mode, the instrument measured waveforms of several components of the electric and magnetic field fluctuations in three frequency bands from 50 Hz to 200 kHz. In the high-frequency band (30–200 kHz) which is relevant to this study, the device recorded waveforms measured at the same time by three magnetic antennas and one electric antenna.

On 28 January 1997, between 1950 and 2125 UT, the Interball 2 spacecraft was located on the night side and moved over the northern auroral region at altitudes of 2.6-3 R_E. During this time interval, the MEMO instrument recorded a highly structured emission of auroral kilometric radiation at frequencies

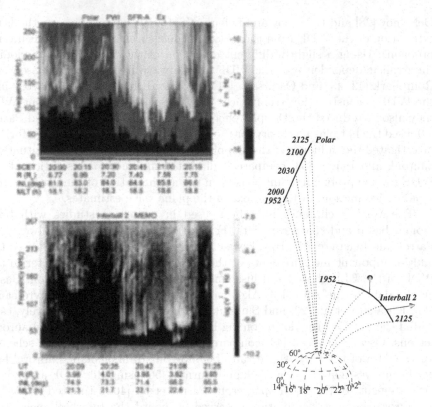

Fig. 12.2. Comparison of AKR observed by the Interball 2 and Polar spacecraft on 28 January 1997 between 1950 and 2125 UT. *Left*: Time-frequency spectrograms recorded by the MEMO instrument onboard Interball 2 (bottom) and by the SFR analyzer of the PWI instrument onboard Polar. The same time interval and frequency range is used for both spectrograms. Coordinates of the two spacecraft are given on the bottom of the spectrograms. Over-plotted white lines show the local electron cyclotron frequency. *Right*: Portions of orbits of the two spacecraft in the corresponding time interval and their projections along the magnetic field lines onto the Earth's surface. The arrows pointing out of the Interball-2 orbit show the average wave-vector directions for two burst-mode intervals at a frequency of 80 kHz (see Fig. 12.5)

between 30 kHz and more than 270 kHz, as reported in [46]. At the same time, the Polar spacecraft also moved over the northern auroral zone but at approximately twice the altitude and on the evening side. The PWI instrument on board Polar [for details of the instrument see, Gurnett et al., 14] observed the same AKR emission. Comparison of observations of the Interball 2 and Polar spacecraft is given in Fig. 12.2 on the left. The strong AKR emission between 2030 and 2125 UT is clearly seen on both spectrograms, demonstrating the global character of AKR. Some detailed time-frequency

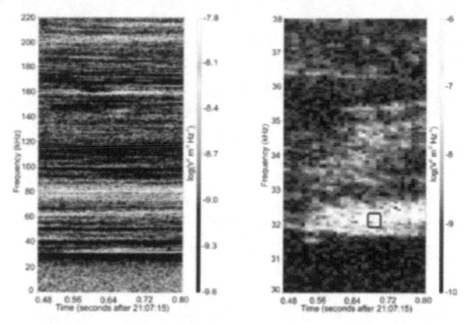

Fig. 12.3. Spectrograms calculated from one snapshot of burst-mode MEMO waveform data recorded on 28 January 1997 after 2107:15.48 UT. The spectrograms show the electric component in the frequency interval 0–220 kHz with a frequency resolution of 260 Hz (*left*) and detailed analysis of pronounced structures around 32 kHz (*right*) with a frequency resolution of 52 Hz. Black rectangle represents data used later for wave propagation analysis (Fig. 12.4). The local electron cyclotron frequency is 23.5 kHz

features are, however, different. Some of those differences may be attributed to different sensitivities of the two instruments as a function of frequency, but most of them are probably owing to different positions of the two spacecraft. As their local time is different, the two spacecraft can observe AKR emitted from different portions of the auroral oval. This also concerns slight timing differences for the onset of intense AKR around 2030 UT. During this orbit, several intervals of burst-mode data were recorded by the MEMO instrument on board the Interball 2 spacecraft. One electric and three magnetic components were always sampled at 533.3 kHz during 0.32 sec. Figure 12.3 shows the power spectrograms calculated from the electric-field data measured during one of those burst-mode intervals. We can see that, at short time scales, the emission still has a complex time-frequency structure with many spectral lines observed at slowly drifting frequencies.

Detailed analysis of the middle portion of this burst-mode interval has been done using the available data of one electric and three magnetic field components. The wave distribution function (WDF), representing the distribution of the wave energy for different wave-vector directions and wave-propagation

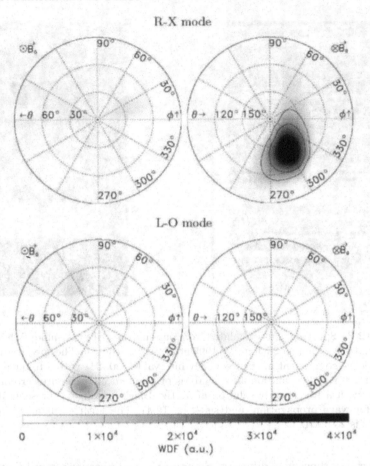

Fig. 12.4. Detailed analysis of the multicomponent MEMO data in the frequency band of 0.4 kHz around 32.1 kHz, between offsets 0.19 and 0.22 sec (see Fig. 12.3). The wave distribution function (WDF) is plotted in the coordinate system connected to the Earth's magnetic field B_0 and to the local magnetic meridian (Fig. 12.1). The polar diagrams represent the energy density as a function of angles θ and ϕ, on the left for downgoing waves, and on the right for upgoing waves, on the top for the R-X mode and on the bottom for the L-O mode

modes, is shown in Fig. 12.4. We use a narrow frequency band around the maximum of the power-spectral density of the intense spectral line observed around 32 kHz. A cold plasma model is used to calculate theoretical wave spectral densities, necessary for simultaneous WDF estimation in both R-X and L-O propagation modes and for both hemispheres of wave-normal directions (upgoing and downgoing with respect to the direction of the ambient magnetic field). The method of "discrete regions" was used to estimate the WDF [63]. It is based on least-squares optimization of the non-negative WDF values in the total number of 824 discrete regions, homogeneously covering

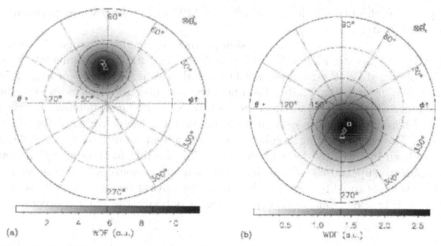

Fig. 12.5. Wave distribution function for the upgoing waves in the R-X mode propagating in the frequency band of 2.2 kHz centered at 80.1 kHz. The same coordinate system as in Fig. 12.4 is used. Two 0.32-s burst-mode snapshots have been used: (**a**) time interval starting at 2037:23.58 UT; (**b**) time interval starting at 2107:15.48 UT. Overplotted are plane wave estimates: triangles for the method of McPherron et al. [35], squares for the method of Means [36], diamonds for the SVD method [64]

the 4π solid angle of wave normal direction in the two propagation modes. The results show that nearly all the wave energy is found in the upgoing R-X mode. Within statistical uncertainties of our analysis, we can neither exclude nor confirm the presence of a small fraction of the L-X mode.

Figure 12.5 shows a comparison of obtained results of wave propagation analysis for two different time intervals separated by 30 minutes. The second of these time intervals is the same as in Fig. 12.3, and both of them were analyzed using different methods by Parrot et al. [46]. For the WDF estimation in Fig. 12.5 we use the model of Gaussian peaks [63]. This model describes concentration of wave energy near one or several wave-normal directions. Information about the WDF can then be reduced to a set of the peak directions and accompanied by the respective energy densities and parameters describing the degree of concentration of the wave energy near the respective directions (peak widths). In this case we model the WDF G as a single peak represented by a Gaussian function in 2D space of wave-normal directions (or, equivalently, on the surface of a unit sphere),

$$G(\theta, \phi) = \frac{\Gamma}{\pi \Delta^2} \exp\left(-\frac{2\left[1 - \cos\theta_0 \cos\theta - \sin\theta_0 \sin\theta \cos(\phi_0 - \phi)\right]}{\Delta^2}\right),$$

$$(12.19)$$

where Γ is the peak energy density, Δ is the peak width, and θ_0 and ϕ_0 define the central direction. These free parameters are optimized by a nonlinear least-squares method to obtain the best model (12.17) for the given experimental spectral matrix.

The results show that the emission propagates at oblique angles from the Earth's magnetic field, at different azimuth angles ϕ for the two time intervals. The obtained WDF estimates show large peak widths, ($\Delta = 20°$ for Fig. 12.5a and $\Delta = 28°$ for Fig. 12.5b). The plane wave results are localized within the 75% intensity level compared to the peak maximum. The differences between the different plane wave estimates are larger for a larger width of the WDF peak. The corresponding angular sizes of the global sources of observed AKR have been investigated by Schreiber et al. [65] leading to the suggestion that we observe superposed waves from many elementary sources. The positions of these sources roughly corresponded to active auroral regions remotely sensed by the Ultraviolet Imager (UVI) instrument onboard the Polar spacecraft.

12.4 Conclusions

We have shown that wave propagation analysis based on multicomponent spacecraft data can be a useful tool for investigation of high-frequency wave phenomena. Several analysis methods have been described, some using the plane wave approach and others based on a general continuous distribution of wave energy for different propagation directions. This second approach has been shown to be useful also for recognition of the propagation mode of observed waves. Observations of AKR by the MEMO instrument on board the Interball-2 spacecraft have been analyzed using these methods. We have shown that the observed highly structured AKR emission propagates predominantly in the R-X mode. The propagation analysis has allowed us to trace the waves back to the source regions which, consistently with previously published results, corresponded to large active regions on the auroral oval. These examples demonstrate the value of using the described analysis methods in future in situ investigations of high frequency waves in geospace and in the solar wind.

Acknowledgments

We thank D. A. Gurnett of the University of Iowa, PI of the Polar PWI instrument for the SFR data used in Fig. 12.2. We thank J. D. Menietti of the University of Iowa for useful discussions. This work was supported by the ESA PECS contract No. 98025 and by the GACR grant 202/03/0832.

References

[1] Alexander, J.K. and M.L. Kaiser: J. Geophys. Res. 81, 5948, 1976.
[2] Arthur, C.W., R.L. McPherron, and J.D. Means: Radio Sci. 11, 833, 1976.ö

[3] Baumback, M.M. and W. Calvert: Geophys. Res. Lett. 5, 857, 1978.
[4] Baumback, M.M., D.A. Gurnett, W. Calvert, et al.: Geophys. Res. Lett. 13, 1105, 1986.
[5] Benediktov, E.A., et al.: Kossm. Issled. 3, 614, 1965.
[6] Calvert, W.: Geophys. Res. Lett. 12, 381, 1985.
[7] de Feraudy, H., et al.: Geophys. Res. Lett. 14, 511, 1987.
[8] Delannoy, C. and F. Lefeuvre: Comp. Phys. Comm. 40, 389, 1986.
[9] Dunckel, N., et al.: J. Geophys. Res. 75, 1854, 1970.
[10] Ergun, R.E., et al.: Astrophys. J. 538, 456, 2000.
[11] Ergun, R.E., et al.: Geophys. Res. Lett. 25, 2061, 1998.
[12] Grard, R.: Ann. Geophys. 24, 955, 1968.
[13] Gurnett, D.A., et al.: Ann. Geophys. 19, 1259, 2001.
[14] Gurnett, D.A., et al.: Space Sci. Rev. 71, 597, 1995.
[15] Gurnett, D.A.: J. Geophys. Res. 79, 4227, 1974.
[16] Gurnett, D.A., et al.: Space Sci. Rev. 23, 103, 1979.
[17] Hanasz, J., et al.: J. Geophys. Res. 106, 3859, 2001.
[18] Hayakawa, M., M. Parrot, and F. Lefeuvre: J. Geophys. Res. 91, 7989, 1986.
[19] Hilgers, A.: Geophys. Res. Lett. 19, 237, 1992.
[20] Huff, R.L., et al.: J. Geophys. Res. 93, 11445, 1988.
[21] James, H.G.: J. Geophys. Res. 85, 3367, 1980.
[22] Kaiser, M.L., et al.: Geophys. Res. Lett. 5, 857, 1978.
[23] Kasahara, Y., et al.: J. Geomag. Geoelectr. 47, 509, 1995.
[24] Kurth, W.S., M.M. Baumback, and D.A. Gurnett: J. Geophys. Res. 80, 2764, 1975.
[25] LaBelle, J. and R.A. Treumann: J. Geophys. Res. 97, 13789, 1992.
[26] Le Quéau, D. and P. Louarn: Planet. Space Sci. 44, 211, 1996.
[27] Lefeuvre, F.: *Analyse de champs d'ondes électromagnétiques aléatoires observées dans la magnétosphère, à partir de la mesure simultanée de leurs six composantes*. Doctoral thesis, Univ. of Orléans, Orléans, France, 1977.
[28] Lefeuvre, F. and C. Delannoy: Ann. Telecommun. 34, 204, 1979.
[29] Lefeuvre, F. and R.A. Helliwell: J. Geophys. Res. 90, 6419, 1985.
[30] Lefeuvre, F., et al.: Ann. Geophys. 4, 457, 1986.
[31] Lefeuvre, F., et al.: Ann. Geophys. 5, 251, 1987.
[32] Lefeuvre, F., et al.: Ann. Geophys. 16, 1117, 1998.
[33] Lefeuvre, F., et al.: J. Geophys. Res. 97, 10601, 1992.
[34] Louarn, P., et al.: J. Geophys. Res. 95, 5983, 1990.
[35] McPherron, R.L., C.T. Russel, and P.J. Coleman, Jr.: Space Sci. Rev. 13, 411, 1972.
[36] Means, J.D.: J. Geophys. Res. 77, 5551, 1972.
[37] Mellott, M.M., R.L. Huff, and D.A. Gurnett: Geophys. Res. Lett. 12, 479, 1985.
[38] Morioka, A., H. Oya, and A. Kobayashi: J. Geomagn. Geoelectr. 42, 443, 1990.
[39] Mutel, R.L., D.A. Gurnett, and I.W. Christopher: Ann. Geophys. 22, 2625, 2004.
[40] Mutel, R.L., et al.: J. Geophys. Res. 108, 1398, doi: 10.1029/2003JA010011, 2003.
[41] Oscarsson, T.: J. Comput. Phys. 110, 221, 1994.
[42] Oscarsson, T. and K.Rönnmark: J. Geophys. Res. 94, 2417, 1989.
[43] Oscarsson, T. and K. Rönnmark: J. Geophys. Res. 95, 21,187, 1990.

[44] Oscarsson, T., G. Sternberg, and O. Santolík: Phys. Chem. Earth (C) **26**, 229, 2001.

[45] Parrot, M. and F. Lefeuvre: Ann. Geophys. 4, 363, 1986.

[46] Parrot, M., et al.: J. Geophys. Res. 106, 315, 2001.

[47] Parrot, M., et al.: Ann. Geophys. 21, 473, 2003.

[48] Parrot, M., et al.: Ann. Geophys. 22, 2597, 2004.

[49] Pinçon, J.L., Y. Marouan, and F. Lefeuvre: Ann. Geophys. 10, 82, 1992.

[50] Press, W.H., et al.: *Numerical Recipes.* Cambridge Univ. Press, New York, 1992

[51] Pritchett, P.L., et al.: J. Geophys. Res. 107, 1437, doi: 10.1029/2002JA009403, 2002.

[52] Pritchett, P.L.: J. Geophys. Res. 89, 8957, 1984.

[53] Roux, A., et al.: J. Geophys. Res. 98, 11657, 1993.

[54] J.C. Samson: Geophys. J. R. Astron. Soc. 34, 403, 1973.

[55] Samson, J.C. and J.V. Olson: Geophys. J. R. Astron. Soc. 61, 115, 1980.

[56] Santolík, O. and D.A. Gurnett: Geophys. Res. Lett. 29, 1481, doi: 10.1029/ 2001GL013666, 2002.

[57] Santolík, O., et al.:   Geophys. Res. Lett. 31, L02801, doi: 10.1029/ 2003GL018757, 2004.

[58] Santolík, O., et al.: J. Geophys. Res. 106, 13191, 2001.

[59] Santolík, O., et al.: J. Geophys. Res. 106, 21137, 2001.

[60] Santolík, O. and M. Parrot: J. Geophys. Res. 101, 10639, 1996.

[61] Santolík, O. and M. Parrot: J. Geophys. Res. 103, 20469, 1998.

[62] Santolík, O. and M. Parrot: J. Geophys. Res. 104, 2459, 1999.

[63] Santolík, O. and M. Parrot: J. Geophys. Res. 105, 18885, 2000.

[64] Santolík, O., M. Parrot, and F. Lefeuvre: Radio. Sci. 38, 1010, doi: 10.1029/ 2000RS002523, 2003.

[65] Schreiber, R., et al.:   J. Geophys. Res. 107, 1381, doi: 10.1029\-/\-2001\ -JA009061, 2002.

[66] Shawhan, S.D.: Space Sci. Rev. 10, 689, 1970.

[67] Stix, T.H.: *Waves in Plasmas.* Am. Inst. of Phys., New York, 1992.

[68] Storey, L.R.O.: Ann. Geophys. 16, 651, 1998.

[69] Storey, L.R.O.: The measurement of wave distribution functions. In: M.A. Stuchly, editor, *Modern Radio Science 1999*, page 249, Oxford University Press, Oxford, 1999.

[70] Storey, L.R.O. and F. Lefeuvre: Theory for the interpretation of measurements of a random electromagnetic wave field in space. In M.J. Rycroft and R.D. Reasenberg (Eds.), *Space Research XIV*, page 381, Akademie-Verlag, Berlin, 1974.

[71] Storey, L.R.O. and F. Lefeuvre: Geophys. J. R. Astron. Soc. 56, 255, 1979.

[72] Storey, L.R.O. and F. Lefeuvre: Geophys. J. R. Astron. Soc. 62, 173, 1980.

[73] Storey, L.R.O., et al.: J. Geophys. Res. 96, 19469, 1991.

[74] Wu, C.S. and L.C. Lee: Astrophys. J. 230, 621, 1979.

13

Phase Correlation of Electrons and Langmuir Waves

C.A. Kletzing[1] and L. Muschietti[2]

[1] Department of Physics and Astronomy, University of Iowa
craig-kletzing@uiowa.edu
[2] Space Sciences Lab., University of California
laurent@ssl.berkeley.edu

Abstract. Multiple spacecraft observations have confirmed the ubiquitous nature of Langmuir waves in the presence of auroral electrons. The electrons show variations consistent with bunching at or near the plasma frequency. Linear analysis of the interaction of a finite Gaussian packet of Langmuir waves shows that there are two components to the perturbation to the electron distribution function, one in-phase (or 180° out-of-phase) with respect to the wave electric field called the resistive component and one which is 90° (or 270°) out-of-phase with respect to the electric field. For small wave packets, the resistive perturbation dominates. For longer wave packets, a non-linear analysis is appropriate which suggests that the electrons become trapped and the reactive phase dominates. Rocket observations have measured both components. The UI observations differ from those of the UC Berkeley observations in that a purely reactive phase bunching was observed as compared to a predominantly resistive perturbation. The resistive phase results of the UC Berkeley group were interpreted as arising from a short wave packet. The UI observations of the reactive phase can be explained by either a long, coherent train of Langmuir waves or that the narrower velocity response of the UI detectors made it possible to capture only one side of the reactive component of the perturbed distribution function for a short wave packet in the linear regime. Future wave-particle correlator experiments should be able to resolve these questions by providing more examples with better velocity space coverage.

Key words: Electron bunching, Langmuir waves, resistive and reactive bunching phases, phase correlators

13.1 Introduction

The precipitation of auroral electrons provides an example of a beam-plasma interaction which generates Langmuir waves from the free energy in the electrons. The resulting waves play several important roles in the Earth's auroral ionosphere. First, the waves Landau damp on the thermal electron population

C.A. Kletzing and L. Muschietti: *Phase Correlation of Electrons and Langmuir Waves*, Lect. Notes Phys. **687**, 313–337 (2006)
www.springerlink.com

and thereby form a direct conduit for energy exchange between the auroral electron beam and the thermal electrons. Several authors have speculated that Langmuir waves play a significant role in establishing the electron temperature in the auroral ionosphere [3, 15]. In addition to heating electrons, the Langmuir/upper hybrid waves radiate away some of their energy into electromagnetic radiation, which can serve for remote sensing of auroral plasma processes from ground level and from satellites. For example, auroral roar is an EM emission observed near 2–3 and 4–4.5 MHz at ground level [14, 30] and from satellites [1, 12].

Understanding these auroral wave emissions is important not only to fully understand terrestrial aurora and related phenomena, but also because they shed light on analogous emission processes elsewhere in the solar system and beyond. For example, the generation of auroral roar is similar to that of terrestrial continuum radiation, which is generated via mode conversion of upper hybrid waves at the plasmapause, and possibly continuum radiations at other planets as well. Solar type III radiation results from mode conversion of Langmuir waves in the solar wind, and recent observations of structured type III emission [Reiner et al., 23, 24] indicate the significance of frequency structure in the causative Langmuir waves.

High frequency (HF) electric field observations in the topside auroral ionosphere, for which the plasma frequency is typically greater than the electron cyclotron frequency, have revealed plasma waves in the range $f_{pe} \leq f \leq f_{uh}$ ever since the earliest measurements [2, 29]. Simultaneous wave and electron distribution measurements have shown that the waves are excited by Landau resonance but that temporal variation of the distribution function or wave refraction from vertical density gradients can limit wave amplitudes as shown by McFadden et al. [19]. Many examples of waves near f_{pe} were observed using Aureol/ARCAD 3 satellite wave receivers and the free-energy source was identified as the electrons [Beghin et al., 3]. Although rocket-borne receivers have detected HF waves at E-region altitudes attributed to generation by secondary electrons [Kelley and Earle, 13], most observations pertain to altitudes from 300 km and up, where the waves have amplitudes ranging from less than 1 mV/m to as large as 1 V/m [4, 6, 8, 17, 18, 28] These waves are highly bursty in time, sometimes lasting as little as 1 ms but sometimes appearing continuous for ~1s. Recent high-resolution experiments reveal that they can have complex frequency structure [17, 18, 25]. Other investigators have detected plasma waves through particle-particle correlator techniques. Using a rocket-borne electron detector of large geometric factor, ~5% modulation at frequencies 4.2–5.6 MHz was found during a 7-second interval when the rocket passed the boundary of two oppositely directed Birkeland sheet currents [27]. Strong modulation (~30%) was observed at 2.65 MHz of 4–5 keV electrons, corresponding to energies at which a positive slope was observed in the perpendicular and parallel velocity distribution functions [Gough and Urban, 11]. Rocket measurements in which 1.4 MHz fluctuations were detected in the 7.5 keV electrons, just below the electron beam energy, were reported to occur simultaneously with a positive slope in the electron distribution function [Gough et al., 10].

Linear theory explains the existence and parallel polarization of auroral Langmuir waves [see, e.g., Nicholson, 22]. Theory and simulations have shown that Langmuir waves in the auroral plasma most likely do not develop into strong turbulence but that observed non-linear features in amplitudes and modulations of the waves are consistent with nonlinear interactions between the Langmuir waves and ion acoustic waves [Newman et al., 21]. More recently, the difference between Langmuir wave-electron interactions both at rocket altitudes and at the altitude of the Freja spacecraft have been studied, finding that standard quasi-linear diffusion theory does not hold for large amplitude Langmuir waves at Freja altitudes, but should hold at lower altitudes [Sanbonmatsu et al., 26].

Correlating waves and particles directly probes the physics of their interaction by providing a superior picture of the microphysics compared to statistically associating an unstable feature of the distribution function with the presence of waves. Elementary theory implies that if the electrons and the waves are exchanging energy, the electrons will have an oscillatory component at a velocity equal to the phase velocity of the waves. Identifying the velocity at which the electron distribution function has this oscillatory component determines the wave phase velocity and therefore the wave number. Measuring the phase of this oscillatory component of the electron distribution function relative to the wave electric field yields further information. The phase bunching splits into two pieces: 1) the resistive component, in phase with the electric field indicating wave-electron energy exchange; and 2) the reactive component, 90° out of phase with the electric field indicating trapping [Nicholson, 22].

Detailed treatment of phase relationships by Muschietti et al. [20] for Gaussian Langmuir wave packets, show that the linear perturbation in the distribution function may be considered as the sum of a resistive and reactive components. For both components, the perturbation narrows and increases in magnitude as the wave packet length increases. Electron detectors with broad energy resolution can only detect the resistive component because it has only positive polarity and adds over the entire energy range. Detecting the bipolar nature of the reactive component requires narrow energy response $(\Delta v/v \leq 5 - 6\%)$ detectors.

Only three high frequency (MHz) wave-particle correlation experiments have been reported in the literature [7, 9, 16]. Very similar experiments have been tried on FREJA by Boehm et al. [5] and on FAST by Ergun et al. [7] but were limited in phase resolution. The first two of these experiments used a correlator that worked by binning individual detected particles according to the phase of the strongest wave detected by a broadband (0.2–5 MHz) wave receiver [Ergun et al., 7, 9]. The most recent experiment used a correlator which had higher phase resolution and detected electrons bunched 90° out-of-phase with respect to the electric field, suggesting non-linear evolution and trapping of the electrons.

In what follows, we discuss the theory of linear perturbations to the distribution by a Gaussian wave packet and also present the theory of extended wave packets via a BGK analysis. We then describe in detail two of the three high frequency wave-particle correlator measurements referred to above. We then conclude with a discussion of these results in the context of theory.

13.2 Finite-Size Wave Packet in a Vlasov Plasma

Consider a coherent, Langmuir packet propagating in the x direction with a phase $\psi = kx - \omega t$ where the frequency ω is close to the plasma frequency ω_{p}. The wave packet is assumed to be localized, which we specify by a form factor $\eta(x,t)$ which is piecewise continuous, bounded, and vanishing for $x \to \pm\infty$. Its slow time dependence describes the drift due to the group velocity of the Langmuir wave or the growth (or damping) due to the interaction with the electrons. Explicitly, the wave electric field is written as

$$E(x,t) = E_0\eta(x,t)e^{i\psi} + c.c. \tag{13.1}$$

where c.c. is the complex conjugate.

The electrons are assumed to consist of two populations: a dense background and energetic, streaming particles. In the problem that we treat, the density of the energetic electrons is orders of magnitude smaller than the background density, so that only the latter determines the real part of the dispersion relation. The electrons are described by a homogeneous distribution function $F(v)$ which includes velocities around the phase velocity of the Langmuir wave, $v_{\mathrm{p}} \equiv \omega/k$. Under the influence of the wave field, the streaming population develops a time-dependent, inhomogeneous component $f(x,v,t)$ which satisfies the Vlasov equation

$$\frac{d}{dt}f \equiv \left(\frac{\partial}{\partial t} + v\frac{\partial}{\partial x}\right)f = \frac{eE}{m}\frac{\partial}{\partial v}(F+f) \tag{13.2}$$

where $-e$ and m are the electron charge and mass, respectively. Our goal is to find explicit expressions for $f(x,v,t)$ and to analyze their phases versus the wave phase ψ in view of correlator applications.

In the linear approximation, one neglects the perturbation $\partial f/\partial v$ on the right-hand side of (13.2) and formally integrates the equation,

$$f_L(x,v,t) = \frac{eE_0}{m}\int_{-\infty}^{t}\eta(x',t')e^{i\psi(x',t')}\frac{\partial}{\partial v'}F(v')\,dt' + c.c. \tag{13.3}$$

The integration is carried out along the trajectories $x'(t')$, $v'(t')$ that have x and v for end points at $t' = t$, namely $x'(t) = x$ and $v'(t) = v$. The perturbation at large negative times when $x' \to \pm\infty$ is ignored. From (13.3) one sees that the linear perturbation f_L oscillates in time with the same frequency as the electric field. One also notices that the perturbed distribution depends upon the structure of the form factor η along the past trajectories of the particles, which leads to a phase shift relative to the present wave phase $\psi(x, t)$.

13.2.1 Linear Perturbation of the Electrons

Taking as characteristics the straight trajectories $v'(t') = v$ and $x'(t') = x + v(t' - t)$, where v is constant and positive, one rewrites (13.3) as

$$f_L(x, v, t) = A_0\, e^{i(kx-\omega t)}\frac{\partial F}{\partial v}\int_{-\infty}^{0}\eta(x + v\tau, \tau + t)e^{i(kv-\omega)\tau}d\tau \;+ c.c. \quad (13.4)$$

where $A \equiv eE_0/m$. The integral explicitly relates the distribution f_L to the motion of the particles at earlier times, so that the effect of spatial gradients in the electric field amplitude will be directly seen in the resulting phase relation.

Let us for instance consider the simple profile of a square window with a width L,

$$\eta(x) = \begin{cases} 0 & : \text{ if } x < 0 \\ 1 & : \text{ if } 0 < x < L \\ 0 & : \text{ if } x > L \end{cases} \quad (13.5)$$

From (13.4), one readily obtains

$$f_L(x, v, t) = \frac{iAe^{i\psi}}{\omega - kv}\frac{\partial F}{\partial v}\left[\begin{matrix} 1 - \beta^x & : \text{ if } 0 < x < L \\ \beta^{(x-L)} - \beta^x & : \text{ if } L < x \end{matrix}\right] + c.c. \quad (13.6)$$

where $\beta \equiv \exp i(\omega/v - k)$. In the square bracket to the right, the exponents reflect the presence of the boundaries and modify the phase of the perturbation relative to the wave phase ψ. They also yield a ballistic term $\beta^{(x-L)} - \beta^x$ downstream of the interaction region. An important point is that these boundary terms keep the perturbed distribution function of resonant electrons bounded. The usual expression for a plane wave [(13.6) with the square bracket equal to unity] shows the perturbation to have a singularity at resonance, where the assumed linear solution, therefore, breaks down. Instead, the right-hand side for $0 < x < L$ can be expanded for velocities close to the phase velocity v_p. The resulting expression is finite. One can then differentiate it with respect to v and thus evaluate at $v = v_p$ the derivative we neglected in the linearization of Vlasov equation (13.2). Using the notation $F' \equiv \partial F/\partial v$, the result can be written as

$$\frac{\partial f_L}{\partial v}\Big|_{v_p} = F'(v_p)\frac{2A}{\omega v_p}\left[(v_p\frac{\partial \ln F'}{\partial v} - 1)kx\,\cos\psi + (kx)^2\sin\psi\right] \quad (13.7)$$

The term proportional to $\sin \psi$ (thus out of phase with the electric field) is seen to grow quadratically with the distance into the packet. Therefore, the packet must have a finite extent to assure the validity of the linear perturbation, which we write $(kL)^2 \ll \omega v_p/(2A)$. This inequality introduces an important parameter to the problem at hand

$$\mu \equiv \frac{2eE_0 k}{m} \left(\frac{kL}{\omega}\right)^2. \tag{13.8}$$

This quantity measures the effect of the localized electric field on the electrons and can be used as a small expansion parameter. We can think of it as the square of the bounce frequency times the transit duration of a resonant particle.

13.2.2 Case of a Gaussian Packet

A more realistic envelope of wavepacket is provided by a Gaussian. In fact, since the dispersion relation of Langmuir waves is quadratic, a Gaussian packet moving at the group velocity describes well a propagating Langmuir packet. Its dispersion time is given by $t_d \approx (L/\lambda_d)^2 \omega_p^{-1}$ with λ_d the Debye length. This time is very long compared to, for example, the transit duration of resonant electrons, $t_t \approx kL\,\omega_p^{-1}$. For simplicity, we choose here a static form factor

$$\eta(x) = \exp[-x^2/(2L)^2], \tag{13.9}$$

which is justified since for many applications the group velocity u is much smaller than the velocity of resonant electrons, $u/v_p = 3\,(k\lambda_d)^2 \ll 1$. After substituting (13.9) into (13.4) and some algebra, one obtains

$$f_L(x,v,t) = A\,\eta(x)e^{i(kx-\omega t)}\frac{\partial F}{\partial v}\frac{L}{v}(-i)Z(\xi) + c.c. \tag{13.10}$$

where Z is the usual plasma dispersion function, yet has here a completely different argument:

$$\xi = (\omega - kv)\frac{L}{v} - i\frac{x}{2L}. \tag{13.11}$$

The real part of the argument is the Doppler-shifted frequency seen by a traversing electron times its transit duration through the wavepacket. It determines the proximity to resonance during the interaction. The imaginary part of the argument describes the position with respect to the center of the packet.

Let us evaluate now (13.10) at the center of the packet. We can define a resonance function that represents the response of the electron distribution to the wave field:

$$R \equiv \frac{-iL}{v}Z(\xi_r) = \frac{L}{v}\left(\sqrt{\pi}e^{-\xi_r^2} - i\,2e^{-\xi_r^2}\int_0^{\xi_r} e^{y^2}\,dy\right) \tag{13.12}$$

The real part of R, associated with the Gaussian $e^{-\xi_r^2}$, represents the resistive contribution (in phase with the electric field). The imaginary part, associated

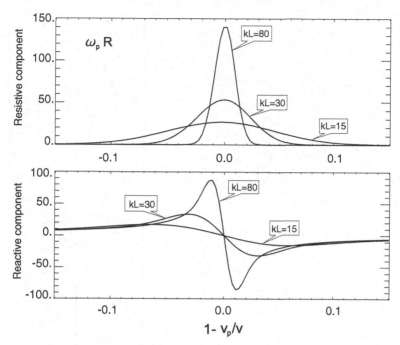

Fig. 13.1. Resonance function defined by (13.12). Real and imaginary parts of R are shown for various packet sizes (Reprinted with permission from American Geophysical Union)

with the Dawson integral, represents the reactive contribution (out of phase with the electric field). When the packet is large so that $\omega L/v \to \infty$, $Re(R)$ tends to the usual delta function of the Plemelj formula, $\pi\delta(kv-\omega)$, and $Im(R)$ tends to the principal part of $1/(kv - \omega)$. Plots of real and imaginary parts of R are displayed in Fig. 13.1 for various sizes of packets. One clearly sees the tendency to a delta function for large kL and, conversely, the broadening of the resonance for a more localized wavepacket.

The perturbed oscillating distribution has components in and out of phase with respect to the electric field. Let us split f_L of (13.10) into resistive and reactive terms, in a way that emulates the procedure performed by the wave correlator on rocket data. One obtains

$$f_L = \frac{\mu}{kL}\frac{v_p^2}{v}F'(v)\,\eta(x)\,[Z_i\cos(kx - \omega t) + Z_r\sin(kx - \omega t)] \qquad (13.13)$$

where the acceleration factor $A = eE_0/m$ has been rewritten in terms of μ using (13.8). The perturbation is thus either rather resistive or reactive depending upon the relative weight of Z_i and Z_r. Its amplitude is proportional to the wave amplitude, to the packet size, and to the slope of the distribution F. Figure 13.2 displays the perturbation in phase space by means of contours. Solid contours with gray shadings indicate a positive value, i.e. an

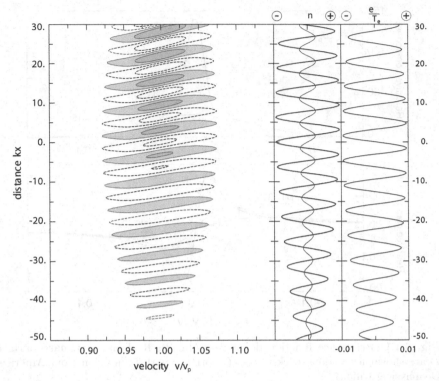

Fig. 13.2. Linear perturbed distribution as described by (13.13) with power-law model $F(v) \sim v^{-4}$. Contours with gray shadings indicate enhancements at levels 0.006 (light) and 0.018 (dark). Dashed contours denote depletions at levels -0.006 and -0.018. The perturbation (normalized to $F(v_p)$) features a chain of bunch-ellipses centered at the phase velocity v_p. Panel marked δn at right displays the resonant density perturbation, split into its resistive (*thick line*) and reactive (*thin line*) components. Rightmost panel shows the potential ϕ of the Langmuir packet with characteristics: $kL = 30$, $\mu = 0.09$

accumulation of electrons. Dotted contours indicate a negative value, i.e. a dearth of electrons. The bunching of the particles yields a chain of ellipses centered on the phase velocity v_p and zeroes of the potential, which is shown in the rightmost panel. The panel marked δn displays both the resistive (thick line) and reactive (thin line) components of the density perturbation. It is clear that the linear perturbation is mostly resistive. In addition, from the maxima of δn being in phase with $\partial\phi/\partial x > 0$ we can conclude that the wave accelerates the electrons, whereby it is being damped. This is consistent with the power-law model, $F(v) \sim v^{-4}$, we have chosen for drawing the plot.

Note the slight tilt of the ellipses in Fig. 13.2. A consequence is that the result of a velocity-integration of the perturbed distribution critically depends on the integration window's width and centering. For computing δn shown in the mid-panel we used a 7% width on either side of v_p. However, if the window is narrower and not centered on v_p, the integration will emphasize the reactive component. Furthermore, the sign of the latter depends upon whether the window happens to be centered a little above or below v_p.

In Fig. 13.2 the parameter μ is small, $\mu = 0.09$, hence the interaction is justifiably in the linear regime. However, when either the wave amplitude is larger or the packet is more extended, the linear solution to the Vlasov equation loses its validity and one must account for nonlinear corrections. These corrections to the orbits of resonant electrons δv and δx are difficult to compute analytically. Instead, we resort to particle-in-cell simulations, where consistent interactions between field and particles are automatically taken into account. The results of these simulations are shown in Fig. 13.3 which display the bunch-ellipses in a weakly nonlinear regime with $\mu = 1.2$. Solid contours and shaded areas indicate a positive value, or an accumulation of electrons. Dotted contours indicate a negative value, or a dearth of electrons. Two points must be noted. First, the wave is here driven by a bump on the tail of the distribution function. Accordingly, resonant electrons are decelerated

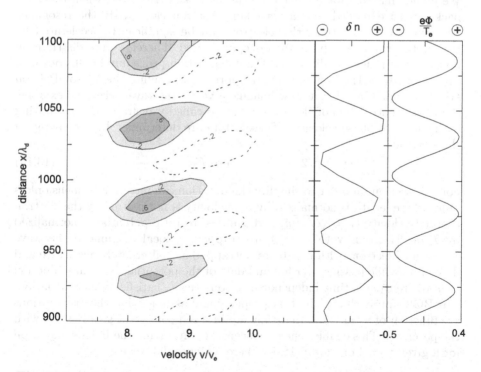

Fig. 13.3. Bunch-ellipses in a weakly nonlinear regime, from a particle-in-cell simulation where the Langmuir packet is amplified by a bump on tail located at $v = 10v_e$. Here $\mu = 1.2$ and $v_p = 8.7v_e$. Accumulation locales (*shaded in gray*) have decelerated while dearth locales (*dashed contours*) accelerated. More in the text

by the wave and maxima of δn are in phase with $\partial\phi/\partial x < 0$. Second, since bunches of accumulated electrons are decelerated, the nonlinear correction to their orbit has $\delta v < 0$ and $\delta x < 0$. In contrast, bunches associated with a dearth of electrons are accelerated and thus have $\delta v > 0$ and $\delta x > 0$. Hence, the two types of bunch-ellipses are no longer aligned at $v = v_p$ as in Fig. 13.2. One can see in the plot that the solid contours have moved to the left while the dotted contours have moved to the right. In addition, their position in x drifts and slowly converges toward maxima of the potential, whereby their phase becomes more reactive.

13.3 Extended Wave Packet: A BGK Analysis

The only nonlinear term in the Vlasov-Poisson system of equations is the acceleration term on the right hand side of (13.2). In Sect. 13.2, we chose to linearize this term by splitting the distribution in a large, homogeneous part F and a small, inhomogeneous part f, and then neglecting $\partial f/\partial v$ versus $\partial F/\partial v$. We found that this was justified as long as $\mu \ll 1$. However, when the wave packet is so extended that a traversing electron can satisfy the resonance condition for a long time, this electron can be significantly accelerated by the electric field. Its orbit is then deeply altered. Therefore, the distribution function is strongly modified and the linearization procedure breaks down.

The BGK method offers another approach to solving the Vlasov-Poisson system of equations. Let us now imagine a very long wavepacket and examine the phase-space orbits of electrons in the so-called waveframe (frame moving with the wave phase velocity). These orbits are determined by the energy of the electrons

$$w = v^2/2 - \phi(x) \quad \text{where} \quad E(x) = -\frac{\partial\phi}{\partial x} \ . \tag{13.14}$$

For this section, in order to simplify the notation, we introduce dimensionless units where length is normalized by λ_d, velocity is normalized by the electron thermal velocity $v_e = \sqrt{T_e/m}$, and the electrostatic potential is normalized by T_e/e. Electrons with $w < 0$ are trapped in local maxima of the wave potential. Electrons with $w > 0$ are untrapped and alternately accelerate and decelerate while passing over hill and dale of the potential. Now, any function of w where $\phi(x)$ is time-independent automatically satisfies Vlasov equation. The BGK approach exploits that property and assumes that the distributions are function of w only and that they are in a self-consistent steady state with the potential. This enables one to concentrate on solving the Poisson equation for a given model of wave field, e.g. here a sinusoidal wave

$$\phi(x) = \Psi \sin kx \ . \tag{13.15}$$

Let $F_e(w)$ and $F_t(w)$ be the distribution functions of, respectively, the passing and the trapped electrons. Poisson equation reads

$$\frac{d^2\phi}{dx^2} = -n_i + \int_\Psi^\infty \frac{F_e^+(w) + F_e^-(w)}{\sqrt{2}(w+\phi)^{1/2}} dw + \int_{-\phi}^\Psi \frac{\sqrt{2}F_t(w)}{(w+\phi)^{1/2}} dw \ . \qquad (13.16)$$

Terms on the right hand side represent, in order, the density of the ions, which are supposed to form a constant background, the density contribution from the passing electrons, and that from the trapped electrons. The passing electrons have been split into those moving to the right, $F_e^+(w)$, and those moving to the left, $F_e^-(w)$. The trapped electrons, by contrast, must be symmetric with the same flux of right and left moving particles, $F_t^+(w) = F_t^-(w) = F_t(w)$, in a stationary situation.

13.3.1 Passing Electrons

A model of distribution for the ambient electrons that is practical for computational purposes and representative of the observed distributions is given by

$$F_e^\pm(v) = \frac{2}{\pi} \frac{1}{[1+(v\pm v_p)^2]^2} \ , \qquad (13.17)$$

where v is an absolute number measuring the velocity from the wave frame, and the direction is selected by the $\pm$ sign. Note that this distribution is normalized to unity and becomes a power-law in v^{-4} at large velocities. In the presence of the wave it translates into

$$F_e^\pm(w) = \frac{2}{\pi\left[1+\left(\sqrt{2}(w-\psi)^{1/2}\pm v_p\right)^2\right]^2} \qquad \text{with } w > \Psi . \qquad (13.18)$$

After integrating this expression for the densities of right and left moving particles, one obtains the total density of passing electrons as

$$n_p = 1 - \frac{2}{\pi} \frac{\sqrt{\varphi}\,(1+\varphi-v_p^2)}{v_p^4 + 2v_p^2(1-\varphi)+(1+\varphi)^2} - \frac{1}{\pi}\arctan\left(\frac{2\sqrt{\varphi}}{v_p^2+1-\varphi}\right) \qquad (13.19)$$

where the notation $\varphi \equiv 2(\Psi+\phi)$ is introduced. The density is maximum where the potential is minimum, at $\varphi = 0$, and monotonically decreases toward a minimum where the potential is maximum, at $\varphi = 4\Psi$. We will assume that $4\Psi \ll 1$, which enables us to expand the complicated expression above into the simpler

$$n_p(\phi) = 1 - \frac{4\sqrt{2}}{\pi(v_p^2+1)^2}(\Psi+\phi)^{1/2} - \frac{16\sqrt{2}(5v_p^2-1)}{3\pi(v_p^2+1)^4}(\Psi+\phi)^{3/2} \qquad (13.20)$$

The most important term, in $\varphi^{1/2}$, has a coefficient that is related to the value of the ambient distribution at the wave phase velocity ((13.17) with $v = 0$). This is in agreement with the intuitive idea that electrons which move slowly along the separatrix (located at $v = \sqrt{2\Psi(1+\sin kx)}$) are strongly affected by the potential and thus contribute the most to the density variations. By contrast, fast-moving electrons far from the separatrix hardly "notice" the potential and pass by quasi undisturbed.

13.3.2 Trapped Electrons

We formally define the density of trapped electrons n_t as an unknown function of ϕ through the expression

$$n_t(\phi) \equiv \int_{-\phi}^{\Psi} \frac{\sqrt{2} F_t(w)}{(w+\phi)^{1/2}} dw . \qquad (13.21)$$

This integral equation can be inverted for $F_t(w)$, which yields

$$F_t(w) = \frac{1}{\sqrt{2}\pi} \int_{-\Psi}^{-w} \frac{1}{(-w-\phi)^{1/2}} \frac{d}{d\phi} n_t(\phi) \, d\phi . \qquad (13.22)$$

The potential $\phi(x)$ is given by (13.15) and requires a net density perturbation

$$n_s(\phi) = -k^2\phi \quad \text{with} \quad n_s = n_p + n_t . \qquad (13.23)$$

Substituting n_s and n_p for n_t in (13.22), one obtains after integration

$$F_t(w) = \frac{2}{\pi(v_p^2+1)^2} - \frac{\sqrt{2}}{\pi v_p^2}(\Psi - w)^{1/2} + \frac{4(5v_p^2 - 1)}{\pi(v_p^2+1)^4}(\Psi - w) \quad (13.24)$$
$$\text{with} \quad -\Psi \le w < \Psi .$$

This is an explicit expression for the distribution of trapped electrons that is consistent with a sinusoidal wave. Due to the v_p^8 dependence in the coefficient's denominator, the third term is considerably smaller than the second. Thus, the sign of the second term indicates that the distribution is punctuated with periodic holes located where $(\Psi - w)$ maximizes, namely at maxima of the potential ϕ.

An illustration of the perturbed distribution is provided in Fig. 13.4. It has been recast in terms of x and v in the plasma frame and normalized to $f_e(v_p)$. The display is similar to Figs. 13.2 and 13.3, including the potential in the right panel. Gray shadings indicate the phase space density with light grays denoting depletion. Clearly, a velocity integration by means of a window centered on v_p will provide a density perturbation δn which is (1) reactive and (2) characterized by $\delta n < 0$ where ϕ is maximum.

13.4 Electron Phase Sorting Measurements

The predictions of the theory of Langmuir wave interaction with electrons raise the question of experimental tests of the theory. In particular, measurements of the portion of the electron distribution which is resonant with the waves along with measurements of the waves themselves permits detailed comparison of theory and data which probes the success of these models..

The past 15–20 years have seen the development of wave-particle correlators which sort individual electron counts into bins corresponding to the phase

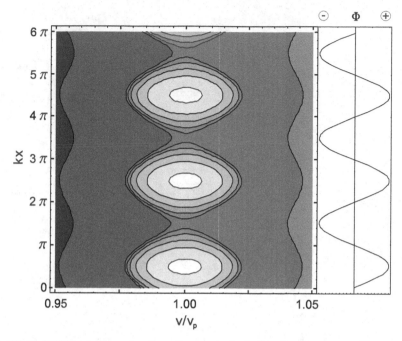

Fig. 13.4. BGK solution showing the perturbed distribution in a self-consistent state with an infinite wave train $\phi(x) = \Psi \sin kx$. Phase space density is indicated by gray shadings where darker gray means more particles. Clearly, integrating over a velocity window around v_p produces a reactive signal with density minima in phase with maxima of the potential and vice versa

of the observed wave as each electron arrives at the payload. The advantage of instrumentation of this type is that by accumulating the sorted electron counts over many wave periods, the average phase relation of the electrons relative to the wave can be determined for a relatively modest usage of spacecraft telemetry bandwidth. In this way, key aspects of the detailed physical interaction of Langmuir waves and electrons can be studied as never before.

13.4.1 Measurements of the Resistive Component

The first electron phase sorting correlator which measured the electron to Langmuir wave phase relation was developed at UC Berkeley in the late 1980's for use on auroral sounding rockets. The principle of operation is illustrated in the schematic shown in Fig. 13.5.

The incoming waveform from the parallel electric field antenna is first passed through a bandpass filter to remove signals outside the frequency range expected for Langmuir waves. At rocket altitudes in the auroral zone, this corresponds to a 200 kHz-5 MHz band. By removing frequencies outside this range, any significant signal remaining has high likelihood of being that from

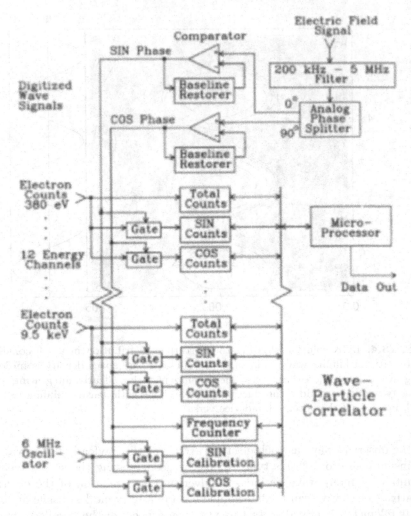

Fig. 13.5. Schematic diagram of the UC Berkeley electron phase sorting correlator from the work of Ergun et al. [9]. The instrumentation sorted electron into two phase bins of 180° width, one centered on 0° phase and one centered on 90° phase. By combining these bins with total counts over the entire interval, the phase of electrons could be determined within four 90° intervals (Reprinted with permission from American Geophysical Union)

Langmuir waves. After this filter, the signal was then passed through an analog phase splitter to produce one signal with no phase shift and an identical signal shifted by 90° in phase. Each of these two signals was then passed through a comparator to produce a one-bit digitization of the signal. Because the count rate of phase-bunched electrons is typically a small fraction of the total electron count rate (of the order of 1%), it is essential that this digitized signal have a precise 50% duty cycle so as not to bias the results. This was ensured through the use of a baseline restoring circuit which monitored the integrated digitized signal as a function of time and adjusted the baseline of the incoming analog signal to yield an accurate 50% duty cycle.

The two digitized signals, dubbed SIN for the 0° phase signal and COS for the 90° signal, were used to sort electrons at several fixed energies as they arrived at the rocket payload. When a given signal was high, the counter was gated open and allowed to accept incoming electron counts. When the signal was low, the counter was gated closed and would not accept electron counts. In this way, counts were accumulated over the 180° phase interval centered on 0° phase (the COS counter) and over the 180° phase interval centered on 90° phase (the SIN counter). In addition to these two counters, a third counter accumulated counts over the entire wave period. The set of three counters was duplicated 12 times to analyze electron counts from 12 fixed energy channels of a fast electron spectrometer that was oriented to detect electron precipitation parallel to the magnetic field.

Under the assumption that the phase correlation of the electrons is sinusoidal, the square root of the sum of the squares of the SIN and COS channels yields the amplitude of the correlation. The phase of the correlation is then the opposite of the inverse tangent of the ratio of the SIN and COS counters.

In analyzing correlator results, it is important to consider how random fluctuations in counting rate can influence the results. Poisson statistics can give significant variations away from the average count particularly for small numbers of counts. Thus it is essential to analyze correlation data in terms of standard deviations σ away from the expected background. In terms of standard deviations, the probability of 2σ variation is once in 22 samples and is generally not considered significant. A 3σ variation is expected to occur once in every 370 measurements. However, a 3.5σ variation should occur only once every 2149 samples and 4σ events occur only once every 15783 samples. Thus to be considered truly significant, events with $\sigma > 3.5$ are preferred.

The correlator was flown on a rocket launched from Poker Flat, Alaska on March 4, 1988 which crossed several auroral arcs during the expansion phase of substorm. During the flight, the parallel component of the electric field reached amplitudes of 100 mV/m in the frequency band from 200 kHz to 5 MHz with a dominant frequency of 1.4 MHz indicating the presence of strong Langmuir oscillations. The output of the correlator showed several events with significant correlation with electrons from the eight lowest energy channels which covered a range of energies from 380 eV to 3.2 keV. The correlations appeared first in the higher energy channels and then moved to lower energy.

Analysis of the data from this rocket flight yielded five events with fluctuations greater than 4σ. These five events, plotted as function of wave electric potential, are shown in Fig. 13.6 taken from Ergun et al. [9].

Each event is plotted with the associated error bar as determined from experimental uncertainties as well as the 1σ statistical error of the counts. All five events are located near 90° or 270° with respect to the wave electric

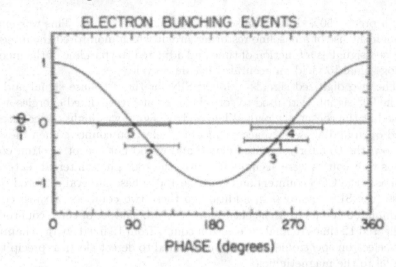

Fig. 13.6. Phase of bunched electrons for five events of correlations at levels greater than 4σ from Ergun et al. [9]. The phase of each event is plotted along with error bars derived from experimental uncertainties (Reprinted with permission from American Geophysical Union)

potential. This corresponds to 0° or 180° with respect to the wave electric field. The frequency of all 5 events was 1.4 MHz which, when combined with the electron energies of 380 eV to 3.2 keV, yields wavelengths of 8 m to 20 m. Events 2 and 5 were such that more electrons were being accelerated by the wave than were being decelerated and thus corresponds to wave damping. They occurred during periods of weaker amplitudes of 10–20 mV/m. Events 1, 3, and 4 had more electrons decelerating than accelerating corresponding to wave growth. At these times the wave amplitudes were near 100 mV/m or greater.

Following the analysis of Sect. 13.2, the value of μ can be calculated using the observed electric field amplitude of 80 mV/m, the inferred wavelength $\lambda = 12.8$ m, and observed the plasma frequency of 1.4 MHz along with a value for kL. For these events, kL was estimated to be of the order of $kL = 30$ which yields a value of $\mu = 0.09$. This is consistent with the linear analysis for a short wave packet which predicts the observation of the resistive component of the electron perturbation because in this regime, the resistive perturbation is generally larger than the reactive perturbation as illustrated in Fig. 13.2.

13.4.2 Measurements of the Reactive Component

Further improvement in the measurement of the phase of electron bunching in Langmuir waves in the auroral zone was achieved using a new phase correlator developed at the University of Iowa (UI) in the late 1990's. Figure 13.7 shows

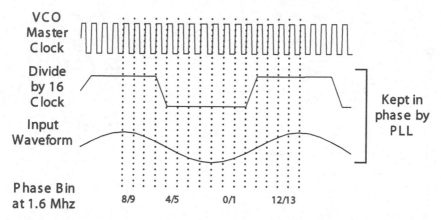

Fig. 13.7. The UI wave-particle correlator uses a phase-locked loop (PLL) locked to the measured waveform with a clock derived from master voltage-controlled oscillator (VCO) running at 16 times the frequency. This master clock subdivides the wave in 16 phase bins. Electron counts are sorted into the bins as they arrive to produce a map of phase bunching (Reprinted with permission from American Geophysical Union)

the principle of operation. The correlator used a voltage-controlled oscillator (VCO) running at 16 times the expected frequencies for Langmuir waves.

The VCO clock signal was divided by 16 to provide a signal which was then aligned to the measured AC parallel electric field through the use of a phase-locked loop (PLL). The PLL output a signal which indicated when the loop was properly locked and also the frequency of the locked signal which is then compared with the analog waveform data to ensure that the PLL was locked to signals of interest. Because the VCO master clock runs at 16 times the frequency of the wave to which the PLL is locked, it provides a highly accurate means of sub-dividing the measured wave into 16 phase bins. As electrons are counted by the detectors, they are sorted into the appropriate phase bin associated with their arrival time. Calibration of the detector, wave, and correlator electronics verified the accuracy of the phase bins and determined the timing delays through the system. The timing delays cause the absolute bin number which corresponds to 0° phase shift (as well as other phase angles) to shift as a function of input wave frequency. For example, at 1.6 MHz, 0° phase shift between the wave and the electrons occurs between correlator bins 8 and 9. The bottom of Fig. 13.7 shows the calibrated phase angles of the bins at a frequency of 1.6 MHz.

The UI correlator was flown on a rocket flight which was launched from Poker Flat, Alaska in February 6, 2002. Strong Langmuir waves were observed as the rocket traversed an auroral form near the poleward boundary of the auroral precipitation. The Langmuir waves were associated with a burst of field-aligned electrons at energies below 1 keV and well below the inverted-V

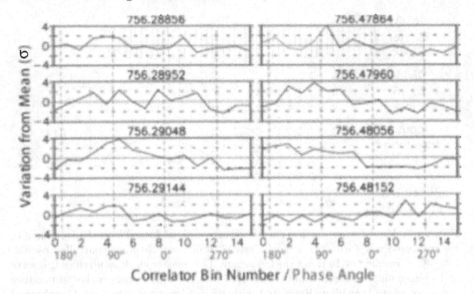

Fig. 13.8. Two examples of significant correlation of Langmuir waves with electrons at 468 eV. For each of the 16 phase bins, the count level is shown in terms of standard deviations away from the average number of counts that would be expected for a bin for the total number of counts received (Reprinted with permission from American Geophysical Union)

peak energy of 5 keV. During this period of Langmuir emission, two intervals of stronger emission with amplitudes of 60–200 mV/m were observed. Each of these intervals were of the order of 50 ms in duration. During both intervals, the field-aligned wave power was 8.5–11 times that of the perpendicular power indicating field-aligned waves. Because the payload telemetered continuous waveform data, it was possible to determine that the Langmuir waves were remarkably monochromatic with little or no variation in frequency. During each of the two larger amplitude bursts, the amplitudes varied slowly over hundreds of wave periods.

During each of the two bursts of strong Langmuir waves, significant wave-particle correlation was found as shown in Fig. 13.8.

Each set of panels shows four consecutive sets of correlator measurements of electrons with energy of 468 eV and which were sampled every millisecond. Data are plotted as standard deviations away from the expected average count rate in each phase bin. To be regarded as significant, we require a correlation level of at least 3.5σ. Although there is frequent variation of the order of 2σ shown in Fig. 13.8, these are not considered significant because the probability of 2σ variation is once in 22 samples as discussed above. With this in mind, in the left set of panels, the first and third panels show notable correlation. The first panel shows a weakly significant correlation with three adjacent phase bins with $\sim2\sigma$ levels of correlation corresponding to a 3.3σ variation.

However, the third panel from the top has two consecutive phase bins with phase bunching above 3σ. The combined probability of two such bins occurring together exceeds that of a single 4σ event. The bins in which the phase bunching occurs are those at 90° with respect to the wave electric field (positive away from the ionosphere) for the wave frequency of 1.6 MHz.

The second example shown in the right set of panels in Fig. 13.8 shows a second event with similar high levels of significance in the top three panels. Initially, a single channel with more than 4σ significance occurs at 90°. In the second panel, this feature broadens and shifts to somewhat larger phase shift with four phase bins above 2σ representing a combined significance of more than 4σ. The calculated probability of random occurrence of the counts represented by the four phase bins centered around 90° is one in 932,068 suggesting that this is a highly significant correlation. The third panel shows further advance to larger phase angles, with most significant set of signals now showing more than -4σ (5 channels at -2σ each) between 270° and 0°.

Figure 13.9 shows three sequential distribution function plots from immediately before, during and immediately after the first correlation event. The times given above each plot correspond to the center time of each 40 ms energy sweep. As can be seen, at the time of the correlation, a small, downward electron beam parallel to the magnetic field is measured at the same energy as the correlated electrons. This is indicated by a small arrow at the velocity corresponding to 468 eV. The isolated beam is not present in the distribution function measured before the correlation or in the distribution measured after the correlation. Although the time resolution of these measurements is much lower than for the correlator, this suggests that the correlation arises from this beam.

The brief increase in electron phase space density at the energy of correlated electrons illustrated in Fig. 13.9 was also seen for the second correlation event. In both cases the preceding and following energy sweeps did not show this feature, suggesting that the resonant electrons were only briefly in the detectable energy range. From the energy of the correlated electrons (468 eV) and the frequency of the waves (1.6 MHz), we can derive the wavelength of the Langmuir waves as 8.2 m, similar to the results of the UC Berkeley measurements.

The waveform data makes it possible to estimate the packet parameters to aid in the interpretation of correlation events. Figure 13.10 shows 15 ms of data which includes the interval of the second set of correlations.

The interval during which the correlations were observed is indicated by a heavy line above the waveform data. As can be seen, the correlations occurred during the largest wave amplitudes. If we assume a background thermal energy of 0.2 eV and use the inferred wavelength of 8.2 m along with the observed Langmuir frequency of 1.6 MHz, then the group velocity for Langmuir waves,

$$v_g = \frac{\partial \omega}{\partial k} \simeq \frac{3v_{th}^2 k}{\omega_{pe}} = \frac{3v_{th}^2}{\lambda f_{pe}}, \tag{13.25}$$

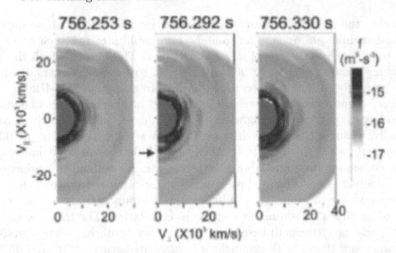

Fig. 13.9. Sequential distribution functions from before, during, and after the correlation event shown in the left panel of Fig. 13.8. The velocity corresponding to the correlated electrons is indicated with an arrow in the middle panel and shows that a field-aligned beam is present at this time (Reprinted with permission from American Geophysical Union)

gives the value of $v_g = 8.02$ km/s $= 8$ m/ms. If the wave is traveling down the field line, then the amount of the wave packet above the correlation observations extends from the time of the correlations in Fig. 13.10 to the end of the packet some 3 ms later. This yields a length of the wavepacket above rocket payload of 24.1 m at the time of the correlations and corresponds to a value of $kL = 18.5$. A rough average for the amplitude of the electric field during this interval is 130 mV/m. Combining this with the estimate of kL gives a value of $\mu = 0.12$, consistent with the linear theory and similar to the value used in Fig. 13.2.

It should be pointed out, however, that the usual limitations of single point spacecraft measurements apply to this interpretation. With the data at hand, we cannot rule out a scenario in which the electron beam has suddenly appeared and the wave packet has grown over time at the spacecraft location. Indeed, such a scenario would be consistent with the quasi-exponential increase in amplitude of the packet in the 10 ms preceding the correlation observations. In this case, there is no way to ascertain the amount of wave packet above the payload, but the shape of the packet suggests that the correlations were observed shortly after the linear growth phase ended and some type of non-linear saturation began to operate. The group velocity calculation still applies, but now becomes a lower bound on the length of the packet. By observing the packet for 3 ms we know that it extended at least 24 m above the rocket, but it may have extended much further, and then disappeared due to temporal effects in the driving electron distribution. A third alternative is that the payload may have moved into and then out of a pre-existing region

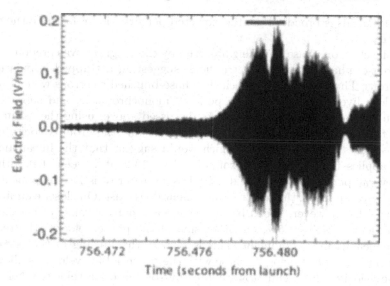

Fig. 13.10. Langmuir waveform data which includes the interval of the second set of correlation events. The correlation interval is indicated by the solid bar above the waveform

of Langmuir waves such that the observed wave envelope is determined by the spatial structure of the Langmuir waves along the rocket track. However, given the rocket velocity of roughly 1 km/s, this latter scenario would require a rather small wave packet and does not seem likely.

13.5 Discussion

The electron correlator observations reported by the UC Berkeley were predominantly resistive, that is, in-phase or 180° out-of-phase with the wave electric field. Although they had expected to see the reactive component, the observations of the resistive component prompted the analysis developed in Sect. 13.2 and led to the conclusion that they had observed electrons which were bunched by relatively short wave packets with kL of the order of 50–100. This result was also consistent with the value for μ that was determined which suggested that thee observed waves were in the linear regime.

As shown in the middle panel of Fig. 13.2, for $\mu = 0.09$, the perturbation to the density and hence, to the electron distribution, is dominated by the resistive component for small values of μ. Although the reactive component is present, as shown in Fig. 13.1, its bipolar character requires relatively narrow response electron detectors so as not to average the positive and negative perturbation together, yielding no perturbation. The detectors which made these resistive correlator measurements had $\Delta v/v \simeq 16\%$ and as shown in

Fig. 13.1, this is broader than the expected perturbation relative to the phase velocity.

The electron phase bunching observed by the UI group was predominantly 90° out-of-phase with the electric field, suggesting a trapped population of electrons. The wave field in which the phase-bunched electrons were observed was long-lived in terms of wave periods, monochromatic, and showed slow modulation of the wave envelope. As discussed above, using the assumption that the wave packet moved past the payload due the group velocity of the waves yields a low value of μ which would suggest that the linear analysis also applies to this case. In examining Fig. 13.2, it is seen that although the overall perturbation to the density has a greater resistive component than reactive component, the reactive component does exist. On closer examination of Fig. 13.2, however, it can be seen that for a narrow range of velocities of the order of $\Delta v/v = 5\%$, on either side of the phase velocity, the reactive component is more dominant. This is seen in the shifting of the perturbed distribution toward being in phase with the potential for velocities above the phase velocity and a shift toward 180° out-of-phase for velocities below the phase velocity. Because the UI electron detectors had a comparably narrow velocity response $\Delta v/v \simeq 5\%$, they are capable of capturing one or the other side of the perturbation and could measure a reactive perturbation even for a short wave packet.

In the case that the envelope of the observed waves is predominantly due to temporal evolution, then it is likely that the wave packet extends a significant distance above the rocket payload. This is suggested by the fact that the electron distribution will be unstable over a wide range of altitudes and would be expected to grow waves over a region of many wavelengths in extent. The character of the wave envelope also suggests a transition away from the linear stage of growth. Taken together, these two arguments suggest that an alternative explanation is that the correlations indicate trapping and thus the BGK analysis is appropriate. As shown in the right hand panel of Fig. 13.3, when the electron-Langmuir wave interaction becomes nonlinear, the density perturbation shifts toward a purely reactive phase as the electrons become trapped. Indeed the bunching of the electrons observed by the UI group is such that the positive perturbation was that of trapped electrons.

To resolve this ambiguity will require future experiments which provide adjacent energy channels with narrow response so that the full character of the perturbed distribution function can be revealed. This would allow one to see if both reactive pieces, that is the positive perturbation below the phase velocity and the negative perturbation above the phase velocity shown in Fig. 13.1, are observed side-by-side as the linear model would suggest or if only a single reactive perturbation is observed, consistent with the BGK analysis.

13.6 Conclusions

Multiple spacecraft observations have confirmed the ubiquitous nature of Langmuir waves in the presence of auroral electrons. Early observations have shown clear evidence that the electrons show variations consistent with bunching at or near the Langmuir frequency. Linear analysis of the interaction of a finite Gaussian packet of Langmuir waves shows that there are two components to the perturbation to the electron distribution function, one in-phase (or 180° out-of-phase) with respect to the electric field called the resistive component and one which is 90° (or 270°) out-of-phase with respect to the electric field. For small wave packets, the resistive perturbation dominates. For longer wave packets, a non-linear analysis is appropriate which suggests that the electrons have interacted long enough to become trapped and the reactive phase becomes dominant.

Rocket observations of the phase bunching of the electrons using wave-particle correlators have measured both components. The UI observations [Kletzing et al., 16] differ from those of the UC Berkeley observations [Ergun et al., 7, 9] in that a purely reactive phase bunching was observed as compared to a predominantly resistive perturbation. The resistive phase results of the UC Berkeley group were interpreted as arising from a short wave packet. The UI observations of the reactive phase can be explained by either a long, coherent train of Langmuir waves or that the narrower velocity response of the UI detectors made it possible to capture only one side of the reactive component of the perturbed distribution function for a short wave packet in the linear regime. Future wave-particle correlator experiments should be able to resolve these questions by providing more examples with better velocity space coverage.

References

[1] Bale, S.D.: Observation of the topside ionospheric mf/hf radio emission from space, Geophys. Res. Lett. 26, 667, 1999.
[2] Bauer, S.J. and R.G. Stone: Satellite observations of radio noise in the magnetosphere, Nature 218, 1145, 1968.
[3] Beghin, C., J.L. Rauch, and J.M. Bosqued: Electrostatic plasma waves and hf auroral hiss generated at low altitude, J. Geophys. Res. 94, 1359, 1989.
[4] Boehm, M.H.: Waves and static electric fields in the auroral acceleration region, PhD thesis, University of California, Berkeley, 1987.
[5] Boehm, M.H., G. Paschmann, J. Clemmons, H. Höfner, R. Frenzel, M. Ertl, G. Haerendel, P. Hill, H. Lauche, L. Eliasson, and R. Lundin: The tesp electron spectrometer and correlator (F7) on Freja, Space Sci. Rev. 70, 509, 1995.
[6] Bonnell, J.W., P.M. Kintner, J.E. Wahlund, and J.A. Holtet: Modulated langmuir waves: observations from freja and scifer, J. Geophys. Res. 102, 17233, 1997.

[7] Ergun, R.E., C.W. Carlson, and J.P. McFadden: Wave-particle correlator instrument design, In: R.F. Pfaff, J.E. Borovsky, and D.T. Young (Eds.), *Measurement Techniques in Space Plasmas: Particles*, volume 102 of AGU Geophys. Monogr. Ser., p. 4325, AGU, Washington, D.C., 1998.

[8] Ergun, R.E., C.W. Carlson, J.P. McFadden, J.H. Clemmons, and M.H. Boehm: Evidence of a transverse modulational instability in a space plasma, Geophys. Res. Lett. 18, 1177, 1991.

[9] Ergun, R.E., C.W. Carlson, J.P. McFadden, D.M. TonThat, J.H. Clemmons, and M.H. Boehm: Observation of electron bunching during Landau growth and damping, J. Geophys. Res. 96, 11371, 1991.

[10] Gough, M.P., P.J. Christiansen, and K. Wilhelm: Auroral beam-plasma interactions: particle correlator investigations, J. Geophys. Res. 90, 12287, 1990.

[11] Gough, M.P. and A. Urban: Auroral beam/plasma interaction observed directly, Plan. Space Sci. 31, 875, 1983.

[12] James, H.G., E.L. Hagg, and L.P. Strange: Narrowband radio noise in the topside ionosphere, *AGARD Conf. Proc.*, AGARD-CP-138, 24–1–24–7, 1974.

[13] Kelley, M.C. and G.D. Earle: Upper hybrid and Langmuir turbulence in the auroral e-region, J. Geophys. Res. 93, 1993, 1988.

[14] Kellogg, P.J. and S.J. Monson: Radio emissions from the aurora, Geophys. Res. Lett. 6, 297, 1979.

[15] Kintner, P.M., J. Bonnell, S. Powell, and J.E. Wahlund: First results from the freja hf snapshot receiver, Geophys. Res. Lett. 22, 287, 1995.

[16] Kletzing, C.A., S.R. Bounds, J. LaBelle, and M. Samara: Observation of the reactive component of langmuir wave phase-bunched electrons, Geophys. Res. Lett., 32, L05106, doi:10.1029/2004GL021175, 2005.

[17] McAdams, K.L. and J. LaBelle: Narrowband structure in hf waves above the electron plasma frequency in the auroral ionosphere, Geophys. Res. Lett. 26, 1825, 1999.

[18] McAdams, K.L., J. LaBelle, M.L. Trimpi, P.M. Kintner, and R.A. Arnoldy: Rocket observations of banded stucture in waves near the langmuir frequency in the auroral ionosphere, J. Geophys. Res. 104, 28109, 1999.

[19] McFadden, J.P., C.W. Carlson, and M.H. Boehm: High-frequency waves generated by auroral electrons, J. Geophys. Res. 91, 12079, 1986.

[20] Muschietti, L., I. Roth, and R. Ergun: Interaction of Langmuir wave packets with streaming electrons: phase-correlation aspects, Phys. Plasmas 1, 1008, 1994.

[21] Newman, D.L., M.V. Goldman, and R.E. Ergun: Langmuir turbulence in the auroral zone 2. nonlinear theory and simulations: J. Geophys. Res. 99, 6377, 1994.

[22] Nicholson, D.R.: *Introduction to Plasma Theory*, Wiley, New York, 1983.

[23] Reiner, M.J. and M.L. Kaiser: Complex type iii-like radio emissions observed from 1 to 14 mhz, Geophys. Res. Lett. 26, 397, 1999.

[24] Reiner, M.J., M. Karlicky, K. Jiricka, H. Aurass, G. Mann, and M.L. Kaiser: On the solar origin of complex type iii-like radio bursts observed at and below 1 mhz, Astrophys. J. 530, 1049, 2000.

[25] Samara, M., J. LaBelle, C. A. Kletzing, and S.R. Bounds: Rocket observations of structured upper hybrid waves at $f_{uh} = 2f_{ce}$, Geophys. Res. Lett., submitted, 2004.

[26] Sanbonmatsu, K.Y., I. Doxas, M.V. Goldman, and D.L. Newman: Non-Markovian electron diffusion in the auroral ionosphere at high Langmuir-wave intensities, Geophys. Res. Lett. 24, 807, 1997.

[27] Spiger, R.J., J.S. Murphree, H.R. Anderson, and R.F. Loewenstein: Modulation of auroral electron fluxes in the frequency range 50 kHz to 10 MHz, J. Geophys. Res. 81, 1269, 1976.

[28] Stasiewicz, K., B. Holback, V. Krasnoselskikh, M. Boehm, R. Boström, and P.M. Kintner: Parametric instabilities of langmuir waves observed by freja, J. Geophys. Res. 101, 21515, 1996.

[29] Walsh, D., F.T. Haddock, and H.F. Schulte: Cosmic radio intensities at 1.225 and 2.0 mc measured up to and altitude of 1700 km, Space Res. 4, 935, 1964.

[30] Weatherwax, A.T., J. LaBelle, M.L. Trimpi, and R. Brittain: Ground-based observations of radio emissions near $2f_{ce}$ and $3f_{ce}$ in the auroral zone, Geophys. Res. Lett. 20, 1447, 1993.

Index

Lecture Notes in Physics

For information about earlier volumes
please contact your bookseller or Springer
LNP Online archive: springerlink.com